ENERGY FUTURE

ENERGY FUTURE

Report of the
Energy Project at the
Harvard Business School

ROBERT STOBAUGH
& DANIEL YERGIN, Editors

I.C. Bupp, Mel Horwitch,
Sergio Koreisha,
Modesto A. Maidique
and Frank Schuller

RANDOM HOUSE NEW YORK

Faculty research at the Harvard Business School is undertaken with the expectation of publication. In such publication the authors responsible for the
research project are also responsible for statements of fact, opinions, and
conclusions expressed. Neither the Harvard Business School, its faculty as a
whole, nor the president and fellows of Harvard College reach conclusions or
make recommendations as results of faculty research.

Library of Congress Cataloging in Publication Data

Main entry under title:
Energy future.

   Includes index.
   1. Energy policy—United States. 2. Power resources
—United States. I. Stobaugh, Robert B. II. Yergin,
Daniel. III. Harvard University. Graduate School of
Business Administration.
HD9502.U52E4914   333.7   78–21329
ISBN 0–394–50163–2

Manufactured in the United States of America
9 8 7

To Beverly and Angela
who contributed a great deal

ACKNOWLEDGMENTS

In the course of researching and writing this book, we two communicated with over three hundred business executives, government officials, labor union leaders, analysts, academics, and other specialists. With some, it was a matter of only a small point. Others gave enormously of their time and attention. Our coauthors had similar exchanges with many hundreds more. Obviously, because of the space required to list a thousand or so names, but also because some spoke to us candidly on condition of anonymity, we do not thank them by name. But we do express our deepest appreciation to all of them. Our undertaking would not have been possible without such assistance.

Fortunately, there are also those we can thank by name. We start not with a person, but with an institution, the Harvard Business School, and in particular its Division of Research and its Associates, which made this entire effort possible in a way that few other institutions could—or would—have. It provided the resources so that we had ample time to explore these complex matters critically and pragmatically. It was Dean Lawrence Fouraker who asked Robert Stobaugh to organize the Energy Project in 1972, and then gave it his continued support. Professor Richard Walton, formerly head of the Division of Research, provided the necessary funding and encouragement for its launching. His successor, Associate Dean Richard Rosenbloom supported and encouraged us in a manifold of ways as the research project took shape as a book, and to him all of us have a special and lasting gratitude. Our colleagues in the Production and Operations Management Area at the school were willing to be in the classroom, enabling us to carry on with the required research and writing. For this, we thank them, as well as former Associate Dean Walter Salmon and Associate Dean John McArthur, Area Chairmen Wickham Skinner, Robert Hayes, James Healy, and Philip Thurston, all of whom handled the difficult task of working out faculty schedules. As the title of the book suggests, this is a report of a project located at the Harvard Business School, and not a report of the school. And, of course, the conclusions and opinions are those of the authors, for, as the school's policy statement says: "Neither the Harvard Business School [nor] its faculty as a whole . . . reach conclusions or make recommendations as results of faculty research."

We would also like to thank the Center for International Affairs at Harvard, and its directors, Raymond Vernon, Benjamin Brown, and Samuel Huntington, for it was in the Center's International Energy Seminar that important parts of this book took shape.

Nancy Estes served as combination administrative aide and secretary for the Project. Whether scheduling meetings at chaotic moments or typing at the oddest hours, she kept the Project running. Her efforts and skills, plus her kind nature, were invaluable. Nancy Armstrong weekly took time away from the further development of her magnificent soprano voice to bring order out of chaos and a whirlwind of pieces of paper. She, too, has kept the Project ticking, and has been a pleasure to work with. Gay Auerbach has been involved with this book in many different ways, mixing enthusiasm and competence with an always useful irony. Carmen Vaubel, although working away on other assignments, also chipped in on weekends and late at night. Stacy Miller, as she has for other publications of the Energy Project, spent many hours in the library helping us to find information that did not want to be found. On behalf of our coauthors, let us also thank those who worked with them and saw too many drafts of their chapters—Regina Collingsgru, Mary Day, Muriel Drysdale, Althea Martinez, Rose Giacobbe, and other members of the Word Processing Center.

We speak often in this book of the transition from imported oil to a more balanced energy system. There is also the transition from raw research to a finished book, for what value are years spent researching if the fruits remain unknown? Jason Epstein, editor-in-chief at Random House, saw merit in what we were trying to do, chose to take us on as his authors, encouraged us, but also challenged us many times with tough and blunt questions. His colleague Grant Ujifusa also brought great dedication to this book, and gave it much more time and concern than the rules of publishing allow. As has been written about him before, he "brought insight, commitment, and patience to this project." Carolyn Lumsden worked diligently as copy editor. Helen Brann believed that this was an important book that should be published, and made that possible. We thank her deeply for her continuing support, and also express great appreciation to her colleague, John Hartnett.

Late in this process, but fortunately not too late for the book, we had the invaluable experience of working with William Bundy, editor of *Foreign Affairs,* who has been editor of some of the truly significant energy articles of the 1970's. We express our appreciation to him and to his thoughtful colleague, James Chace.

Four other people must be mentioned: Sidney Robbins, this time not a coauthor with Robert Stobaugh, provided a careful reading of the entire manuscript. Max Hall provided editorial guidance in the early phases for a number of us. Herbert Holloman, director of MIT's Center for Policy Alternatives, provided financial support for our earlier studies of nuclear power, a careful reading of several chapters, and moral support for the entire undertaking. Elmer Funkhouser provided much encouragement and help. In addition, we thank the members of the Resources for the Future Project, "Energy: The Next Twenty Years" (of which Robert Stobaugh is a member), for discussion of our manuscript.

Only those who have participated in this kind of research project know the costs exacted on family life. Taking one away night after night, weekend after weekend, vacation after vacation, the work never ends. Our families deserve special medals for understanding.

As editors, we were extremely fortunate to be working with an unusual group of colleagues, who put aside many other obligations in pursuit of that intangible goal, a first-rate chapter. Strong-willed and determined, they have worked closely with us and with each other, not only on their chapters, but on ours, and never surrendered their independence of judgment. For such a group to be so harmonious throughout this demanding project is a tribute to their abilities and scholarly commitments, and also to their understandings of the frailties of the editors.

As we look through this manuscript, we are reminded of what has occurred to us so often before: The "energy problem" is so fragmented, in so many different pieces. We never doubted the importance of trying to make sense out of those pieces; sometimes we doubted the feasibility. We were willing to forgo much in order to try, for we regarded it as extraordinarily exciting to have the opportunity to try to make sense out of the pieces. It was also a great responsibility. We approached this work in that double spirit.

Robert Stobaugh
Daniel Yergin

Soldiers Field
1979

CONTENTS

ENERGY FUTURE

ROBERT STOBAUGH
AND DANIEL YERGIN

1

The End of Easy Oil

In 1968, the State Department sent the word to foreign governments—American oil production would soon reach the limits of its
capacity. Friendly governments needed to know that the cushion of
the U.S.'s extra capacity, which could be called into production during an emergency, was about to disappear. The end of an era was
at hand.

But few people anywhere thought seriously about the implications of losing the cushion, for the industrial world had grown
increasingly comfortable using oil to fuel the unprecedented economic growth of the 1950's and 1960's. Western Europe and Japan
relied mainly on the Middle East, and the United States also was
beginning to import from that region. Middle Eastern oil was the
world's favorite fuel—easy to produce in very large volumes (a dime
or two a barrel), easy to transport, and easy to burn—certainly easier
than coal.

In 1970, some 111 years after the birth of the American oil industry, domestic production peaked and began to decline. But the
demand for oil continued to surge, and that demand could be met
only by more and more oil from the Middle East, which meant increasing dependence—and increasing vulnerability. The idea that
there was something threatening in the growing dependence was

an idea better ignored. Even if one recognized a potential problem, what to do about it was hardly clear when the general momentum to use more oil was regarded as unstoppable.

The first oil shock, in late 1973 and early 1974, definitely marked the end of the era of secure and cheap oil. Arab oil producers embargoed the United States and reduced overall output and shipments to other nations. For the first time, OPEC countries stopped negotiating a price with the oil companies; they instead unilaterally set the price on a take-it-or-leave-it basis. The oil buyers had no choice, and they took it, paying the higher price, eight times higher by the end of 1974 than five years earlier.[1] And so the petroleum exporting countries defined a new era for the rest of the world—one of insecure supplies of expensive oil.

Yet today, as we enter the 1980's, even after the second oil shock that accompanied the fall of the Shah of Iran, the cause and consequences of the new era of oil have yet to be taken seriously in the United States. The key contradiction is this: While the declared aim of American policy is to reduce the use of imported oil, the United States is in fact becoming more and more dependent upon it. Between 1973 and early 1979, U.S. oil imports almost doubled, and had begun to provide half of the nation's oil. By current trends the United States will be even more dependent on imported oil in the 1980's.

Does this matter?

We think it does. The oil crisis of 1973–74 constituted a turning point in postwar history, delivering a powerful economic and political shock to the entire world. It interrupted or perhaps even permanently slowed postwar economic growth. And it set in motion a drastic shift in world power, in the very substance of international politics. Curiously, however, in the aftermath it became fashionable to discuss the crisis as though it were a unique event, a freak storm that had been weathered.

We have consistently disagreed, viewing it instead as a warning of a fundamental and dangerous disorder, for the basic conditions that allowed the first shock, and now the second, have continued to prevail. Indeed, higher real oil prices seem assured for the future, with the only questions being how soon and how high. Any price increase has immediate, undesirable effects on all oil-importing nations, causing a direct loss in national income. If the price rise is very gradual over a period of many years, thereby allowing the oil-importing nations a gradual adjustment, then the direct effects might be the main ones.

A large, sudden increase in oil prices would have serious indirect effects. It would exacerbate inflation, place further strains on the international monetary system, and sharply contract the demand for goods and service, further reducing national income. In short, the economic consequences would likely be a major recession, or possibly even a depression.

The political consequences are potentially no less serious. Slower economic growth and high inflation intensify conflicts not only within Western nations, but also among them. As the world's largest oil importer, the United States would bear much of the blame for higher oil prices. A bitter competition for oil could ensue, damaging the Western security and trading systems. Putting aside American relations with the industrialized nations, greater reliance on Middle East imports would certainly mean that U.S. foreign policy would be increasingly constrained by its oil suppliers.

Political instability in the Middle East, supply interruptions, the extension of Soviet influence—such factors only make a very bad situation much worse. *This point must be underlined.* For the industrial nations to continue to depend on Middle Eastern oil in the way current trends indicate means heavy reliance on a region of high political tension and risk. In the last three decades, the Middle East has been subjected to a half-dozen wars, a dozen revolutions, and innumerable assassination and territorial disputes. Dependence reinforces the twin vulnerabilities—interruption of supplies and major price increases.

When we began this book, the prospects just sketched seemed unlikely until the 1980's—perhaps even the late 1980's. Economic activity, responding to the strains of the first shock, had slowed down, tempering for a time the demand for oil, and thus postponing the reemergence of a tight petroleum market. But political and psychological factors were left out of many of the most prominent forecasts of the international energy scene, for they could not easily be integrated into equations. It became quite fashionable to talk complacently of a glut on the world market.

Beginning in 1977, such forces did begin to make their influence felt, underscoring how crisis-prone is the current energy system. Increases in U.S. oil imports helped trigger the sharp decline of the dollar, which began in 1977 and reached almost panic proportions by the end of 1978, spilling over into the U.S. stock markets. As the economics minister of one of the major European countries told us, "I regard U.S. oil imports and the weakness of the dollar as two sides of the same coin." The selling against the dollar, with all the attendant instability that brings, reflects a discounting by non-Americans of the United States's ability to control its oil imports. One former chairman of the President's Council of Economic Advisers warned of "the rising probability of a serious collapse," and another former chairman said that the financial markets "were clearly signaling the onset of a major financial crisis." [2]

Then, in the latter part of 1978 and in early 1979, Iran exploded in a crisis that threw that country—the supplier of 10 percent of the world's oil—into chaos, choked off production, and finally drove the Shah from the Peacock Throne. The interruption in the flow of Iranian oil helped set the stage for the 15-percent price hike at the December 1978 OPEC meeting. This single increase was greater in dollar value than the total price of oil in 1970. Furthermore, the high prices in the spot markets for world oil in early 1979, the drive by OPEC producers to capture a significant part of that increase, and the policies of the new Iranian government—these make it likely that world oil price in December 1979 will be several dollars more per barrel than that agreed upon at the December 1978 OPEC meeting. But the Iranian crisis itself, more than just the increase in the price of petroleum, dramatized the dangers for Western nations of depending so heavily on traditional regimes going through a rush of modernization.

True, the United States has taken some steps to slow oil imports. But the results seem rather minor compared to the scale and urgency of the problem. Even the much embattled National Energy Act of the Carter Administration, as it finally emerged in October 1978, will contribute considerably less to meeting the country's energy problem than all the controversy might have led one to think.

Of course, one would hardly expect that it would be easy, even under the most favorable circumstances, to frame a program that could stop the growth of oil imports in a fashion that would not disrupt the economy and society. After all, a wide range of complex engineering and other technical problems are involved, from the difficulties of offshore drilling operations to the way in which electricity is priced. Beyond the technical problems, there are a host of competing economic, political, and regional interests, all of them warring over the allocation of vast resources. To put it bluntly, serious talk about energy involves a very great deal of money. There may never have been so rich a sweepstakes in American history, for the OPEC price revolution of 1973 and 1974 increased the world market value of proved U.S. reserves of oil and gas alone by $800 billion.[3] And then, there are uncertainties, which stretch from the revolutionary dreams of some now-unknown Saudi Arabian colonel to the musings of a specialist in photovoltaic technology.

Given these complexities, it is no wonder that people prefer to dream of some simple solution. First it was nuclear power. Then coal. Then fusion. Then Mexican oil. But none of these is a singular remedy. An easy fix is unlikely to offer itself, and a prudent society does not count on it happening. And the task of policy formation is made even more difficult by the vocal and sometimes quite bitter debates that now becloud the American political process.

The disagreements can be striking. At a recent conference, one "expert" described the country's nuclear establishment as "weak and demoralized" because of the bleak outlook for the building of nuclear power plants, while another equally reputable "expert" said that the United States and other governments should get on with the business of selecting sites for the ten-thousand nuclear power plants that he insisted *would* be built to generate the world's electricity. A news article in the *New York Times* was headlined "Oil Report Optimistic"; it reported that a global shortage before the late 1980's was unlikely, and that one before the twenty-first century was only a possibility, "not a probability." Yet elsewhere in the same newspaper on the same day a story warned that the Western world could "face a damaging scramble for oil in the mid-1980s, combined with sharply higher prices." [4] How can governments make reasonable choices when confronted with such uncertainty?

Indeed, making a move away from ever-increasing imports of oil has proved far more difficult than was imagined in 1973 and 1974.

What has gone wrong? We try to answer that question in this book. We also suggest what can go right by sorting out some of the basic issues and by setting out in comparative fashion the different sources of energy and potentials of each.

We began the book with three premises that framed our entire undertaking. First, we see the crises of 1973–74 and 1978–79 not as isolated phenomena, but rather as part of a major transition for both energy producers and users. Second, we believe that healthy economic growth is essential and that reliance on the free market is the best way to achieve it. Third, we feel that thinking about energy raises important questions about income generation and distribution: What are the total costs and benefits involved in any decision, who profits, and who pays? We believe that some attempt, however rough, has to be made to assess the total "social costs" embedded in the problem.

Our primary focus is the United States. This country is the world's largest producer, consumer, and importer of energy. It has a wider range of choices than most other industrial countries, and yet has surprisingly little to show for six years of vociferous debate. In addition, what happens in the United States, whatever shape it takes, has a profound importance for the rest of the world, both through direct effects and as a model to follow.

Our perspective is what is possible and reasonable in the near and middle term—that is, what can get the United States most smoothly through the last two decades of the twentieth century. Fundamentally, we discuss the managing of a transition away from a system of heavy reliance on imported oil to a more balanced one. We have therefore tried to apply what might be called a managerial perspective, broadly defined. We attempt to assess priority and potential, evaluate costs and risks, compare returns on different forms of investment, examine how people and resources can be effectively mobilized, and suggest what can be done to mediate conflict among competing interests.

We explore energy questions pragmatically, from the bottom up, examining not only the technical and economic obstacles and possibilities, but the political and institutional realities as well. This means trying to understand the sometimes humble and often nitty-gritty reality of each energy source. "Coal" to us is not merely a solidified hydrocarbon that exists in abundance in the United States, nor is it a simple coefficient in a mathematical model. "Coal" is in fact a system that involves, among other things, labor-management strife, uncertain environmental effects, and the doubts posed by those who make investment decisions for utilities. Similarly, a serious, not just rhetorical, energy conservation program depends on a set of incentives framed by the government that would satisfy energy users and help create businesses that can "deliver" conservation to the homeowner.

In the chapters that follow, we present our findings and recommendations. But a quick overview of the book follows.

To begin with, imported oil poses too many risks to be calmly accepted. Knowing that it would be controversial, we have tried to put a figure on the "real" cost of a barrel of imported oil. In early 1979, the average price of U.S.-produced oil, kept down by price controls, was $9 a barrel. At that same time, the world market price for oil delivered to the United States was $15 a barrel, and was scheduled to rise another dollar by October 1979—although the Iranian shutdown intervened to signal the real possibility of even higher prices. To compute a price for incremental amounts of imported oil, we have tried to establish the total "social cost"—that is, market price plus the costs to the society of future price hikes and economic disruptions that might result from increasing demand. Although in this calculation we did not include the political and the more dire economic consequences, our best estimate of the cost still came out at $30 to $40 a barrel in 1979 dollars. For convenience we use $35 a barrel, or about three times the average price paid by domestic refiners for crude oil—a mix of domestic and imported. Obviously, this is a very tentative figure, and the assumptions behind it are explained in Chapter 2. Yet, the magnitude of the number is such that the United States should make a much greater attempt to stop the growth of its oil imports.

But can the United States do this? There are four conventional sources of domestic energy: oil, natural gas, coal, and nuclear power. But all four are likely to deliver less energy than projections by their advocates would lead one to believe.

In the debate about oil, three main domestic oil "solutions" have been put forward as alternatives to imported petroleum: break up the industry, decontrol oil prices (and lease more land), and use unconventional technology. The first two are important and controversial political questions, affecting as they do the distribution of income and power in America. But the "solutions" have little to do with augmenting production. Whether the industry is or is not broken up, whether prices are or are not deregulated, the physical production of oil from conventional sources will continue to decline. At best, enhanced recovery and other unconventional means can help to keep domestic supplies flowing at current levels. To be sure, outside the United States new oil fields in Mexico and China are important and will augment world supplies, but, claims to the contrary, they are unlikely to make a substantial change in the world oil balance.

Natural gas, which accounts for over a quarter of America's energy needs, has also been entrapped in a great domestic debate: Should it continue to be regulated, with price based on cost of production, or should it be deregulated, with price based on value? As in the case of oil, the debate is about money and who gets it. But the distributional questions aside, the best that one can expect is that a deregulated price will enable natural gas production to remain at current levels.

A major goal of the Carter National Energy Plan has been to foster the substitution of domestically produced coal for imported oil. Given America's great coal reserves, the proposition appears feasible on paper, but in practice it probably is not. For coal to do what the Carter Administration wants it to do, a traditionally backward industry must be suddenly transformed into a modern, technologically advanced one. Potential users are reluctant to commit themselves to coal, especially because of the uncertainty about meeting environmental requirements. Coal's contribution, in the rest of this century, therefore, is likely to prove more limited than the Administration plans, although its importance is still likely to grow, particularly for utilities.

Nuclear power is the other conventional alternative in which high hopes are placed. Yet, the further development of nuclear power is stalemated by controversy that has passed beyond the boundaries of the technocratic community to become a substantive matter in the political process. It is too soon to assess the long-term consequences for the nuclear power industry of the nuclear accident at Harrisburg—except to say that it hardly improves the industry's prospects. Even without a Harrisburg, however, the problem of what to do with nuclear waste is so confused, and so far from politically acceptable resolution, that it could result in an absolute decline in the energy produced by nuclear power in the next decade. Moreover, one should remember how limited the potential of atomic power is under the most bullish of circumstances: If nuclear power capacity *doubled* in ten years, it would still be providing less than 7 percent of America's total energy.

In short, there is little reason to expect conventional alternatives to make a sizable contribution to reducing our dependence on imported oil. These energy sources—domestic oil and gas, coal, and nuclear power—as a group can increase their contribution to cover, at most, one third to one half of the nation's additional energy needs over the next decade.

On the other hand, the unconventional alternatives, which tend to be played down, can make a much greater contribution than is normally assumed. The unconventional as well as the conventional alternatives should be given a fair chance. To date, they have not received anything like that. According to one estimate, conventional energy sources have received more than $120 billion in incentives and subsidies, while the unconventional sources have received virtually nothing by comparison. And some subsidies for producers continue, while subsidies for consumers to use conventional energy sources run into tens of billions of dollars yearly because of controls that keep prices far below replacement costs.[5]

A major reason for the imbalance between conventional and unconventional sources since 1973 has been the persistent acceptance by policymakers and the media of misleading conclusions drawn from prominent econometric and technological models. Our work has convinced us that while such models can be valuable, they have been accepted too uncritically. Such models are abstractions and, therefore, simplifications of reality. Too often, the assumptions that govern the models are overlooked. Indeed, we believe that unexamined technicalism has helped to create some of the impasses and stalemates in U.S. energy policy. Decisions about energy issues ignite intense bargaining and competition for resources, raise major distributive questions, and threaten strongly held values. The world of human institutions in which the real choices have and will be made is in fact a world of power and politics, one not easily captured in mathematical models. Indeed, without acknowledging the genuine distributional problems, and without looking for ways to mediate conflict among competing groups, stalemate, not coherent energy policies, will persist.[6]

Among the unconventional sources of energy, conservation presents itself as the most immediate opportunity. It should be regarded as a largely untapped source of energy. Indeed, conservation—not coal or nuclear energy—is the major alternative to imported oil. It could perhaps "supply" up to 40 percent of America's current energy usage, although we do not predict that it will. Moreover, the evidence suggests that there is much greater flexibility between energy use and economic growth than is generally assumed, and that a conservation strategy could actually spur growth. Conservation does not require technological breakthroughs. But it has been difficult to tap, because a consistent set of signals—price, incentives, and regulations—is not in place. Moreover, the decentralized character of energy consumption means that decisions to conserve, unlike decisions to produce energy, have to be made by millions and millions of often poorly informed people.

The range of energy possibilities grouped under the heading "solar" could meet one fifth of U.S. energy needs within two decades. Like conservation, solar energy faces a problem of decentralized decision-making. Moreover, the most promising near-term solar energy applications, which use existing, relatively simple technologies, are receiving less support than the more uncertain, more distant high-technology solar applications. Low-technology solar energy can make a significant contribution, but like conservation, it needs a more consistent framework of price, incentive, and regulation.

Those with a stake in conventional wisdom about conventional energy sources may charge that the conclusions of this book are unrealistic, that the authors are romantics opposed to economic growth, or that we advocate basic changes in the way the society is organized. We are not and we do not. Indeed, we do not side with those romanticists who have a vision of the national life decentralized in many spheres through the mechanism of the energy crisis to the point where it becomes a post-industrial pastoral society. For we wish to see the system prosper and in a vital way, and this means greater reliance on the free market. Still, we do not subscribe to the views of the other set of even more powerful romanticists—industrial romanticists, who believe that it is possible to return to an era of unlimited production and that production alone can be the nation's salvation. For genuine alternatives for energy do exist and we want to contribute to the clarification of the choices. We also believe that it is unwise to ask conventional sources to do more than they really can and by so doing to block the transition to a more balanced energy system. To deny the need for a transition does not prevent change, it only postpones it, ensuring that when change does come, it will cause intense disruption, pain, and conflict.

No easy remedy will solve the energy crisis. Solutions, however, will emerge from a recognition and comparison of benefits and risks, possibilities and obstacles, across a wide range. Political choices are therefore involved, which is why the energy crisis is a crisis of our political system.

Some kind of general clarification and understanding is required if the nation is to move beyond stalemate. We hope that we can, to some degree, redefine the terms in which the American energy debate is being waged. We would not by any means insist that we have all the answers, or that we have correctly assessed all aspects of the many intricately interconnected problems involved. But we believe that we are pointing in the right direction. The easy days of easy and cheap oil are truly over, but there is no need that the process of change and adaptation should prove as difficult as it now appears.

We have yet to meet an energy specialist who can keep all the barrels of oil, trillion cubic feet of natural gas, tons of coal, gigawatts of electricity, and quads of energy in his (or her) head. We certainly can't. In order to keep the energy sources straight, and to make clear the comparisons, we have used barrels of oil per day as our basic unit—though, of course, following convention and common sense as the energy source dictates. Thus, the aim of the facing Table I–1 is to lay out the equivalences in barrels per day, in addition to showing U.S. energy consumption in 1977.

CONVERSION FACTORS

One million barrels of oil daily are the equivalent of

Natural gas: 2.08 trillion cubic feet (tcf) per year
Coal: 94 million tons per year
Nuclear: 199 billion kilowatt hours (kwh) per year
        [1 million kwh equal 1 gigawatt hour (GWH)]
Hydro: 203 billion kilowatt hours (kwh) per year
Electrical generating capacity at historical operating rates (about 50 percent): 50 GW
Any energy: 2.12 quadrillion (10^{15}) BTUs per year (quads)

Table I–1

U.S. ENERGY CONSUMPTION, 1977

SOURCES	Millions of Barrels per Day of Oil Equivalent	Percentage of Total	Quantity in Units Commonly Used for Each Source
Petroleum	18.4 [a]	50%	18.4 million barrels per day
Natural Gas	9.2 [b]	25	19.2 trillion cubic feet per year
Coal	6.7 [c]	18	625 million tons per year
Nuclear	1.3	4	251 billion kilowatt hours per year
Hydro	1.1	3	230 billion kilowatt hours per year
Total	36.7	100%	

[a] Includes imports of 8.7 million barrels per day (mbd), or 47 percent of total oil consumption; excludes 0.2 mbd of exports.

[b] Includes imports of 0.5 mbd oil equivalent, or 5 percent of total natural gas.

[c] Excludes exports of 0.6 mbd oil equivalent.

Source: Department of Energy, Energy Information Administration, *Annual Report to Congress, Volume III, 1977* (Washington, D.C.: Government Printing Office, 1978), pp. 5, 23, 51, 145.

2

After the Peak: The Threat of Imported Oil

Oil, the prime mover of industrial life, has become one of its greatest problems. The current patterns of production and consumption, and the political alignments that accompany them, are leading the United States—and the world—into a future laden with uncertainty and danger. Although the nation has had an oil industry for 120 years, the United States has, in just a few short years, become heavily dependent on imported oil. This dependence means ever greater reliance on an unstable part of the world —the Middle East, and especially Saudi Arabia. It would therefore be very much in the nation's interest to reverse the trend toward more oil imports.[1] But contrary to popular impression, domestic oil production is unlikely to be the way to do it. Geology denies that possibility.

THE RISE AND FALL OF AMERICAN OIL POWER

The oil industry had its origins in the middle of the nineteenth century, when kerosene began to replace expensive whale oil as a source of interior lighting.[2] Whereas most of the kerosene was made from coal, some was extracted from crude oil. But the crude oil could be obtained only where it naturally seeped to the surface or from brine water wells. With kerosene priced at $42 a barrel, a considerable opportunity obviously existed for anyone finding a cheap and easily available supply of crude oil. A group of New Haven investors, fortified with a report from a Yale chemist on the economics of refining kerosene from crude oil, formed a company to drill for oil in the vicinity of brine wells in Pennsylvania. The investors hired a railroad conductor, "Colonel" Edwin Drake, to head the venture. The Colonel had neither business nor drilling experience, nor was he a colonel, but he possessed two useful attributes: He was available because he was out of work; and his services came cheaply because he had a railroad pass to get to Pennsylvania. Once there, near Titusville, he struck oil on August 27, 1859. And so the modern oil industry was born.

For more than a century thereafter, the production of crude oil in the United States steadily increased. In 1909, the fiftieth year of the industry, U.S. production reached 500,000 barrels a day,* more than the rest of the world combined.[3] And except for several years after World War I, the United States remained one of the world's leading petroleum exporters. As time went on, what became the largest industry in the world—dominated for most of its history by American companies—produced, refined, and distributed oil to an ever wider and more diverse market. Cheap oil gradually shoved aside coal, and so became the basic source of power for an industrial civilization.

A little-noticed but significant change occurred in 1948. Imports —mostly cheap Venezuelan crude—exceeded American oil exports, which meant that the United States had become for the first time a permanent net importer of oil. Nevertheless, the United States continued to produce half of the world's oil in the early 1950's. Furthermore, it had sufficient unused capacity to produce for export markets in an emergency, as happened during the Suez crisis of 1956. But American oil, which was expensive to produce, could not compete with the low-cost crude that had begun to flow in large quantities from the Middle East. Thus, for the stated reason of "national security," but also to protect politically powerful domestic

* For this chapter, *crude oil* includes natural gas liquids, which were 17 percent of the total in 1977.

producers, the U.S. government placed restrictions on imports in the late 1950's.[4] So protected, domestic oil production continued to climb.

The historic turning point came in 1970, when U.S. spare capacity vanished and U.S. production reached what proved to be its peak—an average of 11.3 million barrels a day. From then on, the level of oil production began to decline. But as demand continued to increase, cheap imported oil took a larger and larger share of the U.S. market as ever-larger cracks began to appear in the oil import barrier. In response to sporadic shortages that began to develop around the country, the Nixon Administration abandoned import quotas in 1973, and imported oil poured in. As Table 2–1 shows, foreign oil now accounts for nearly half of American consumption: [5]

Table 2–1

FOREIGN OIL CONSUMPTION IN THE U.S.

Year	Consumption (millions of barrels a day)	Production (millions of barrels a day)	Imports (millions of barrels a day)	Imports (as a percentage of consumption)
1960	9.7	8.0	1.8	19%
1962	10.2	8.4	2.1	21
1964	10.8	8.8	2.3	21
1966	11.9	9.6	2.6	22
1968	13.0	10.6	2.8	22
1970	14.4	11.3	3.4	24
1972	16.0	11.2	4.7	29
1974	16.2	10.5	6.1	38
1976	17.0	9.7	7.3	43
1979 (est.)	19	10	9	47

What the table demonstrates is a degree of dependence that is novel in American history, which means that the United States has now become vulnerable to a wide range of economic and political threats. The high cost of oil, the resulting effects on the dollar and the international payments system—and supply interruptions—can and have hurt the American economy, as well as those economies of other Western industrial nations and the non-oil developing countries. On top of that are political dangers that would result from changes in the regimes in the key oil-producing countries. Thus, American imports are not a subject that can be understood in terms of conventional economic analysis alone.

Backed by the American government, U.S. oil companies once dominated the international oil system. But as the American market became more dependent on foreign imports, American influence over the world market has diminished. How did this happen?

The U.S. government itself has never owned foreign oil operations. But at key moments it played a decisive role in making it possible for a handful of the U.S.-based companies to establish foreign operations. These companies, along with two foreign competitors, Royal Dutch/Shell and British Petroleum, dominated the world oil market for about a century.*

Around the turn of the century, the American companies, which had already been exporting oil from the United States, began to search abroad for crude oil. Early on, they found it in Mexico, Rumania, and even Japan. At the end of World War I, the U.S. industry produced two thirds of the world's oil supply from domestic sources and another sixth from Mexico. But because wartime demands had strained world reserves, many in the U.S. government were worried about the oil outlook. The major new fields in Texas had not yet been discovered, and so for a few years after World War I the United States was a net importer of oil. "The best technical authorities seem to believe that the peak of petroleum production in the United States will soon be reached," wrote U.S. Secretary of State Robert Lansing in 1919. "The U.S. position can best be characterized as precarious," stated the director of the U.S. Geological Service in 1920. Officials were afraid that the United States would not have adequate foreign supplies. But many in the government, especially in the State Department, felt that foreign oil controlled by U.S. companies was as reliable as if the foreign oil were owned by the U.S. government itself. Thus the U.S. government and the big American oil companies, especially Standard Oil of New Jersey (the largest old Standard Oil descendant, now known as Exxon), pursued a common goal—to place as many foreign sources of oil as possible in American hands.[7]

As the concern over oil supplies became commonplace, the French managed to obtain Germany's share of the oil rights in the

* The world oil market is defined so as to exclude North America and the communist countries, because these two regions have been insulated from time to time from the general market.[6]

Ottoman Empire. But the British controlled most of the oil rights in the Ottoman Empire and, indeed, most of the oil outside North America. Royal Dutch/Shell had aggressively developed production in Latin America and Asia. More importantly, British Petroleum owned the first of the giant oil fields to be discovered in the Middle East, which was found in Iran in 1908.

The interests of the government and the American oil companies were so intertwined that it was hard to tell who was leading whom. It appears, however, that in the very early days, the government's desire for an American presence in the Eastern Hemisphere, especially in the Middle East, was at least as strong as the companies'. In 1920, Congress passed a law that prohibited foreign-owned corporations from acquiring oil leases on U.S. public lands if their own governments did not allow U.S. firms to explore for oil in territories they controlled. The law, aimed at Royal Dutch/Shell, got the Dutch to open up the Netherlands' East Indies, formerly an exclusive Shell preserve, to exploration by American companies.[8]

The biggest prize of all, of course, was the Middle East. There the U.S. drive succeeded admirably. The United States pushed the British aside, prevented the French from expanding their small interests, and then defended the region (or most of it) from Russian encroachment. In a thirty-year period, from the mid-1920's to the mid-1950's, U.S. companies, starting with no base, gained a dominant position in the area that was to hold the future of the world's oil supply. It was one of the most stunning examples of economic expansion in history. The Middle East nations themselves (at the time, some were scarcely nations) were often little more than spectators to Great Power competition.[9]

The U.S. government first obtained an "open door" for American investment in the British and French zones of influence, including Iraq (part of the Ottoman Empire). The result was that Exxon and Mobil ended up owning about a quarter of the Iraq Petroleum Company. In 1934, Iraq became the second major Middle East exporter, after Iran. A pipeline was completed, and oil began to flow in large quantities.[10]

The U.S. government next persuaded the British government to allow Gulf Oil to enter Kuwait, a British protectorate, in 1934. British Petroleum and Gulf thereupon decided to become partners and so avoid competing against each other in negotiations with the Sheik of Kuwait. The U.S. government also obtained British approval for Standard Oil of California (Socal) to enter Bahrain, a small island off the coast of Saudi Arabia that was a British protectorate.[11] The discovery of oil in Bahrain in 1932 encouraged Socal to begin exploration in that vast desert just a few miles across the water—Saudi Arabia. And there it found an oil field to stagger the imagination.

In 1933, Socal easily outbid the Iraq Petroleum Company (IPC) for oil exploration rights in Saudi Arabia, paying $300,000 in gold, compared with IPC's offer of $60,000 worth of Indian rupees. A few years later, Socal sold half interest in its rights in Saudi Arabia to Texaco, thus forming the Arabian American Oil Company—Aramco.[12]

Aramco sought U.S. government aid on several occasions in order to protect its interests in Saudi Arabia. And the United States, after struggling for so many years to secure a place for its companies in the Middle East, was not about to abandon them. World War II interfered with opportunities for travel and cut into Saudi Arabia's revenues from the annual pilgrimages to Mecca. An oil company executive in 1941 warned President Roosevelt that King Ibn Saud "is desperate" and needed $6 million quickly. Washington responded with lend-lease assistance. In 1945, the forerunner to the U.S. National Security Council reaffirmed Saudi Arabia's importance, stating, "It is in our national interest to see that this vital resource [Saudi Arabian oil] remains in American hands." [13]

In 1946, the companies again needed help from Washington. When Socal and Texaco decided to sell part of Aramco to Exxon and Mobil, antitrust problems were raised. The U.S. government thereupon provided antitrust clearance, saying that the enlarged consortium was in the national interest, as it would result in faster development of the Arabian oil fields and hence more money for the King. The U.S. government also pressured the British and French governments to overlook clauses in the Iraq Petroleum Company agreement that would have prevented Exxon and Mobil from buying in.[14]

In 1950, the State Department felt that the Middle East was "highly attractive and highly vulnerable" to communism. When Ibn Saud again demanded more money, the companies faced a quandary. They did not want to accede to the State Department's suggestion that they give up rights to territories they had not yet developed and that the Saudi government could auction off. On the other hand, the companies could not raise prices to European customers, because to do so would undercut postwar recovery and the Marshall Plan. Nor did the companies want to raise royalty payments to Saudi Arabia, because this would mean lower profits. The National Security Council resolved the issue by devising a policy under which Aramco could make its payments to the Saudi Arabian government in the form of income taxes, and then deduct equivalent amounts of money from taxes owed the U.S. Treasury— a tax credit. The principle of tax crediting was common practice for other U.S. industries abroad, and oil companies operating in several other countries had already used it. Still, the U.S. State Department expressed some "concern over what in effect could amount to a subsidy of Aramco's position in Saudi Arabia by U.S. taxpayers." [15]

With Saudi Arabia stabilized by the substantially higher revenues from Aramco, the Americans, for the second time in five years, were forced to turn their attention to Iran, which was then still the largest Middle Eastern producer. In 1946, in the first crisis of the Cold War, the United States had put pressure on the Russians to evacuate northern Iran to keep them from threatening Middle East oil interests. In 1951, the Iranians, led by Prime Minister Mohammad Mosaddegh, nationalized British Petroleum's Iranian properties in an attempt to wrest effective control of their oil industry away from the British. The "seven majors," * controlling 98 percent of the world oil market, responded by instituting a boycott of Iranian oil, and they easily supplied the world markets by increasing production in the other Middle East nations—primarily Saudi Arabia, Kuwait, and Iraq. While the U.S. government made some conciliatory statements toward Iran, it sided with the majors for two reasons: First, Russian influence may have otherwise increased in Iran; and second, if Mosaddegh succeeded, other major oil-producing countries might follow suit.[16]

With a push from the CIA, Mosaddegh fell and the Shah was reinstalled. The U.S. government then took the lead in settling the dispute, encouraging the U.S. majors to join a new consortium of companies being formed to operate Iran's oil fields and market its

* The seven international majors are five U.S. companies—Exxon, Gulf, Mobil, Socal, and Texaco—and the two British-based companies, British Petroleum and Royal Dutch/Shell (60-percent-owned by the Dutch).

crude. Some majors actually needed prodding because they were afraid that a requirement to market Iranian crude would prevent them from increasing production sufficiently in Saudi Arabia to please the King. But the U.S. government wanted the majors to participate in Iran because they were the only U.S. companies considered capable of moving large-enough quantities of Iranian crude to provide the income deemed necessary for the new Iranian government. To avoid antitrust problems, the U.S. majors asked for and received clearance from the Justice Department.[17] The final ownership in the consortium was apportioned as follows: American companies, 40 percent; British (BP and Royal/Dutch Shell), 54 percent; and French (Compagnie Française de Pétroles), 6 percent. The Iranians resented the settlement because foreigners still dominated their oil industry, the main difference being that American influence had diluted that of the British. Although this was an improvement, in Iranian eyes, the resentment was to flare openly a quarter century later when the Iranians once again made a major effort to wash away the vestiges of foreign control.

The Iranian settlement of 1954 produced two changes in the ground rules, which also later helped to undermine the American position throughout the oil-producing world. The first was that Iran became the unquestioned proprietor of its own oil fields, then something novel for an oil-exporting country, but within a few years the norm. Second, the U.S. government persuaded the majors to allow some American "independents" to own a small share of the consortium. (Some of the independents, of course, were large companies, such as Standard Oil of Ohio, Getty Oil Company, and Atlantic Refining.) Howard Page, the Exxon executive whose persistence was crucial in forging the agreement, recalled later that the majors agreed because "people were always yacking about it." So, he explained, some independents were allowed in as "window dressing."[18] The independents soon learned that Iranian oil could be very profitable, although their production level was quite limited under the terms of the agreement. Thus their Iranian experience encouraged them to explore elsewhere overseas, and their subsequent growth made certain that the majors would never again be strong enough to organize an effective boycott like the one against Mosaddegh.

The formation of the Iranian consortium represented the zenith of American oil power in the world. In 1955, the five U.S. majors produced two thirds of the oil for the world oil market, the two British majors almost one third. The French had a mere 2 percent.

In the late 1950's, the competition to sell crude to refineries owned by independents caused oil prices to decline in the world oil market. In the meantime, the "posted price"—that is, the official price used to determine host-government revenues—remained constant. This started the chain of events that led to the birth of the Organization of Petroleum Exporting Countries (OPEC). Although the OPEC story started in the marketplace, it moved to the Exxon boardroom. After a 1959 cut in posted prices, Exxon chairman M. J. "Jack" Rathbone persuaded his fellow directors to cut prices again in August 1960, thereby once more reducing the revenues received by the exporting nations. As was the habit in those days, Exxon did not negotiate with the oil-exporting nations; it simply announced the price cuts. The company said that the lower prices were necessary because sales were occurring at sizable discounts. It added, "The pressure on market prices has also been accentuated by new and widespread offers of Russian oil at low prices." [19]

The producing countries were outraged. The Petroleum Affairs director of the Arab League scoffed at Exxon's explanation that Soviet oil was responsible. Venezuelan Oil Minister Perez Alfonzo blamed the cuts on "pressures" from the world's big consuming countries, which he proceeded to identify: "The United Kingdom is the greatest consumer of Western Europe the same as the United States is the biggest consumer of imported oil in the world." [20]

Outside the oil industry, the price cuts scarcely made a ripple. The entire *New York Times* report was contained in seven sentences on page 41.[21]

The helpless oil-exporting nations had to accept the cuts, and about all they could do was meet among themselves. In September 1960, Iraq invited the five other major oil-exporting nations to a conference in Baghdad. It was at this meeting that OPEC was formed. Years later, eloquent words were to be written about OPEC's beginning: "United in OPEC, they set out to redress the imbalance between cheap oil and imports, and also, in the psychological sense, to redress centuries of colonialism and exploitation." [22] At the time, however, the founding of OPEC was almost a nonevent. Although it was formed on September 9, 1960, the story was not reported in the *New York Times* until September 25. Here is the entire story:

MIDEAST OIL LANDS SEEK PRICE STABILITY

BAGHDAD, Iraq, Sept. 24—(Reuters)—A five-nation oil conference held here earlier this month voted to demand that oil companies try to restore prices to their former level and keep them steady, it was announced here today.

The meeting, held after the major companies cut Middle Eastern crude oil prices, also voted to form a permanent Organization of Petroleum Exporting Countries to unify their oil policies and promote their individual and collective interests.

Iraq, Iran, Saudi Arabia, Venezuela and the Persian Gulf sheikhdom of Kuwait sent delegates to the conference. The sheikhdom of Aqtar [sic] and the Arab League sent observers.

These states hold 90 per cent of the world's oil reserves, according to the chief Saudi Arabian delegate.

Faud Rouhani, OPEC's first Secretary-General, later said that the oil companies initially pretended that "OPEC did not exist." Of course, OPEC eventually gained considerable recognition. Initially it functioned primarily as a clearinghouse for information and as a forum for representatives of member nations. Perez Alfonzo, one of the founding leaders of OPEC, tried to shape the organization so that it would allocate production; but in fact this has not yet been done.[23]

The process set in motion by the Iranian settlement and the establishment of OPEC converged in Libya. In 1956, Libya had granted fifty-one concessions to seventeen companies, including many independents who formed their own groups and whose output was not controlled by the majors. In the early 1960's, the independents began to export large quantities of oil into large European markets, which heretofore the majors had controlled. During the rest of the 1960's, OPEC was still unable to prevent the continuing decline in the price of world oil, which reached a low of from $1.00 to $1.20 a barrel in the Persian Gulf by the end of 1969. The companies and OPEC haggled endlessly over a few pennies per barrel. But by 1970 the situation had changed. Not only had U.S. production reached its limit, but also a number of oil-exporting nations, such as Libya, Algeria, and Kuwait, were approaching production peaks. As a result, Saudi Arabia and Iran were the only nations in OPEC with the ability to expand production substantially.[24]

In May 1970, an accident triggered an historic sequence of events. A bulldozer broke a pipeline carrying oil from the Persian Gulf to a Mediterranean seaport. This pipeline was especially important because the 1967 Arab-Israeli war had closed the tanker route through the Suez Canal. Libya, just across the Mediterranean from Europe, was suddenly in a key position. Colonel Muammer el-Qaddafi, Libya's new dictator, exploited the opportunity by ruthlessly squeezing just the right company—the American-based independent, Occidental—which had no substantial source of crude elsewhere, and thus was almost solely dependent on its Libyan output. Occidental capitulated quickly. At the conclusion of the negotiation, an Occidental executive uttered a few prophetic words: "Everybody who drives a tractor, truck, or car in the Western world will be affected by this." Other companies quickly began to fall.[25]

The overall capitulation of the West occurred in February 1971 at a conference in Tehran. The adroit maneuvering of the OPEC nations, led by the Shah and aided by the vacillation of the American government, gave OPEC their first clear-cut victory over the West. They obtained what was considered at the time to be a large price increase: fifty cents a barrel. America's reign over world oil production was swiftly coming to an end. At that very time began the vast increase in U.S. oil imports.[26]

By the summer of 1973, the stage was set, and center stage was occupied by one country—Saudi Arabia. The demand for petroleum was surging throughout the industrial world. Ordinarily Aramco maintained capacity about 10 percent higher than its actual production requirements, to deal with unforeseen contingencies; another 10 percent was usually available in the summer, at which time demand slackened. These cushions had been adequate in the past. But by the summer of 1973, the only spare capacity in the world belonged to Aramco, and that was equal to just 3 percent of total world production. But world demand was so much greater than expected that even the extra capacity was called into use. The result—no cushion at all. It was just at this time that the United States began to integrate its own voracious need with that of the world market by lifting oil-import quotas—so adding a powerful new demand on Eastern Hemisphere oil. The world oil market was a very tight one indeed by mid-1973.[27]

Simultaneously, the politics of the Middle East were becoming very unstable. In May 1973, President Sadat warned the late King Faisal of Saudi Arabia that Egypt might soon attempt to retake the Arab lands occupied by Israel. At a meeting held in Geneva on May 23, 1973, King Faisal spoke of Sadat's plans to four American oil executives, each representing one of Aramco's parents. Faisal, fearful of the appeal of Arab radicals, warned them that he would not let Saudi Arabia become "more isolated in the Arab world." Unless the United States gave more support to the Arab cause, the King said, American interests in the area "will be lost." [28] The warning, carried to the Nixon Administration by a convoy of alarmed oil executives, was ignored. The Nixon Administration was more concerned at the time with Watergate.

On October 6, 1973, Egypt attacked the Israeli forces along the Suez Canal. On October 12, the very epitome of the American Establishment, attorney John J. McCloy, sent the White House a personal note from the heads of the four Aramco parents, making two basic points: (1) The Arabs would cut back oil production if the United States increased its support to Israel, and (2) such a cutback would have dire consequences on the European and Japanese economies. On October 17, the Arab oil ministers met in Kuwait. They agreed to cut exports by 5 percent and recommended an embargo against unfriendly nations. On October 19, after learning of Nixon's decision to provide $2.5 billion of arms to Israel, King Faisal ordered a 25 percent reduction in Saudi oil output and an embargo against the United States and several other nations. Aramco obeyed immediately. To do otherwise, stated one American oil executive, "would have resulted in even greater cutbacks, and some oil was better than no oil." Most other Arab nations quickly followed the Saudi lead.[29]

The cutbacks dictated by the Arabs were less than 10 percent of the world oil supply, but they caused widespread panic, and some governments pressed the oil companies to deliver extra oil to them at the expense of their allies. According to a British civil servant, Britain's Prime Minster Edward Heath had a "temper tantrum" when British Petroleum, 48-percent owned by the British government, refused to comply. BP had contractual obligations that prevented them from doing so.[30] And some refiners who lacked an assured supply of crude oil were forced to bid for it at auctions run by the various OPEC countries. In Iran on December 14, a barrel went for $17.34. On December 23 in Nigeria, the price of a barrel reached $22.60. As one refiner explained, "We weren't bidding just for oil; we were bidding for our life." With bids of this magnitude reinforcing the Shah's determination to raise prices substantially, representatives of the OPEC nations met in Tehran on December 22 and 23. They announced a price hike. Effective January 1, 1974, OPEC's take was to be $7.00 a barrel, compared with $1.77 before the October War. As oil production in Arab countries began to creep upward, it became clear that the real crisis was one of price, not supply.[31]

The embargo ended on March 18, 1974, six months after it had begun. But the damage had been done: The oil-producing countries had seized control of the world's basic energy source, and during 1974, Saudi Arabia showed what that control meant. Encouraged by demand from independent refineries for its crude, Saudi Arabia took the lead in raising the OPEC take to about $10 a barrel for the entire year—even though Saudi Arabia was usually considered an advocate of price moderation.

Between 1974 and 1978, the world oil market was quite orderly. Demand for OPEC oil dipped temporarily because of slow economic growth, a build-up of production in the North Sea and Alaska, and conservation. Thus, OPEC production in 1978 was only slightly higher than in 1974. From time to time, the oil cartel raised prices, but not enough to keep pace with inflation.

As 1978 began, more and more people in the United States and other Western countries were becoming complacent about the world of imported oil and optimistic about its future prospects. Increasing publicity was given to those analysts predicting that a "glut" of oil would force prices to continue to decline in real terms, perhaps through the 1990's or even to the end of the century. The United States was coming to rely on Iran as a stable ally, as the regional power that would maintain order around the Persian Gulf. This close relationship was expressed in the extensive and expensive transfer of highly sophisticated military technology from the United States to Iran. It was also expressed in the toast that President Carter raised to the Shah on New Year's Eve, December 31, 1977. "Iran under the great leadership of the Shah is an island of stability in one of the more troubled areas of the world," said the President. "This is a great tribute to you, Your Majesty, and to your leadership, and to the respect, admiration, and love which your people give to you." But each passing month in 1978 demonstrated more clearly that the people's love was minimal. Domestic unrest and protests increased, eventually encompassing almost every element of society. The Shah's position weakened, although as late as September 1978 few doubted his ability to weather the storm. Within months, of course, the scale of domestic protest forced the Shah into exile. The oil workers played a key role in his downfall when they cut off oil exports late in 1978. This loss of five million barrels per day on the export market turned a small surplus of world production into a shortage.

In December 1978, OPEC announced price hikes higher than generally expected—5 percent, effective January 1, 1979, with further increases scheduled to make the total 1979 price rise equal to 14.5 percent (or 10 percent for the yearly average increase). But with five million barrels daily of Iranian oil withdrawn from the world market, the consumers were drawing upon oil stocks at a rate that was two million barrels a day greater than normal. That meant that three million barrels daily of missing Iranian production was being made up by increased production elsewhere, notably in Saudi Arabia. But it also meant that the world was two million barrels a day short. The market was suddenly very tight. The price for crude oil not already covered by long-term contracts jumped in early 1979 by as much as $8.00 a barrel above the OPEC price of $13.34 for "marker" crude. The initial spot-market sales were made by oil brokers, but as the price soared OPEC members began to increase their prices for spot-market oil. Sometimes they canceled a long-term contract by invoking the *force majeure* clause, and then immediately offered an identical quantity of identical quality oil to the same customer.

It was obvious that OPEC members would not allow for long the oil companies to reap easy profits on contracted oil bought at prices substantially lower than those of the spot market. Different OPEC members quickly began adding surcharges on all their oil, even if covered by long-term contract.

After the new government took over in Iran, production sufficient for the needs of the domestic market was resumed. After a few weeks, production rose above two million barrels a day, putting Iran once again into the export market. As Iranian oil began to come back onto the market in early March, other OPEC countries cut back on their production to keep the market tight and the spot price high. Few OPEC countries could resist the temptation—and the internal pressures—to capture some of the advantages of the spot-market price increases in the official OPEC price. On March 26, 1979, the OPEC ministers met in Geneva and announced a general increase of several dollars a barrel above the price schedule set just three months earlier. The increase in spot-market prices and the OPEC official prices would have been greater if Saudi Arabia had not allowed Aramco to increase production beyond the level previously scheduled. Still, this meant that on an annual basis, $40 billion was added to the Western world's oil bill.

Iran reentered the world oil market in a spirit quite different from that under the Shah. The religious leader, the Ayatollah Khomeini, as well as the new civilian government, made clear that production would never be allowed to return to its old levels. Better to sell three million barrels per day at a higher price than five million at a lower price. Also, the consortium—along with westernization—was to be banished from Iran. At last, thought many Iranians, they had rid the nation of foreign domination of the oil industry for the first time since its beginning over half a century earlier. Indicative of Iran's new policy, the first tanker of oil was sold to a Japanese company, which did not hesitate to turn around and sell at a higher price to an affiliate of Royal Dutch/Shell, a consortium member.

The cut-off of Iranian exports in the winter 1978–79 focused attention on the short-term market impact. But the long-term implications of the Iranian revolution are no less important. The cut-off revealed that the margin of surplus in the world oil market was more narrow than most experts thought. The "glut"—and talk of the glut—disappeared almost overnight. Moreover, the upheaval in Iran has meant a permanent reduction in the availability of world oil. For not only are Iranian exports unlikely to reach prerevolution levels, but other key producers, particularly Saudi Arabia and Mexico, fearing strains of too-rapid economic development, became more cautious about expanding oil output. In addition, it was realized that tensions other than those involving Arab-Israeli relations could result in supply interruptions. The important role that control of the oil fields played in the Shah's downfall meant that the oil fields in other countries would become, even more so than before, a prime target of political dissidents. A revival of Islamic fundamentalism, coexisting with radicalism, will ensure a rich supply of political dissidents in Iraq, the Gulf sheikdoms, even Saudi Arabia, the mother lode of oil on which the Western world depends for political moderation and increased supplies. Large volumes of oil exports were increasingly seen as a sign of political subservience to the West. As the Sheik of Kuwait said, responding to Queen Elizabeth's compliments on his lavish welfare and development spending: "But, Your Majesty, this kind of spending wins me fewer friends these days." [32]

SAUDI ARABIA: THE NEW CONTROLLER

The member nations of OPEC, accounting for over 90 percent of the world exports of oil, by and large have taken over ownership of the oil concessions within their territories and continue to set prices unilaterally. Meanwhile, the international oil companies have become a combination of contractor and sales agent for OPEC members. But OPEC as an organization still has no mechanism to regulate output in order to control price; it has continued to be primarily a gatherer, analyst, and disseminator of information, and a forum for members, who meet to agree on price.

One country dominates OPEC—Saudi Arabia. It is favored by a unique conjunction of huge reserves, extraordinary ease of exploitation, and a population so tiny (5 or 6 million people) that domestic revenue needs at current prices have no practical effect on the level of oil production. Saudi Arabia is the largest producer in OPEC, with 30 percent of total production and 34 percent of total reserves. In fact, more reserves have been found each year in Saudi Arabia than oil produced during that year. Thus, estimates of oil reserves have risen dramatically from an estimated 5 billion barrels during World War II to more than 150 billion today. And during the four years following the oil-price explosion, Saudi Arabia's financial holdings have grown from a few billion dollars to well over $50 billion, exceeding those of such rich nations as West Germany and Japan. The pricing system for OPEC oil itself shows the importance of Saudi Arabia, for all other OPEC oil is priced in relation to Saudi "marker crude." As one U.S. oil executive summed it up in 1978, "OPEC is a front for Saudi Arabia, which makes the decisions alone, but doesn't want to take the full responsibility for those decisions." Thus, Saudi Oil Minister Yamani could say offhandedly at the conclusion of the 1977 Caracas OPEC meeting, "Saudi Arabia is taking decisions to regulate the market." [33]

Producing nations outside OPEC are not likely to undercut Saudi dominance.[34] The Soviet Union, the world's largest producer (11 million barrels daily), exports about 2.5 million barrels daily, but most goes only to Eastern Europe. Thus, the Soviet Union has little impact on the world market. But Russia is likely to become a smaller exporter, or possibly even a net importer, by 1985. Britain and Norway's combined North Sea production, which was about 1.5 million barrels daily by the end of 1978, is unlikely to rise above 4 million during the 1980's.

Over the next decade, Mexico and possibly the People's Republic of China represent the only likely sources of significant new supplies for the world oil market outside of OPEC.

Mexico, of course, has been an oil producer since the early part of this century, and by the end of 1978 its production had climbed to 1.5 million barrels daily, one third of which was exported. Quite a bit of optimism has been generated in the United States for Mexican oil as estimates of reserves have reached astronomical proportions. Three weeks after taking office at the end of 1976, President Lopez Portillo announced that proven oil reserves were not 6 billion barrels as previously supposed, but 11 billion. By the beginning of 1979 the figure had risen to 40 billion barrels (of which 12 billion represented the oil equivalent of natural gas, although this fact did not appear in some press releases). "Probable" reserves were put at an additional 45 billion, and potential reserves at 200 billion. (Both estimates include about one-third gas.) For comparison, Saudi Arabian proven reserves are some 166 billion barrels of oil.[35]

The euphoria in the United States is not justified. True, the Mexican economy will benefit handsomely, but the Mexican oil is quite unlikely to make any important change in the world oil picture. Mexican exports will be constrained by a ballooning domestic oil consumption brought on by a doubling of population by the end of the century—perhaps to over 125 million. Only a relatively small reserve base has yet been proven, and developing major new oil fields requires considerable time. Bureaucratic and political problems associated with the Mexican oil industry will further slow the development.

Furthermore, the current official policy of Mexico is to limit exports to the pace required by domestic development rather than to the size of its oil reserves or the pressure of foreign demand. "The capacity for monetary digestion," explained President José Lopez Portillo, "is like that of the human body. You can't eat more than you can digest or else you become ill. It's the same with the economy." Mexican leaders want to hold a tight rein on development, because they are concerned that too much oil money too quickly will result in so much inflation that Mexico's other exports and its tourism will be priced out of world markets. If this happened, the country's critical unemployment problem would be made worse. Mexican leaders further reject the alternatives of piling up dollars in bank accounts and other investments outside of Mexico, prepaying Mexico's foreign debt, or paying for massive imports of consumer goods. Thus, production is unlikely to be higher than 3.9 million barrels daily in 1985, with exports of about 2.6 million—less than 4 percent of expected world production. Also, one should remember that even reserves of 100 billion barrels represent less than five years of the world's consumption.[36]

The oil output of the People's Republic of China has been growing rapidly for the past decade, but so has consumption, leaving less than a quarter million barrels daily for the export market. By the mid-1980's China's exports might be one or two million barrels daily, but hardly large enough to challenge Saudi Arabia's dominance. Moreover, all these new suppliers are likely to be near their capacity, thus providing no cushion for sudden surges in demand or cutbacks by Arab producers. Nothing on the horizon suggests that the centrality of Saudi Arabia in the international petroleum market will be challenged in the foreseeable future.[37]

On what kind of country has the world become so dependent?

Saudi Arabia is a family business; not an ordinary family business, but one with 10,000 members, for whom the penalty for attempting to marry without family approval can mean death. In 1901, Ibn Saud, penniless and exiled in Kuwait, set out with some forty men on camels to recapture his family's traditional homeland from other Arabs and the Turks. The stuff of high adventure and epic history, his struggle succeeded in 1932, with the unification of his conquests into the Kingdom of Saudi Arabia.[38] King Ibn Saud was succeeded as king by his sons—Saud in 1952, Faisal in 1964, and after Faisal's assassination, Khalid in 1975.

By 1979, the wealth controlled by the Saud family made the fortunes of legendary Western "oil barons," and even the monetary reserves of the richest industrial nations, look puny by comparison. Indeed, it is reasonable to assume that the Saudi royal family is so wealthy that its primary objective is not maximizing the wealth of the nation it governs, but rather assuring its own survival as the ruling family.[39] Yet there is so much complacency in the United States about the stability of the present regime of Saudi Arabia that until the Iranian crisis, one was not aware of threats to its existence. But threats certainly exist, and an understanding of them is necessary for an analysis of the energy problem.

Like all families, the Sauds do not always agree amongst themselves. In 1977, as the press reported signs of sharp family disagreement, the 250 senior princes met in the capital, Riyadh, to discuss the line of succession. There was no dispute that Crown Prince Fahd would replace King Khalid (then sixty-four) when he stepped down or died. The ailing Khalid had little desire or training for the monarch's job; he had been raised as a bedouin and still preferred the desert life to Riyadh. But the princes could not agree on Fahd's successor. Fahd himself proposed one of his full brothers, which would have meant passing over the prince next in line—Prince Abdullah, who is Fahd's half brother, head of the National Guard, and thought to be less friendly to the West than is Fahd. Abdullah objected and had enough family support to block a decision at the time.[40] The question apparently remains unresolved.

Succession crises are not the only internal threat to the family, as the recent histories of neighboring countries suggest. In 1977 and 1978, political assassinations occurred in Syria, South Yemen, and North Yemen. And in the last quarter-century, dissidents have overthrown ruling groups in Iran, South Yemen, Iraq, Syria, Egypt, and Libya—albeit some of the rulers were kings put in by colonial powers, and none had the family base of the Sauds. The press has reported some attempted coups in Saudi Arabia, the most publicized being one in 1969 by military officers. Any new regime in Saudi Arabia would quite likely have political and economic goals different from those of the present rulers.[41]

Too little is known about internal relations in Saudi Arabia to make any more solid predictions. But experienced "Saudi watchers," men with years of exposure to different strata of Saudi society, including the highest level of the royal family, are increasingly worried about the corrosive influence of instant wealth. Here are some of their comments: [42] "Five years ago [1973], no one heard words against the government; now one hears, 'This government is intolerable and has to go' "; "I judge the government's chance of survival for a half-dozen years to be quite good, and for a dozen years, fairly good. *But there could be a successful revolution this evening.*"

One of the major external dangers to the Saudi Arabian leaders lies in a possible coalition of radical Arab states supported by Colonel el-Qaddafi, President of Libya. Qaddafi is one of the world's most unusual leaders, to put it mildly. Intensely religious, he has been known to go into the desert alone for a month to meditate. Quite erratic and believed to be behind plots to assassinate some other Arab rulers, he has been called a "nut" by Sadat and "crazy" by the Shah of Iran. But he has received substantial military aid from Russia and Cuba.[43]

A neighbor of Saudi Arabia is South Yemen, which with its Marxist government constitutes a major worry to the Saudis. The Soviets and Cubans have been using the country as a staging area for the fighting in the Horn of Africa. South Yemen's artillery battalions, with Soviet officers, helped Ethiopia in its war against Somalia, gaining experience that the Saudis fear could be put to use against the small oil sheikdoms on the Arabian peninsula. Furthermore, South Yemen has been a sanctuary for the Palestinian terrorists believed to be responsible for the 1977–78 assassinations of three North Yemen leaders—two Presidents and a Prime Minister. And the Saudi leaders fear a Palestinian assassination attempt. In early 1979, the rivalry between the Yemens erupted in military conflict, sending shock waves through the entire Arabian Peninsula and drawing direct American involvement.[44]

Iraq, contiguous to Saudi Arabia in the northwest, also possesses Soviet military presence. The radical Iraqi regime has been consistently at odds with the Saudis, with basic ideological differences separating radical Arabs from conservative Saudi Arabia. No matter what the outcome of the Arab-Israeli conflict, Arab radicals would be opposed to rule by a royal family and would want to see the Saudi regime replaced.[45]

Iran's relationship with Saudi Arabia has traditionally been more important to the Saudis than any other except that with the United States. The Iranians have weighed heavily on the minds of the Saud family in two crucial matters: One is oil policy, the other, survival.

The nation with a larger population (35 million) and smaller oil reserves than Saudi Arabia might have a good reason to seek to escalate oil prices.

Although fear of the Soviet Union has created a strong bond between Saudi Arabia and Iran, whose territory has been coveted by the Russians since the last century, the two countries have sharply disagreed on oil-pricing policy, and for centuries the Arabs and the Iranians have been traditionally suspicious of one another. Iran's current internal condition, of course, means that it does not pose much of a direct military threat at the moment; it does, however, pose a danger that the turmoil will spread to Saudi Arabia. A short time ago, Iran did represent a military threat. Kamal Adham, an advisor to King Faisal, said in 1974 that the Saudis knew that the purpose of the Iranian arms build-up was not to fight the Russians, but that it could only be for one purpose—"control of the Gulf." [46] The threat perceived in the Shah's ambitious dreams and his military might has disappeared, but a new, perhaps even greater danger may now exist in the dreams of the Shiite fundamentalists—and in the presence of a large Shiite population in the area of the Saudi oil fields.

In fact, control of the Gulf and its narrow mouth, the Straits ot Hormuz, are critically important to the noncommunist nations. An almost endless chain of tankers passes through these narrow waters, carrying half of OPEC's oil exports—one quarter of the world's total supply. In early 1978 there were reports that Iraq was training frogmen and desert demolition squads, and that Palestinian terrorists had been found on tankers. It is no wonder the United States regards the Arab/Persian Gulf as a security problem second only to that posed by the Soviet Union. In the words of a recent Department of Defense report, "The survival of NATO is as likely to be decided in the Middle East/Persian Gulf as on the plains of Central Europe." [47] The massive dependence of the industrial world on one fragile regime is a frightening reality of modern life.

What role does Saudi Arabia play in the analyses of the much-discussed projections of an oil "shortage" or "gap" of the 1980's? The terms denote not absolute physical shortage, but rather that, at relatively stable oil prices, the world's demand for petroleum exceeds the supply of oil that producers could or would be willing to produce. The consequence of this demand would be another round of price hikes beyond those of 1973–74 and 1978–79.

But there has been much controversy and confusion over these projections. What is usually omitted from reports about the projections is the way in which their conclusions are affected by assumptions made about a number of very uncertain things. Responsible forecasters can make very different assumptions with very different outcomes. The key uncertainties include: [48]

1. Rate of economic growth in the United States and elsewhere in the world
2. Energy usage per unit of economic output (in other words, amount of conservation)
3. Oil production in Saudi Arabia
4. Oil production in other OPEC nations
5. Oil production in non-OPEC nations
6. Contributions from other "conventional" energy sources—natural gas, coal, and nuclear
7. Contributions by nontraditional sources, such as shale oil, solar, tar sands, geothermal, wave power, and so on

Different assumptions can produce enormously different forecasts. For instance, by making different assumptions on two items—rate of economic growth and energy usage per unit of economic output—specialists have forecast world energy consumption to increase by as little as 60 percent by the year 2000 or by as much as 230 percent.[49]

All forecasts should thus be offered with some modesty. As things stand, it seems reasonable to believe that at current or slightly higher prices, demand for OPEC oil (including consumption within OPEC countries of 4 million barrels daily) could be between 34 and 44 million barrels a day in the 1980's. But in the late 1980's, the capacity of OPEC nations will probably not exceed 44 million barrels per day, assuming that Saudi capacity is 19 million barrels a day. Thus we come to what is perhaps the controlling uncertainty—Saudi Arabian production.

But how much oil will Saudi Arabia choose to make available? It is an open question. As of 1979, the Saudis have the physical capability to produce 11 million barrels daily, and plans have been announced to increase that capacity to 14 million barrels daily during the 1980's. Some engineers believe that it may be physically possible for Saudi Arabia to increase its output by 1 million barrels daily every year up to a level of 16 to 20 million barrels daily, and then to sustain such a rate until the end of the century.[50]

The royal family currently favors an increase in capacity, knowing it will help keep prices relatively stable. These Saudis are concerned about the condition of the economics of the industrial nations, which are both their chief customers and outlets for their savings. The strongly anticommunist Saudis fear that economic stagnation will increase leftist influence in Europe. Further, the rulers want to maintain good political relations with the United States, to which they look as an ally, an arms supplier, a source of technical advice, and a force to help settle the Arab-Israeli dispute.[51]

Yet even if the Saudis want to keep prices at a moderate level, there is a limit to how much they can increase production in order to do so. Hence a policy of high production and price moderation in the nearer term would still mean higher prices at some later date, when Saudi Arabia is closer to its production limit.

But the higher prices could come sooner rather than later. There is a growing pressure within the royal family not to increase oil exports much above present levels.[52] Proponents of this view reason as follows: (1) Saudi Arabia should distance itself from the West and become more friendly with Iraq and other radical Arab elements. (2) Waste and corruption can be better controlled if industrial development is scaled down. (3) The North Yemen labor pool is nearly exhausted, and an increase in migrant labor from other countries could bring about increased political instability. (4) A valuable commodity should not be depleted quickly in order to build up large savings abroad, when the real return on these investments is negative after correcting for inflation. (5) Even after the completion of a massive natural gas project, any oil production in excess of 5 million barrels a day will result in the flaring of natural gas.[53]

The sentiment against high production levels is stronger among some of the younger members of the royal family. They are afraid that the older generation will deplete the oil and gas, spend much of the money, and leave them and their own children with bank accounts shrunk from inflation.[54]

If the present Saudi government were overthrown, the new regime would probably be more radical, less concerned about the effects of its pricing policies on the West, and thus more likely to increase the cost of its oil as rapidly as it could. In any case, the United States would be unwise to base its energy policy on the premise that the Saudis will increase production substantially, thereby moderating any price increases.

We must note, however, that some other observers hold an entirely different view as to how much oil the Saudis will produce. This other interpretation eschews politics in favor of a strictly economic argument. The premise is, "They will produce as much as will maximize their revenues," even if this means producing 20 million barrels daily. This view, of course, is strongly challenged by those who consider other factors, political as well as technical. "I would characterize that as science fiction," said one Department of Energy official. But if the profit-maximizing view is correct, a substantial price rise—perhaps even a doubling—is a reasonable possibility over the next ten years or so; current prices seem to be far below the level that would maximize revenues.[55]

Thus, whatever view one holds about the motivations of Saudi decision-makers, the result seems to be the same—substantially higher oil prices, though no one can now accurately predict how much or how fast. If there are no major "accidents" or conflicts, a likely scenario is that after the 1979 price increases, oil prices will remain essentially stable in real terms for a few years. At some point, demand will begin to run up against even Saudi capacity, and with that the Saudis will lose control of the international petroleum market. Prices will then move up to substantially higher levels— perhaps 50 percent or more above current real prices.[56]

The rate of increase could be gradual. But a number of the past price increases have been triggered by accidents: the 1956 Suez War; the combination of the Suez Canal closure resulting from the 1967 war, the underestimation of European demand by the majors in 1971, and the break of the Trans-Arabian Pipeline by a Syrian bulldozer; the 1973 war, the cutbacks, and the subsequent high prices bid.[57] Who would have thought that an elderly, irate cleric, living in a suburb of Paris, communicating by cassettes with his followers, could have brought Iranian oil production to a standstill? So it is quite possible that some quirk of history will again intervene—an Arab cutback of production in case of another Arab-Israeli war, tankers sunk in the Straits of Hormuz, or a revolution within Saudi Arabia that damages the oil fields, for example. The price rise could then be quite sudden as buyers bid for oil, but it is hoped that the immediate shock could be moderated by the mechanism of the International Energy Agency (IEA), which has developed a plan for allocating oil in case of an emergency. The United States and nineteen other consuming countries are members of the agency, which was established in 1974 to coordinate the energy policies of the noncommunist industrial nations. Along with other IEA members, the United States has embarked on a program to maintain a strategic storage of oil equal to about four months' supply of U.S. imports.[58]

The experience of 1973–74 showed us what we might expect from higher real oil prices. Increased prices meant, first, that the United States had to do more work to buy a given quantity of oil. Second, the payments to foreign governments contracted domestic demand for goods and services, and in effect became an excise tax, a "drag" on the economies of the importing countries. Industrial nations deflated their economies in the face of oil-induced inflation and balance-of-payments deficits, which further reduced growth and further increased unemployment. Finally, there were massive strains on the international monetary system. Some analysts believe that the 1973–74 price hikes will eventually cost industrial nations over half a trillion dollars in lost economic growth.[59]

We might see even greater losses in the 1980's. But even before reaching that point, even with the present "moderate" oil prices, there is an unpleasant reality with an immediate threat looming behind it. The United States, by its current actions, could be leading the world to a major recession, if not a depression. U.S. oil imports are a major contributor to the continuing large deficit in the U.S. balance of payments and to a decline in the value of the dollar in relation to a number of foreign currencies. This obviously strains the international monetary system, creating great uncertainty. But the devalued dollar also gives American-made goods an advantage in foreign markets, which could easily encourage foreign nations to protect their industries with increased tariffs or quotas. That in turn would lead U.S. industries to seek the same. The result is a potential epidemic of folly: a return to the beggar-thy-neighbor policies of the 1930's, with lower volumes of world trade and a world recession.[60] Thus, whether or not U.S. oil imports cause higher world oil prices, they could easily cause losses in economic output.

But higher oil prices in the future would certainly accentuate current trends that are undesirable. High unemployment, inflation, the efforts by various groups to maintain their relative and absolute economic positions—all create internal political problems. These problems in turn increase the likelihood of political conflict among nations. Since 1973, serious clashes have developed among Western allies over oil and nuclear power, which show that energy questions —involving dependence and vulnerability, employment and growth, the stability of governments—have taken on extraordinary sensitivity. In fact, they have become security issues of paramount importance. Thus, higher oil prices in the 1980's—following a decade of tension and suspicion caused by the 1973 price increases—could severely strain relations among the Western nations.[61]

About the most optimistic current forecast that can be made is that the world will drift along with a little slower economic growth and a little higher inflation than usual. Indeed, some of the optimists—those saying not to worry about substantially higher prices in the 1980's—are basing their outlook on continued low economic growth.[62] Hardly a satisfactory solution, for this would extract a high price in the form of unemployment, diminished opportunity, and the inevitable political conflicts.

U.S. DOMESTIC OIL POLICY

If oil imports create so many problems, the obvious thing to do is to lower them, or at least to keep them from growing. Can this be done through a domestic American oil policy? Can the decline in domestic production be reversed and the supply of domestic oil increased? In all the arguments and debates, the charges and countercharges, of the last six years, three different domestic oil "solutions" can be identified. The oil companies generally propound two of the solutions. The first is to deregulate oil prices and speedily grant offshore oil licenses. The second is to produce more oil by unconventional means, such as enhanced recovery and shale oil. Opponents of the oil industry offer the third solution—break up the big companies. Yet none of these provides us with a real solution, for none will increase U.S. oil output by any significant amount over its current level of about 10 million barrels daily. This reality has been obscured in the fight over the $400 billion suddenly created when the OPEC price hikes dramatically increased the worth of proven U.S. oil reserves.[63] Although the various domestic "solutions" involve interesting and important issues, none is likely to result in a higher level of U.S. oil production than now exists.

The elimination of price controls, it is said, would help in two ways. First, the industry claims that controlled prices have reduced oil exploration to below what it otherwise would have been, although it is impossible to tell by how much. An analysis of oil company budgets, however, shows that exploration and drilling have increased substantially, which is not surprising, for the rewards for newly found oil in the United States are still substantial. But in spite of dramatically higher rewards since 1973 for finding oil, U.S. proven reserves have continued to decline. Second, the oil industry argues that price controls have been created and administered in a way to generate great uncertainty. This assertion is certainly true. The whole jerry-built structure—"old" oil, "new" oil, "new new" oil, and "stripper" oil, altogether some seventeen categories in recent legislation—has fostered a chaotic, confusing, and expensive bureaucratic monster. But it is not evident that either price controls or the way they are administered have substantially held back exploration or production.[64]

Even if price controls are completely eliminated, as announced by President Carter in April 1979, production of oil from existing fields would likely decline from the existing level of 10 million barrels daily. *In the late 1980's, only 5 million barrels a day are likely to come from reserves that were known to exist in 1978, including Alaska. It is unlikely that U.S. production in the late 1980's will include more than 4 million barrels daily of oil from new fields found between 1978 and the late 1980's. And even this is a high estimate.* The prospects of finding a big field onshore or offshore in the Gulf of Mexico are quite small because these territories have already been intensively searched. Over 2 million wells have been drilled in the United States—four times as many as in all the rest of the noncommunist world combined.

New territories, especially the outer continental shelf and the Alaskan North Slope, are the main hopes for finding a big field.[65] As one oil company executive put it, "Sure, higher prices will help, but a bigger factor is access to new acreage. Even a price of a hundred dollars a barrel won't give you any oil unless you have someplace to drill." [66] Environmental concerns have slowed and will slow the development of the new territories. And the drilling off the East Coast that has been done so far has been disappointing. But even if a major new field were found and environmental constraints did not exist, finding and developing a major new field in such inhospitable areas could take the better part of a decade.

What about the second proposed domestic solution—new sources from unconventional means? A large quantity of oil—more than half—is left behind in conventional recovery methods. For years, oil companies have been developing increasingly sophisticated methods for extracting a portion of this oil, and not surprisingly, the OPEC price hikes boosted interest in the effort. The newer methods, the so-called enhanced recovery, ordinarily involve the injection of a chemical compound or heat into an oil field. The amount of additional oil that can be obtained depends on the tightness with which the remaining oil sticks to the oil-bearing sands and on the porosity of the sands. Unfortunately, years are needed to determine the effectiveness of any given method for any single field. And as of 1978, enhanced recovery is still more promise than reality. It is unlikely that more than one million barrels daily will be made available in the late 1980's through enhanced recovery methods.[67]

Another potential new source is oil-bearing shale rock. Although considered a new source, oil was actually being distilled from shale in the eastern United States prior to Colonel Drake's oil discovery that effectively started the world oil industry in 1859. Drake's find closed down the shale-oil industry. In Colorado in the early part of the twentieth century, vast shale-oil deposits were discovered. These reserves still dwarf known reserves in conventional oil fields, but the problem has always been to get the oil out. According to a report issued in 1918 by the U.S. Geological Survey, "The production of oil in this country will continue to grow . . . because of the shale resource . . . No one may be bold enough to foretell what tremendous figure of production may be reached within the next ten years." [68]

Subsequently, oil from shale has been said to be within a whisker of economic grasp. Three decades ago, chemical engineering students were taught that if oil prices went up 20 percent, oil from shale would become profitable. Since then, the price of oil has increased by a good deal more than 20 percent in real terms, but shale oil apparently still remains unprofitable. As the price of oil has risen, so has the estimated cost of producing oil from shale. Many engineers and promoters say that the price of oil must go up just another 20 percent for oil from shale to be economic.[69]

Even if existing shale processes were to prove economic, either with or without a government subsidy, the capital and time requirements for them would still be very large. A production level equal to about a half of 1 percent of U.S. oil consumption—100,000 barrels a day—would require a billion dollars and a decade for development. Furthermore, under present technology, water requirements are potentially very large, which would generate opposition from farmers and ranchers. Nor are environmentalists happy about a process that breaks some shale rock into fine particles that float into the sky. Meanwhile, the shale rock that remains expands to a volume greater than it previously occupied, which necessitates filling in some canyons. The so-called *in situ* process, in which the rock is burned underground, involves less environmental damage than conventional means, but shale oil still is unlikely to make any contribution to the national energy budget by the late 1980's and very little by the year 2000.[70]

Thus, the total U.S. oil output in the late 1980's from both known and newly found oil fields and from enhanced recovery will likely approximate 10 million barrels daily, about the same as current production. Even this level is quite speculative and perhaps on the optimistic side. To maintain that production level would require the finding of almost four billion barrels annually; but there has been only one year in the last thirty in which more than three billion barrels of reserves have been found.

What about the third proposal, breaking up the companies? Companies in the oil industry have been large, at least since John D. Rockefeller organized the Standard Oil Trust in 1882. And even after the Trust was broken into thirty-four separate companies in 1911, oil companies remained big. In 1978, four out of the top six companies in the Fortune 500 listing of the largest U.S. industrial companies were petroleum firms, with three—Exxon, Mobil, and Standard of California—direct descendants of Rockefeller's Standard Oil Trust. Far down on the list of the oil companies is Marathon Oil, the eighteenth in size, but still ranking number forty-eight in the Fortune listing.[71]

Divestiture, or "dismemberment," as the companies call it, could take one of two forms. One is a horizontal breakup, in which oil companies would be required to sell off non-oil activities, such as coal, an issue we discuss in Chapter 4. The other is vertical divestiture, which typically would involve breaking up each of the largest dozen or so oil firms into three separate companies—the first restricted to oil exploration and production; the second, to pipelines; and the third, to refining and marketing.

Would vertical divestiture increase the supply of domestically produced oil? This question has traditionally been argued in economic terms: How does one measure the trade-off between increased competition and decreased efficiency? Proponents of divestiture argue that the present pattern of vertical integration makes the firms noncompetitive, and they also say that the great size of firms makes them unwieldy and inefficient. Opponents of divestiture assert more or less the opposite, namely that the industry as it stands is highly competitive, and that the current pattern of vertical integration improves efficiency.

Numerous studies have been conducted on vertical divestiture. A careful review of them leads to a most disappointing conclusion: Economic analyses do not yet enable one to determine whether breaking up the companies would result in any meaningful changes in either competition or efficiency. *What is clear is that there is no evidence that divestiture would lead to greater domestic supplies of oil.* Moreover, the consequences of a breakup might well be inconsequential to the consumer, on the order of a tank of gasoline a year, but whether a tank richer or a tank poorer cannot be determined.[72] As for the stockholders, any pinch would be hardly felt. Finally, the units remaining after divestiture would have no difficulty in obtaining adequate financing to continue operations. An exploration and production company spun off from Exxon would not, as one book alleges, be a "mom and pop wildcatter." It would remain one of the world's largest firms.[73]

There might well be, however, significant political and psychological effects. Eight out of ten Americans place some blame on the oil companies for energy problems. Many of them might well assume that breaking up the most visible of those companies would do much to solve the problems, thus increasing complacency and making American consumers even less willing to accept the higher prices needed to dampen demand and stimulate alternatives. Many citizens, however, might derive considerable psychic income from seeing large companies broken up into somewhat smaller ones, and the number of executives earning $500,000 a year substantially reduced. On the other hand, there would be a large multiplication in the number of senior vice-presidents earning over $100,000 a year in the successor companies.[74]

While some American oil policies have tended to keep production lower than it would otherwise have been, other policies have tended to encourage oil consumption to rise. It is true, of course, that prices have been allowed to go up substantially in the United States, but oil is still being sold far below its economic value on the world market, thereby giving American consumers a subsidy of $15 billion a year. And after slight drops in 1974 and 1975, U.S. oil consumption climbed 13 percent in the following two years.[75]

The important point to remember is that there is no domestic oil solution to the problem of increasing U.S. oil imports, no way that production from American oil wells can close the gap of 9 million barrels daily between what the United States produces and what it consumes. Any hopes rest, to varying degrees, on the other energy sources—natural gas, coal, nuclear, solar—and conservation. Our conclusion, therefore, differs fundamentally from the optimistic projections of the last several years. In November 1974—one year after the embargo—Administration experts, with the aid of large computer models, argued that domestic production could reach 15 million or even 20 million barrels daily, and that U.S. imports thus could be reduced to zero. They said this would require an eleven-dollar-a-barrel oil price (in 1973 prices), which in fact was not much higher than the 1975–78 prices for newly found oil.[76]

With no possible domestic oil solution, it is obvious that U.S. oil policy must come to terms with the undesirable features of imported oil—the possibilities of supply cutoffs and further higher prices. A supply cutoff, although more dramatic than increasing prices, is actually easier to prepare for, as indeed the United States is now doing through its strategic storage program. Current plans call for four months' supply of oil, but it makes sense to expand capacity to at least six months. That kind of reserve on hand would help deter any embargo by oil producers, because many of the producers could not afford to go that long without revenues. Furthermore, it would help provide oil in cases such as the Iranian shutdown of 1978–79. The costs associated with the storing of a six-month supply—estimated at $2 billion a year—sound high, but are small compared with the gains in economic and political security. The United States should also encourage other Western nations to expand their strategic oil reserves as well, which would further diminish the potential effectiveness of a future embargo.[77]

But strategic storage deals only with a part of the problem. The overall pattern of consumption, which leads to ever-increasing oil imports, is clearly unstable, creating "costs" that are not reflected in prices paid by U.S. consumers. Most economists and business-men have recommended the decontrol of oil prices, and indeed at the margin this would help to encourage production and dampen demand. But a number of knowledgeable politicians explain that a free-market solution is simply not acceptable, at least not now, for a variety of reasons.[78] As three senators have said in private inter-views: "Everyone is taken in by the conspiratorial theory of history. The public doesn't believe that there is an energy problem, and I hear even from well-to-do businessmen, 'The oil companies are capping those wells' "; "Inflation is our number-one problem. We don't want higher prices"; "If we had not kept the lid on oil prices [in 1973–74], the oil companies would have been broken up within thirty days." In short, allowing oil prices to rise creates some very difficult political and social questions about income distribution and equity.

But whatever the outcome of U.S. oil-pricing policy, an impor-tant fact should not be lost sight of: *Even world market prices would still be much too low to reflect the real risks caused by oil imports.* These include such things as higher oil prices, slower economic growth, and international political tension. Virtually *all* participants in the debate have ignored the costs associated with these risks,[79] but they need to be addressed and thought through.

What do we mean? We are talking about what are called external costs, or externalities. These are costs induced by oil imports that are not borne by individual oil consumers, as such costs are not reflected in the prices involved in the transaction between producer and consumer—that is, they are external to those prices. For in-stance, the cost of cleaning up an oil spill on a beach on Cape Cod is not included in the price of a gallon of gasoline in Boston. Ex-ternal costs plus market price equal what is called the social cost.[80]

Predicting oil prices and their economic effects involves art as well as science, requiring sensitivity to the political situation as well as to straight economics. Indeed, by its very nature any estimate of the total costs associated with U.S. oil imports must be highly un-certain. But it is so important for policymakers to have some idea of the possible magnitude of such costs that we feel compelled to offer an illustration of the possible, regardless of how tentative and uncertain such an illustration might be. Our illustration is based on the following set of relationships, depicted in Figure 2–1, that illustrate how much it might be worth to the United States to keep its imports at the current 9 million barrels daily, as opposed to what might become a 14-million-barrel level by the late 1980's.[81]

Figure 2–1

ILLUSTRATION OF POSSIBLE EFFECTS
OF INCREASED U.S. IMPORTS OF 5 MILLION BARRELS DAILY
(14 INSTEAD OF 9 MILLION BARRELS DAILY)

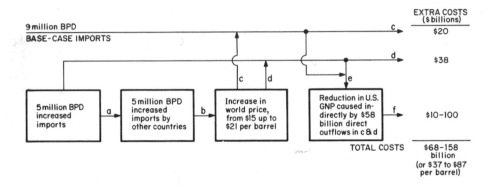

ARROW	DESCRIPTION OF FORCES
a	The U.S. policies that cause 5 million barrels daily of extra U.S. imports affect policies in the other industrial nations so that their imports rise 5 million barrels daily.
b	The additional 10 million barrels daily of world oil imports cause world price to be $21 a barrel instead of $15.
c	The $6 difference in world price causes the 9 million barrels daily of oil imports under the base case to rise in cost by $20 billion yearly ($6 × 365 × 9 million).
d	The increased oil imports of 5 million barrels daily cost $21 a barrel, or $38 billion yearly.
e, f	The outflow of funds from the United States due to the increase in world oil price reduces U.S. national income by contracting demand for U.S. goods and services. The magnitude depends on the amount of funds recycled back to the

United States by the oil exporters for investment or purchases, and on the policies (government purchases, transfer payments, changes in taxes, and changes in money supply) undertaken by the U.S. government that offset—or reinforce —the contraction. Government policies will depend on many factors, including the speed of the price rise and fear of inflation, which in itself is exacerbated by higher oil prices, not only by their direct effects but also because of their indirect effects on wage demands.

Note A: Estimates are needed for each of the above forces for each year, with total costs discounted to the present by use of an appropriate discount rate.

Note B: The range of estimates for the values for each force and the probability of each value occurring are dependent on the estimated values of the prior forces. Thus, if one wanted to estimate, say, ten possible outcomes for each of five forces, the total number of possible scenarios would be 10^5. But since each force also depends on conditions in the prior period for that force, the total number of possible scenarios for a ten-year period would run into the billions. For examples of works that have handled complex, multiperiod problems, see Claude L. Pomper, *International Investment Planning: An Integrated Approach* (New York: North-Holland, 1976); Howard Raiffa, *Decision Analysis: Introductory Lectures on Choices Under Uncertainty* (Reading, Mass.: Addison-Wesley, 1967); and Burton Rothberg, "A Decision Theoretic Model of Eastern Hemisphere Oil Exploration," Unpublished D.B.A. thesis, Harvard, 1974.

1. The first potential cost is a direct one—a higher level of oil imports might cause world oil prices to be higher. Not only do U.S. imports impact directly on the world oil market, they also affect the imports of the other industrial nations. If the relationships shown in a recently published study by the Organization for Economic Cooperation and Development (OECD)* are correct, then the addition of 5 million extra barrels to the current 9 million barrels daily of U.S. imports would result in an additional 5 million barrels daily of oil imports by the other industrial nations. Hence, the additional demand for OPEC oil would be 10 million barrels daily. The reason

* The OECD, of course, is the organization composed of the industrialized, noncommunist countries.

for this tie between U.S. oil imports and the oil imports of other industrial nations is that the United States, with its large market and
by its allocation of resources, can make or break various energy options. Without a stronger U.S. emphasis on conservation, it is more
difficult for other industrial nations to implement additional conservation measures, especially since their consumers already are paying
higher prices than American consumers. A commitment to solar
energy in the United States would make it a much more commercial
option for the others. Environmental restraints on nuclear power in
the United States have had a powerful echo effect on nuclear development in Germany and Japan. In sum, increased American
reliance on oil makes it difficult for the other governments to consider alternatives to oil. With U.S. imports of 9 million barrels
daily, a plausible estimate is that OPEC exports would be 30
million. But with U.S. imports at 14 million, OPEC exports would
be 40 million.[82] (See Table 2–2.)

OPEC oil prices would almost surely be higher if OPEC oil exports were 40 million barrels daily instead of 30. And if the 40
million were near OPEC's capacity, then the upward pressure on
price would be even stronger. A number of persons—especially
consultants and government officials—have made estimates of future prices of OPEC oil. Given the difficulties that we discuss in
the Appendix about making accurate forecasts of oil prices and
volumes, it is not surprising that estimates of prices in real terms
for the late 1980's vary widely. The estimates vary all the way from
no change in price to an increase as high as 100 percent, with a
tendency for the lower estimates of prices to be associated with
lower levels of demand. For our illustration, a range of prices of
40 percent is assumed between the low and high levels of exports.
The matter can be put as follows: If OPEC exports are 30 million
barrels daily in the late 1980's, the price of OPEC oil would be the
same in real terms as in early 1979, before the quick run-up in
world oil prices—that is, about $15 a barrel delivered to U.S. re-
fineries. But if OPEC exports are 40 million barrels daily in the
late 1980's, then the price of OPEC oil is assumed to be 40 percent
higher, or $21 a barrel delivered to U.S. refineries. The difference
between the two OPEC export levels—30 million and 40 million
barrels daily—represents, of course, the difference between U.S.
imports of 9 million and 14 million barrels daily.

Table 2–2

ILLUSTRATION OF POSSIBLE EFFECTS OF U.S. OIL IMPORTS
ON SUPPLY OF AND DEMAND FOR OPEC OIL, LATE 1980'S
(millions of barrels of oil daily)

	Est. 1979*	Late 1980's	
		"Low" U.S. Imports	"High" U.S. Imports
Demand			
U.S. Imports	9	9	14
Imports by Other OECD	21	23[b]	28
Net Imports by Communist Nations	−2	0	0
Net Imports by Other Nations	1	−2	−2
TOTAL IMPORTS OF OPEC OIL	29	30	40
Consumption within OPEC Nations	2	4	4
TOTAL DEMAND FOR OPEC OIL	31	34	44
Supply of OPEC Oil			
Outside Southern Persian Gulf	17.6	18	20
Southern Persian Gulf (excluding Saudi Arabia)[a]	4.4	5	5
Saudi Arabia	9.2	11	19
TOTAL	31.2	34	44

* Estimated from the Department of Energy, *Monthly Energy Statistics,* and British Petroleum Company, *BP Statistical Review.*

[a] Includes Kuwait, United Arab Emirates, and Qatar.

[b] This increase over 1979 reflects an expected recovery from the 1974–77 recession in Europe.

Thus, the increase in the world price of oil of $6 a barrel would raise the cost to the United States of the first 9 million barrels per day by $20 billion yearly. The additional 5 million barrels daily at their market price of $21 a barrel would cost the United States $38 billion yearly. Therefore, the total costs of the additional 5 million barrels daily to the United States would be $58 billion yearly, or $32 for each of the additional 5 million barrels.

To be sure, it is possible to change the assumptions in this illustration and arrive at different numbers for the cost of each additional barrel of oil. But based on almost any reasonable set of assumptions that we tried, the calculated price for the additional oil was substantially higher than the $15 price, and usually it was above $30 a barrel.[83]

2. The second potential cost represents the indirect economic effects of the outflow of the additional $58 billion in direct payments for oil. These indirect costs can vary over an enormously wide range. Whether these indirect costs are low or high depends on a multitude of factors, including the amount of funds recycled from oil exporters back to the United States either for purchases or investments, the speed of any recycle of funds, the gradualness of any increase in oil imports and prices, the economic policies of the federal government to offset the contraction in demand for U.S. goods and services caused by the outflow of dollars, the degree to which workers demand higher wages because of actual or anticipated inflation related to oil imports, the reaction of the foreign exchange market, the reaction of the stock market, and the effect on investor confidence.

In general, a gradual increase in prices, by allowing adequate time for adjustment, would result in relatively low indirect costs— perhaps only a fraction of the $58 billion annual increase in direct outflows. But a rapid increase in oil prices could create economic costs several times those of the direct outflows and could last for a number of years—as occurred because of the oil price increases in 1973–74. Thus, a relevant range for these indirect costs is probably as wide as $10 to $100 billion. In terms of affecting the cost estimated for the marginal 5 million barrels of oil daily, about all we can really say is that these indirect costs might range from about $5 up to $50 a barrel.[84]

3. To these potential costs must be added some social and political risks, difficult to estimate but no less real. Within the United States, increased inflation is socially undesirable; indeed, U.S. leaders have begun to worry seriously about the impact of inflation "on the fabric of American society." Next, U.S. oil imports, even at their

present level, are an important contribution to political tensions within the Western community. In the summer of 1978, for example, the United States sought trade concessions from the Europeans and the Japanese, who in turn insisted that the United States restrain its oil imports. The Europeans and Japanese were alarmed by the effects of U.S. demand on the monetary system and on present and future oil prices. A headline in *The Economist* succinctly captured their worry: "Will American oil greed doom the world?" [85] As one foreign leader wrote us, "The energy question gives me more worry than the 20,000 Soviet tanks on the border. The profligacy with which oil continues to be used in the United States is leading us all into disaster." Finally, as U.S. oil imports increase, U.S. foreign policy will be more influenced by the desires of the oil exporters, which might involve such things as sales of advanced aircraft to Saudi Arabia and policies toward illegal immigrants from Mexico.

This illustration, admittedly crude, suggests that the total potential costs of the additional 5 million barrels daily come, in round numbers, to between $35 and $85 a barrel, not counting some potentially quite serious social and political costs. [86]

Because we are venturing into heretofore unexplored territory and because any number of refinements in the calculations are possible (such as discounting all future payments to the different times that oil-consuming decisions would be made), we believe that the possible order of magnitudes rather than the exact numbers is important. [87] For convenience, we use the lower boundary of our potential range—say, $30 to $40 a barrel—as a reference point in Chapter 8.

It is possible, of course, to construct a much more sophisticated model of the total costs to the nation of incremental oil imports. But whatever the model, we feel safe in concluding that investment decisions by individuals and firms, based on today's cost of oil, are dramatically lower than any reasonable estimate of the true cost to the nation as a whole. And that would be the case even if oil in the United States were at world prices. [88]

One possible solution is to place a tariff on imported oil, thus raising its price to something approaching $35 a barrel. Many economists will recommend a response of this sort when the prices of imports work against achieving some national goal. But such a tariff on oil is simply not politically acceptable, and it is unrealistic to think that a tariff high enough to reflect the true social cost of imported oil would ever be enacted.[89]

This means that U.S. energy policy should give alternative sources of energy, including conservation, an "equal chance" with the social cost of imported oil. In the face of domestic petroleum, which is subject to price control, and the exclusion of external costs from imported oil prices, this means that other energy sources and conservation deserve U.S. government support.

It also means that the production of domestic oil should be encouraged to keep current levels of production from falling. New territories for exploration offshore and in Alaska should be opened, under stringent environmental requirements, with mechanisms for the timely resolution of regulatory and environmental issues.[90] Deregulating the prices paid to producers for new finds of oil and for enhanced recovery oil is reasonable as well. Such deregulation would add some financial incentives and help eliminate political uncertainty for the producers. Financial payments—perhaps in the form of guaranteed market contracts for the output of the facilities —should be given for new technology, such as coal liquefaction and oil from shale, provided a working agreement can be reached with environmentalists and farmers.

Even if it does not become politically feasible to move quickly to raise the general level of domestic oil prices to world levels quickly, the United States should move in this direction. American oil consumers could then make investment decisions with the certain knowledge that oil prices would be steadily rising toward the world market level.[91]

Attention also needs to be given to the effects on competition within the industry of any change in the pricing system. The U.S. government is already heavily involved with payments from one part of the oil industry to another. These payments affect industry competition, and their elimination, unless carefully designed, could bankrupt some companies not vertically integrated. If this should happen, divestiture would again become a major political issue diverting effort and attention away from the crucial jobs of finding oil and reducing oil consumption. To the smaller companies not vertically integrated, life next to a major is like being a small animal lying next to an elephant. If the elephant master, the government, instructs the elephant to roll over, it can do so—though, perhaps, slowly and awkwardly. Of course, some small animals might be squashed in the process.[92]

We have dwelt on two countries—Saudi Arabia and the United States. Continued reliance on oil means that the future of the world economy will, to a high degree, depend on Saudi Arabia, a nation of perhaps 5 or 6 million people—a highly traditional society, but one going through what may be the most rapid and total social and economic transformation in the history of the world. Dependence here puts the rest of the world in a highly vulnerable position. The very foundation of the international economy will be affected not only by conscious decisions in Riyadh about production levels, but also by other possible contingencies—accidents, a Soviet presence, an unresolved Arab-Israeli conflict, tensions in the Arab/Persian Gulf, shifts in the attitudes of Saudi leaders, or the overthrow of those leaders.

The United States is at the center of the world oil problem, having failed to come to grips with the decline of its influence over the world petroleum market and the true costs of its oil imports. By allowing its citizens to receive $15 billion in subsidies to use oil, and by ignoring the even larger external costs associated with imported oil, the United States has been encouraging a form of behavior that will drain the world of the commodity. This is a reckless course, increasing the vulnerability of the entire Western world and undermining the leadership of the United States within it. In short, increasing dependence on imported oil poses a threat to American political and economic interests; that much must now be clear.

Americans should not delude themselves into thinking that there is some huge hidden reservoir of domestic oil that will free them from the heavy cost of imported oil. Of course, reasonable measures should be taken to keep domestic oil production from declining further. But the handwriting is clear. To the extent that any solution at all exists to the problem posed by the peaking of U.S. oil production and the growth of imports, it will be found in energy sources other than oil.

I. C. BUPP
AND FRANK SCHULLER

3

Natural Gas: How to Slice a Shrinking Pie

Natural gas is a premium fuel, the energy prince of hydrocarbons. The 20 trillion cubic feet of natural gas that Americans consumed in 1978 accounted for about one quarter of the energy used by the country.* But unlike oil, natural gas has remained for the most part a domestically produced fuel, with only 5 percent imported. Moreover, unlike coal, natural gas burns clean, without soot and sulphur emissions; nor does its extraction and transportation cause the environmental damage associated with the black solid. Finally, unlike nuclear power, natural gas poses no waste disposal problem. Yet this energy prince of the hydrocarbons is also one of the fuels for which the word "crisis" has seemed particularly appropriate. In the winters of 1976–77 and 1977–78, gas supplies were short, and cutoffs caused the loss of millions of dollars of business, much inconvenience, and many hardships for thousands of Americans.

These cutoffs, following the OPEC oil price increases, exacerbated a controversy that has been brewing for years. As one congressional staff member said, "When the Carter Administration

* Excluding natural gas liquids, which are usually reported with oil production.

brought its energy legislation to the Hill, it stumbled into one of the great religious wars in American politics. The war over government regulation of natural gas prices goes back a full generation. The Administration's talk about energy waste, their econometric models, and their computer printouts have not been very effective against the deeply entrenched positions of the warring camps." The staff person who made these observations in August 1978 did not find it surprising that it took more than a year and a half of deliberation for Congress to agree to a compromise akin to the natural gas provisions of the Carter energy proposal. Nor did he consider this failure to be evidence of "congressional foot-dragging." [1] The fact is that natural gas poses an overwhelmingly difficult political issue, involving an objective conflict of interest among several groups and several regions, very high financial stakes—perhaps as high as $400 billion—and correspondingly high passions.[2]

The current compromise has not by any means ended these deep divisions. Confusion, controversy, and a vast amount of litigation are already evident as the new Federal Energy Regulation Commission, gas producers, pipeline companies, and the states struggle over the interpretation of the Natural Gas Policy Act of 1978. Creating even more uncertainty has been the recognition in late 1978 and early 1979 by the Department of Energy of a surplus of natural gas, commonly called the "gas bubble." Energy Secretary James Schlesinger encouraged industrial users and utilities that had converted from gas to oil to go back to gas. This contrasted sharply with the earlier Administration policy favoring a shift away from natural gas, particularly to coal. "I understand now what hell is," Schlesinger said during the congressional debate over the 1978 bill. "Hell is endless and eternal sessions of the natural gas conference." After the Act's passage, one industry executive observed that Schlesinger had seen only the beginning.

Prices are central to the natural gas controversy. Before the passage of the Natural Gas Policy Act in 1978, approximately two thirds of the gas produced in the United States was under federal price controls. Now all of it is. As one might expect, the controlled price is considered too high by many buyers and too low by many sellers.

What shaped the controversy into its present form was the existence of two very different markets. The interstate market, which historically has been subject to federal price controls, refers to gas produced in one state and transported to another. The intrastate market, which has been subject to federal price controls only since late 1978, means gas consumed entirely in the state where it is produced. The two markets have evolved with entirely different systems of pricing. Through the smoke of the battle, one can see an all-important basic issue: Should the price of gas be based on its cost of production, or should the price be based on its value in the marketplace?

Practically the only thing on which all the warring parties agree is that for the past decade America's proven reserves of gas have been steadily declining. While gas production peaked in 1973 and had fallen approximately 12 percent by 1978, the level of proven reserves peaked in 1967 and by 1978 had fallen some 25 percent, to a volume equal to only ten years at the then-current rate of consumption. Clearly, new discoveries of natural gas have failed to replace what has been consumed. But the recent decline of proven gas reserves does not necessarily mean that the United States is running out of gas in any physical sense. Indeed, many, perhaps even most, informed geologists believe that enough gas exists onshore and offshore under the United States' continental shelf to sustain a national consumption rate about equal to the current 20 tcf level for at least twenty-five to thirty years, but at higher prices than Americans are accustomed to paying. Beyond that, there is doubt that even very considerably higher prices would sustain consumption much above the current rate.[3]

The conflict over price is likely to continue. And to make sense of it requires an understanding of how the price of natural gas came to be regulated in the first place.

TOWARD REGULATION

The first gas company in the United States was established in Baltimore, Maryland, in 1816. It sold synthetic gas produced from coal. The organization of similar companies in other cities followed in the next two decades. The business of all these companies represented the practical application of the discovery by a seventeenth century Belgian chemist that coal could be burned in a way to yield a flammable gaseous substance. The first large-scale use of manufactured gas in the early nineteenth century in America was for street lighting. Large-scale use of gas for cooking did not occur until late in the nineteenth century.[4]

Toward the end of the last century, high-cost manufactured gas met ever sharper competition from cheap kerosene for household gas lighting, and from electric arc lamps for street lighting. Soon thereafter, central electricity generating stations were developed, and electricity began to compete for the entire lighting market. As the competitive situation stabilized in the early twentieth century, the manufactured gas industry was left with only the market for residential cooking and water heating.

Manufactured gas eventually lost even this limited market to a new competitor: low-cost natural gas, which was found primarily in the South and Southwest, and to a lesser extent in the West and Midwest. For many years natural gas remained an essentially local fuel; the earliest natural gas pipelines, built roughly between 1890 and 1925, were rarely more than 150 miles long. Because they were usually within one state, they could be subjected, as public utilities, to state control. It was not until after World War II that the technology became available to allow the economical transmission of natural gas by pipeline over long distances. The $50 billion pipeline network that now connects the gas-producing areas of the Southwest, Gulf Coast, and Appalachia to all of the country's metropolitan areas was almost entirely built during the fifteen years after the end of World War II.

While the gas industry was evolving into a nationwide supplier, the markets for its products also were undergoing an evolution. Natural gas came to be widely used for residential and commercial space heating and for a variety of industrial purposes, especially as boiler fuel.

The transcontinental pipelines that carry natural gas from producing areas to urban consumers are, of course, only one part of the natural gas industry. There are also about 1,500 local distribution companies, mostly privately owned, that buy gas from the pipelines and make it available to consumers. To do this, the companies operate a network of smaller pipelines that serve homes and commercial establishments (Figure 3–1, page 60). As public utilities, the distribution companies come under price regulation as "natural monopolies" from municipal and state agencies.[5]

As soon as pipelines began to cross state borders, regulation by individual states became impossible. Thus, the Natural Gas Act of

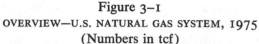

Figure 3–1
OVERVIEW—U.S. NATURAL GAS SYSTEM, 1975
(Numbers in tcf)

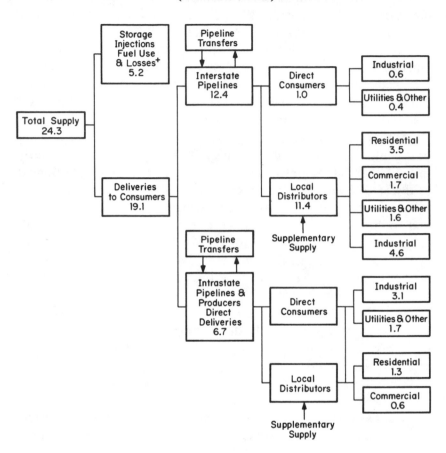

* Supply includes U.S. marketed production, withdrawals from storage, and imports.

† Gas for such purposes as lease and plant fuel, pipeline compressor, fuel, extraction loss, and transmission losses.

Source: Federal Energy Administration, *National Energy Outlook* (Washington, D.C.: Government Printing Office, February 1976), p. 118; based primarily on data from Bureau of Mines, *Natural Gas Production and Consumption: 1974* (Washington, D.C.: Mineral Industry Surveys, 1975).

1938 created a new government agency, the Federal Power Commission (FPC),* with authority to ensure that the prices charged by the interstate pipeline companies to local gas distributors were "just and reasonable." [6]

But the 1938 Natural Gas Act was ambiguous on a key point: whether the FPC was also to regulate the price of gas in the field that producers charged to the interstate pipelines, the so-called wellhead price. In a historic 1954 decision, *Phillips Petroleum Co. v. Wisconsin,* the Supreme Court ruled that it should, by deciding in favor of a suit brought by the attorney general of Wisconsin on behalf of gas consumers in his state. The Wisconsin attorney general argued that gas consumers needed protection from wellhead price increases that were being "passed through" by the pipeline companies. The court accepted the reasoning that gas producers could exact "excessive" prices at the wellhead, ultimately at the expense of the consumer. But the Supreme Court provided the basis for the current controversy by failing to offer specific criteria for deciding exactly what a "just and reasonable" price should be. The same decision laid the groundwork for a related conflict by creating two distinct markets for the fuel. The FPC interpreted this ruling to apply only to interstate gas, thereby leaving intrastate gas free of wellhead price controls.[7]

Congress immediately set out to overturn the Supreme Court decision. In 1956, both the House and the Senate voted to deregulate the wellhead price of natural gas. Had it not been for a quirk, President Eisenhower, who favored deregulation, would have signed the bill. During the debate on the bill, Senator Francis Case, who also favored deregulation, indignantly disclosed that a lobbyist favoring the bill's passage had left an envelope containing twenty-five $100 bills to be given to him. Case ordered the money returned and decided to vote against the bill. The lobbyist protested the "inference that it was some kind of a shady deal" and declared that there were "no strings attached." But President Eisenhower, blaming the "arrogant" gas lobby, vetoed the bill, saying that he could not risk creating "doubt among the American people concerning the integrity of governmental processes." [8]

---

* The Federal Power Commission (FPC) became the Federal Energy Regulatory Commission (FERC) when the Department of Energy was created.

TWO-TIER MARKET: INTERSTATE AND INTRASTATE

What evolved from the regulation of only part of the natural gas industry were two markets with two different approaches to determining price and, of course, two different prices. The key to understanding the entire U.S. natural gas situation is to recognize what these differences are. The federal government's regulation of "just and reasonable" prices for the interstate market was cast in terms of the cost of production, or at least on regulators' opinions about the *cost* of production.[9] In contrast, in the intrastate market, supply and demand determined price; thus price was based on *value,* or at least on the buyers' and sellers' opinions about value.

At the time of the *Phillips* v. *Wisconsin* decision in 1954, gas was so abundant and so cheap that producers considered it a by-product in their search for oil and often flared it. In those days, few wells were drilled solely to find gas. The average price of all natural gas sold in the United States in 1953, for example, was nine cents per thousand cubic feet (mcf), or only one-fifth the price of domestic oil with comparable heat content. The regulation of the interstate market initially had little effect on the price of gas, for interstate prices continued to rise moderately after regulation, just as they had before, and the prices in the interstate and intrastate markets remained essentially identical. In fact, gas remained so abundant and cheap that oilmen who found gas instead of oil would curse their luck.

But this situation began to change in the 1960's, when a decline in the gas reserves signaled the end of the era of abundant and cheap gas. In 1967, for example, in the interstate market, the average price of all gas was 17 cents per mcf, whereas the price of newly discovered gas had been allowed to rise to 19 cents, reflecting the FPC's conclusion that the cost of producing gas was rising (Table 3–1). Furthermore, prices in the interstate and intrastate markets had begun to diverge as the price of newly discovered gas sold into the intrastate market had climbed to 22 cents per mcf. The average price of all gas sold into the intrastate market—although not published—began to exceed the average price of interstate gas by a wide margin.[10]

The fact that intrastate prices rose faster than interstate prices suggests that the regulated cost-based price no longer reflected

Table 3–1

PRICES OF U.S. NATURAL GAS, 1966–77
(cents per thousand cubic feet)

| | Interstate | | Intrastate |
	Average Price	New Contract Price	New Contract Price
1966	17¢	17¢	20¢ (est.)
1967	17	19	22 (est.)
1968	17	19	23 (est.)
1969	18	20	26 (est.)
1970	18	n.a.	29 (est.)
1971	19	27	45 (est.)
1972	21	29	63 (est.)
1973	23	37	80 (est.)
1974	27	46	100 (est.)
1975	34	57	140
1976	48	142	160
1977	69	142	190

Sources: Interstate averages are prices paid to domestic producers by major interstate pipeline companies: 1966–71 from U.S. Senate, Committee on Interior and Insular Affairs, *Natural Gas Policy Issues and Options,* 93 Cong., 1 sess. (Washington, D.C.: Government Printing Office, 1973), p. 220, which also shows that prices paid by interstate pipeline companies are essentially equal to prices paid by major interstate pipeline companies. 1972–76 from Energy Information Administration, Department of Energy, on December 18, 1978. 1977 from Energy Information Administration, DOE, *Monthly Energy Review,* November 1978, p. 77. Annual prices represent an arithmetic average of the monthly prices.

Interstate new: 1966–69 from *Natural Gas Policy Issues and Options,* p. 217. 1971–75 from Subcommittee on Energy and Power of the Committee on Interstate and Foreign Commerce, House of Representatives, *Long-Term Natural Gas Legislation,* part I, 94 Cong., 2 sess. (Washington, D.C.: Government Printing Office, January 20–February 2, 1976), p. 470. Prices include area rate ceilings, optional procedures, limited-term contracts, and small producer sales. 1976 and 1977 figures are national ceiling wellhead prices effective on July 27, 1976, for all wells commenced on or after January 1, 1975; Federal Power Commission, *Opinion and Order on Rehearing Modifying in Part Opinion No. 770 and Granting Petitions for Intervention,* Opinion No. 770A (Washington, D.C.: FPC, November 15, 1976), pp. 12–15.

Intrastate new: 1966–74 from interviews with executives of intrastate pipelines. 1975–77 from arithmetic average of new contracts in Louisiana and Texas, by producers that also sell into the interstate market, rounded to nearest 10 cents. *Monthly Energy Review,* November 1978, p. 78.

buyers' opinions about the value of gas. And, of course, producers, having a choice between selling gas to one or another of the two markets, preferred to sell where they could get the higher price. The year 1970 marked a turning point in the sales of gas into the two markets. In that year, the new reserves dedicated to the intrastate market jumped from the 1969 level of one third of the total up to two thirds.[11]

True, the FPC permitted the price of newly discovered interstate gas, as well as gas previously committed under long-term contracts, to rise substantially after 1970. In 1976, the price of newly discovered natural gas sold into the interstate market was set at $1.42 per mcf. But the intrastate market had risen even more, to $1.60 per mcf. Accordingly, the intrastate market continued to capture the bulk of the new supplies.[12] During the winter of 1976–77, consumers with gas curtailments along interstate pipelines felt the damaging impact of the two-tiered gas market. The damage would have been much greater except that long-term contracts with the pipelines had committed approximately two thirds of total U.S. gas production to the interstate market.

How does a buyer of natural gas determine its value? In principle, one does it by assessing the cost of some alternative energy source other than gas. The problem is that when the main alternative to gas is high-priced oil, a *value-based* price for domestic natural gas will be much higher than the *cost-based* prices that the government had enforced for interstate gas. In 1977, for example, the average price of all gas sold into the interstate market was 69 cents per mcf, or about one third of the price of oil with comparable heat content; but the price of intrastate gas approximated the price of oil. This means that a shift to a value-based free-market system —the abandonment of all price regulation—would cause an enormous transfer of real wealth from gas consumers to gas producers.

Who profits from higher prices has been the nub of the dispute over natural gas policy. On one side, for example, the governor of Texas, a gas-producing state, can plausibly assert that "for over twenty-three years, federal regulation of natural gas prices has *undervalued* natural gas in the marketplace, stimulated artificially high demand for natural gas, and provided little incentive for the development and production of additional supplies." On the other side, a public service commissioner from New York, a consuming state, can ask with equal plausibility "whether it is logical as far as the American consumer is concerned to allow the OPEC nations to establish a value of commodity price for our domestic reserves of natural gas." [13]

The controversy goes beyond the basic issue of whether price should be determined by value or by cost. It extends to a question concerning the methods of determining cost.

How much does gas cost to produce? In theory, one can determine a precise production cost for gas. In reality, no one can do much more than guess. Specialists have even disagreed on whether costs of production have really been rising or falling. Jules Joskow of the National Economic Research Association, a respected economic consulting firm, has argued that while drilling costs have been increasing, the unit cost of newly discovered gas has been declining. Professor Henry Steele of the University of Houston, another respected specialist, has maintained the opposite.[14] The prospects for resolving such arguments are very dim, for several difficult issues present themselves.

First, a company that is a going concern will commit a more or less continuous stream of expenditures for exploration and development, much of which it cannot assign to a specific gas reserve. These costs include initial expenditures to determine prospective sites as well as expenditures associated with dry holes.

Second, about one quarter of U.S. gas production comes from wells that also produce oil. In these cases, it is impossible to calculate a meaningful cost of production for gas because of the dominance of joint costs that are related to the production of both oil and gas. Often the two resources are discovered either in the same geological formation or in separate formations penetrated by the same well. Cost analysis for any given producer requires an inherently arbitrary allocation of costs between natural gas and oil. Hence a very wide range of estimates is readily obtainable, depending on the methods and assumptions one chooses to invoke. Different estimates do not necessarily reflect different "real" costs; they reflect instead the differences in the analytic techniques used. As a result, studies placed in evidence before the FPC have shown differences in estimated average unit costs of gas of 500 percent or more for a single company.[15]

Even without the joint costs, attempts to compute cost per unit volume of gas reserves quickly break down. Theoretically, a calculation of discovery costs should be based on the volume of gas discovered by a given expenditure. To do so, two numbers are needed: the total "proven" reserves of gas and the total costs incurred in finding and developing these reserves. But a company encounters a major problem in trying to ascertain the quantity of reserves actually discovered during any given period since reservoir engineers are likely to differ broadly in their initial estimates. Then, during production, unanticipated developments may sharply raise or lower the expected volumes of ultimate recovery as well as sharply raise or lower expected costs.[16]

Of course, after a gas field is depleted and the wells are abandoned, it may be possible, on the basis of some assumptions about the allocation of overhead costs, to reckon the total cost of what was produced. But because few fields are similar enough to support very precise cost comparisons, such information would not be particularly valuable in estimating costs in other fields. In fact, there is no reasonably predictable relationship between money spent on exploration and the amount of gas (and oil) discovered. The disparity reflects differences in producers' business judgment, technical proficiency, and a strong element of luck.

The truth is that calculating the production cost of natural gas, even under the best of circumstances, is highly imprecise and ultimately arbitrary. Most experts would probably agree with Senator Henry Jackson, chairman of the Senate Energy Committee, who said that it is very difficult to have "a good strong feeling in your stomach that you know exactly what the hell you are doing." [17]

HOW MUCH GAS DO WE HAVE
AND AT WHAT PRICE?

The pricing issue leads into the baffling question of supply. Opponents of price regulation have steadily maintained that control of wellhead prices would inevitably cause shortages, and they point to the curtailments during the winter of 1976–77 as an example. They reason that producers will look for new reserves only if the price they are permitted to charge for new gas is higher than the costs of exploration, drilling, and production. But a government policy of setting "just and reasonable" prices on the basis of historic production costs will diminish or eliminate incentives to find new gas reserves in a time of sharply rising costs. Few of those persons who favor price regulation reject the basic logic of this argument. But many question whether increased reserves come at too high a price. For example, John O'Leary, then the administrator of the Federal Energy Agency, said in 1977, "You have to ask yourself, 'What more do you get out of raising the price of natural gas' . . . every indication we have is that you get very little." [18]

Experts use two different approaches to answer that supply question. Economists typically estimate the supply that would be forthcoming at various price levels. Geologists, on the other hand, typically ignore price and relate supply to the size of recoverable reserves. Within both groups of experts, there is deep disagreement.

Some economists contend that supply is not very responsive to price; others state the opposite. In 1976, for example, the General Accounting Office declared that few additional reserves would likely be discovered at prices above $1.75 per mcf. At the same time that the GAO was painting its pessimistic picture, a task force within the Energy Research and Development Administration estimated that a rise in the price of natural gas from $1.75 to $2.50 per mcf would increase U.S. recoverable reserves by 20 percent.[19]

To add to the confusion, the differences among geologists are at least as large as those among economists. Between 1972 and 1974, the U.S. Geological Survey issued three optimistic reports. At a consumption rate of 20 tcf per year, the different USGS studies projected sufficient gas supply to last anywhere from forty-four to a hundred years. And in 1977, the Central Intelligence Agency estimated that the United States could continue to consume natural gas at a rate of 20 tcf per year for fifty to sixty years.[20] Other equally respectable "stock" estimates contradict all of these relatively optimistic outlooks. For example, in 1974 Shell and Mobil projected total gas reserves sufficient to last only twenty to thirty years at contemporary consumption rates; and in 1976 Exxon estimated an even more meager stock, good for only fifteen to twenty years.[21] A range of estimates that varies between fifteen and a hundred years is hardly a sure guide for policy.

True, the United States can supplement its supply by importing natural gas from Canada and Mexico via pipelines. Indeed, since the early 1970's, imports from Canada have been providing about one tcf annually, or some 5 percent of the total natural gas used in the United States. We expect Canadian imports to remain at about that level for an indefinite period, with the price based on world energy values.

As yet, the United States has imported only meager volumes of Mexican gas. Negotiations for much greater volumes of gas broke down in 1978 over price. Although the Mexicans then announced plans to use all their gas internally, many observers expect the United States eventually to import up to one tcf yearly. In the face of recent enthusiasm about Mexican hydrocarbons, it is worth noting that, at most, this would only be 5 percent of total natural gas consumption in the United States.[22]

Almost all of the gas that has been produced and consumed until now comes from highly porous sedimentary rock, such as sandstone, at depths less than 15,000 feet. But gas is known to occur in other more "unconventional" geologic formations. This makes the question of gas supplies even more bewildering.

Below 15,000 feet, gas is found in two kinds of formation. One is very deep porous sandstone. But the deeper one drills a well, the more it costs per foot. Hence, attempts to develop deep gas wells are very costly. For example, one company told us of spending $5 million to drill and complete a gas well deeper than 15,000 feet, compared with $100,000 for a well only 3,000 feet deep. Uncertainty is also greater; less is known about the geology of the deep reservoirs because there has been less drilling at these depths.

A second unconventional source typically found below 15,000 feet, geopressured brine, faces an additional cost problem because of the need to handle very large volumes of water.

Another unconventional source of gas is in sedimentary rock with low porosity, such as Devonian shale and coal. Because of the low porosity of the rock, it must be fractured to allow the gas to migrate to the well. Sometimes the fracturing does not work, but even if it does, the output per well is very low relative to that typically found in conventional wells. Although limited experience does not allow projections to be made with confidence, experienced observers believe that gas from these two sources is unlikely to make an important contribution to U.S. energy supply any time in the foreseeable future.[23]

Technology heretofore not widely used promises two additional sources of gas. One is synthetic gas (SNG), manufactured from coal. This process was developed in Germany during World War II. In the 1960's, the United States borrowed and improved on the thirty-year-old German technology to produce SNG in several pilot plants established to prove commercial feasibility. But the pilot plants showed the process to be uneconomical at then-current natural gas prices. Later, as gas prices rose and supplies became tighter, several oil companies and gas pipeline firms announced that they would construct commercial SNG plants, mostly in the West, where huge strip mines provided inexpensive coal. As yet, none have been built, and many companies have abandoned their plans in the face of inflation, uncertain operating and construction costs, and environmental concerns. In addition, the federal regulatory authorities have hindered the commercialization of SNG by refusing to let the applicants charge what they consider an adequate price. Such a price would be considerably higher than gas from conventional sources. In short, SNG is a long-range possibility at best.[24]

The bright orange flares in overseas oil fields where natural gas is being burned off as waste have caught the eyes of natural gas companies and policymakers. If this gas could be shipped to the United States, the imports could supplement domestic supplies. Since construction of an undersea pipeline from the Middle East, North Africa, or Indonesia to the United States is an obvious economical and technical impossibility, U.S. firms have turned to the proven technology of liquefying the gas at very low temperatures and pumping it aboard specially designed tankers for delivery to the United States. Here the liquid is regasified and fed into pipelines. As proven reserves of domestic gas declined in the 1970's, imported liquified natural gas (LNG) became the promised salvation for many companies. As Harvey Proctor, executive vice-president of Pacific Lighting Companies, explained, "The Pacific Lighting Companies, as distributors of gas to southern California, are having to cut back on gas service to low-priority, large-volume customers. These cutbacks will deepen progressively as our supplies decline, until by the early 1980's we will be forced to curtail service to the smaller industries and businesses which have no alternative fuel capabilities. . . . Our two LNG projects, one involving gas from Indonesia, the other from the Cook Inlet in south Alaska, are the only feasible projects which could supply gas to southern California within the critical time period needed to head off the prospect of curtailment of gas service to high-priority industrial customers and the resulting mass unemployment. . . . The impact of delays on these LNG projects is almost too great to be calculated." [25]

The Distrigas Company of Boston became the first importer of

LNG, importing LNG from Algeria into Boston in 1974. By 1977 its imports had built up to an annual rate of about .04 tcf, less than one quarter of one percent of total American gas consumption, but locally important. In the spring of 1978 two other much larger LNG projects began operating. The El Paso Natural Gas Company began to import Algerian LNG to two terminals on the East Coast, one in the Chesapeake Bay, the other near Savannah, Georgia, at an annual combined rate of about one-third tcf per year.

An additional ten projects for LNG large-quantity importation were being planned in 1978. But in December, the administrator of the Energy Regulatory Agency of the Department of Energy denied applications by Tenneco and El Paso Natural Gas for two large LNG import projects. These decisions signaled the Carter Administration's deep reluctance to see LNG imports increase beyond their current modest levels.

Even if federal government policy should change, new LNG projects face at least three additional obstacles.

The first is a dispute over safety. Some scientific experts have claimed that the handling and transportation of LNG poses great hazards to the public. Accidents to LNG tankers entering port or while unloading, accidents at LNG storage facilities, and motor vehicle accidents involving trucks that transport LNG—all threaten the public with the risk of potentially catastrophic explosions and fires. Equally reputable scientists support the gas industry's view that such risks are under control. They cite the special design features and operating requirements of LNG ships and storage facilities as protection against catastrophe, and argue that highway transport of LNG is no more hazardous than the transportation of gasoline.

As with the nuclear safety controversy, there seem to be serious gaps in government-sponsored research and development programs to answer the issues that are in dispute. There is, for example, no good experimental basis for predicting the consequences of a collision between an LNG ship and another vessel. Gaps in scientific knowledge such as these create so much uncertainty that a wide range of risk assessments is possible.

Cost is a second obstacle to the wide-scale use of LNG. To build facilities for a typical project requires over $2 billion, most of which is needed for the liquefaction plant and the specialized ocean-going ships. If absolutely nothing were paid to the exporting country for the value of the gas, a price of at least $2.50

per mcf would be required to cover operating expenses and amortization of this capital investment. Because of such high handling and delivery costs, the Algerian government, in its initial contracts with U.S. firms, accepted prices for the value of the gas (exclusive of handling and delivery) that were equivalent to only a small fraction of the OPEC price for oil. But virtually all gas specialists expect that gap to narrow in the future. Some predict prices for LNG delivered to the United States to be in the range of $5 to $7 per mcf by the early 1980's, a view reinforced by indications that OPEC members are considering ways to control the world market prices for LNG.

A third obstacle is that importation of LNG puts the United States in virtually the same position of foreign dependence that characterizes the importation of oil. The difference is that radical Algeria rather than conservative Saudi Arabia would dominate the imports.[26]

THE ARGUMENT OVER NATURAL GAS POLICY

It is hardly surprising that the argument over natural gas policy has been so bitter and drawn out. Consumers, producers, the people of gas-rich geographic regions, and regulatory authorities—all have something to gain as well as something to lose; no one is neutral.

At the risk of oversimplification, four broadly based factions can be identified: Natural gas producers, who want deregulation of all wellhead gas prices, arguing that this would prevent a shortage of domestic natural gas in the decades ahead. Interstate pipeline companies, which also favor wellhead price deregulation, but believe that it would not materially relieve the nation's long-term chronic shortage of domestic gas. The government, which also believes there is a gas shortage, but sees this as a reason to maintain wellhead price controls. And the public, the majority of which favors price controls, because it doubts that there really will be a gas shortage.

Each of these groups can point to some evidence that appears to support its particular combination of beliefs. Each can find "experts" to argue its case in public or before legislative committees, and each, quite naturally, has patrons in either the executive or legislative branches of government, or both. Indeed, much of the debate that the public hears is actually among specialists representing the various factions, drawing upon well-stocked arsenals of competing theories and contradictory data.

Natural gas producers. In 1978, there were some 6,000 producers selling to the pipelines. These producers can be separated into four categories: [27]

1. Two dozen large companies, mostly integrated petroleum firms, accounting for over one half of total production
2. A handful of producing affiliates of certain pipeline companies
3. Approximately 500 large "independent" producers
4. Approximately 5,400 small independent producers

Ben Cubbage, president of the Independent Oil Producers Association, summarized the producers' position: "What we are running out of is very cheap natural gas. The vast remaining supplies will not be discovered at current regulated prices. Developing this gas requires immediate deregulation." [28]

Regulated prices, according to producers, often are not high enough to cover production costs. In the case of existing gas, producers state that in many instances the regulated price is below the costs of expanding the potential reserves in existing fields; for newly discovered gas, the price is frequently below the cost of exploration and production. Any study that indicates that a shortage would still exist at higher prices, producers say, is superficial, because it does not account for the distorting effects caused by price regulation.

The intensity of belief among producers is by no means uniform. It varies roughly according to the size of the company. At one end are the small independents, operating in relatively low-quality, small-margin gas fields, which contribute approximately half the U.S. supplies. Jim Daugherty, an independent drilling contractor and gas producer in western Kentucky, captured the independent producers' position: "Hell, yes, there's gas out there, and there's folks right over the state line in Illinois that will pay for it at whatever price it takes to produce it, but that price to produce it ain't the price that those fellas back in Washington say you can sell it for. So we don't plan to drill for gas until it's profitable." [29] In fact, the independents want more than just higher prices; they want complete and instant deregulation of newly discovered gas.

As far as independents are concerned, alternative estimates of price and supply of natural gas are beside the point: The point is that big risks should mean big rewards. So any public policy that tries to moderate such rewards in the interest of other objectives is unjust. In 1977 and 1978, the independents mounted a well-organized and well-financed lobbying effort to end government price regulation once and for all. Their intransigence helped delay a compromise on the natural gas bill.

For the large, fully integrated oil companies at the other end of the producer spectrum, life is not so simple. These companies are struggling to rescue their primary business, oil, from government regulation, and to stave off possible "dismemberment." Thus, taking a low profile on gas, they have generally limited themselves to saying that deregulation would benefit the public at large by stimulating exploration and drilling.

The interstate pipeline companies. The executives of many of these companies face a stark future. Over a period of years their business could gradually die because they would have no gas to transport. One of the few points of general agreement in the natural gas debate is that division into a two-tier market has ultimately reduced the volume of gas transported by the interstate pipelines. The twenty-five companies in this business, in the words of President Carter's National Energy Plan, have been systematically starved for gas because they "are effectively excluded from bidding on gas in the strong intrastate markets." [30]

Pipeline companies argue that even if higher wellhead prices are necessary to increase gas supplies, the public could still enjoy lower prices because of pipeline economies. If increased gas supplies enabled the interstate pipelines to operate nearer to capacity, their fixed costs would be spread over larger volumes, thereby lowering the cost of transportation. Since the prices that the pipelines charge for transportation, as distinct from the prices they pay at the wellhead, would remain under government control, substantial savings could accrue to consumers who live far from the gas fields. For example, New York City residents paid about three dollars per mcf for gas in 1976. About 80 percent of that charge represented transportation costs. Proponents of wellhead price deregulation say that the price of natural gas delivered to residential customers could fall by 10 percent or more if the interstate pipelines were able to operate at full capacity. [31]

But many in the pipeline business are not optimistic about increased gas supplies. Howard Boyd, chairman of El Paso Company, has said, "Our founder, Paul Kayser, told me before he retired as chairman of the board in 1965 that someday in the next twenty-five years we would run out of gas, and to be prepared for it." El Paso has sunk over a billion dollars in a fleet of tankers for LNG importation as one hedge against a shrinking domestic gas supply.[32]

Since 1968, many other large pipeline companies concerned about dwindling supplies have also diversified to reduce their dependence on natural gas transmission. Houston Natural Gas, an intrastate pipeline, for example, acquired a worldwide manufacturer and distributor of carbon dioxide and a coal company, while Southern Natural Gas has become a global offshore drill-rig operator. "We are striving to become a diversified energy company," says Robert Herring, Houston's chairman. "Gas's role will likely diminish in importance to the nation, but who can say what the next big energy source will be? So Houston Natural Gas Company will be in them all." [33]

The pipelines' drive to diversify stems from two related causes. Within the industry itself, there is real doubt that large undiscovered reserves exist. In addition, the companies are skeptical that after twenty-three years of regulation, public policy will be able to adjust to the present circumstances. As Ed Najaiko, a vice-president of El Paso, has said, "At higher prices, we believe, based on historical trends, that more gas will be discovered. If you ask me if deregulation will provide us with enough gas at prices competitive with alternative fuels to meet the needs of our customers, I don't know, but I would be unwilling to gamble El Paso's future on that uncertainty." [34]

The government. A variety of usually shifting positions toward natural gas policy exists within both the executive and legislative branches of the federal government. During the 1976 campaign, Jimmy Carter pledged to work for deregulation; but once in office, Carter reversed himself and opposed deregulation and value-based pricing, saying that supply, after all, was not very elastic with respect to price. In the summer of 1978, a coalition of very strange political bedfellows formed as both liberal and conservative members of Congress allied to try to defeat a natural gas compromise, but for very different reasons. The legislation seemed to promise prices that would be too high for the liberals' constituencies and too low for the conservatives'. The conservatives also opposed any regulation of intrastate gas.

However, the Carter Administration generally adhered to a policy of continued regulation to control the price of a commodity it believed to be in limited supply. In September 1977, Carter explained his position: "Deregulation will only increase the price which American consumers must bear. Therefore, we must continue regulating the price of natural gas." In like fashion, a central theme of the Department of Energy in 1977 and 1978 was that gas supplies were relatively inelastic with respect to price.[35]

The public. If the producers are wrong, if gas supply really is inelastic with respect to price, then the end of wellhead price regulation on newly discovered gas would indeed mean a transfer of wealth from consumers to producers without any compensating economic or social benefit. As Lee White, one of the more eloquent spokesmen for consumer interests, put it,[36] "If there were reason to believe that these excessive drains on the economy and on family income would produce greater volumes of gas than would otherwise be produced, one would be willing at least to consider such an alternative. However, this is not the case. Every econometric model developed to demonstrate this relationship has been picked to pieces by opponents. All we have to go on is the general gut reaction of producers. . . . This is really not good enough."

The public at large seems to doubt the reality of gas shortage, to suspect corporate manipulation, and to question the rationale for higher gas prices.[37] The public feels that it is unfair for domestic gas producers to get windfall profits on proven reserves simply because OPEC's price increase for oil has increased the value of gas as well. OPEC's action has abruptly changed the worth of gas, a commodity that was already under great pressure in the energy market. In such circumstances, so dramatic an increase in values would create enormous stakes and correspondingly high passions, and would almost inevitably assure the classic political controversy over who gets what.

Further heightening the passions are regional differences of interest. The public in New England and the Middle Atlantic states, areas with essentially no gas production, have historically paid a higher price for natural gas than the national average because of greater transportation costs. Furthermore, the 1970's shift of the new gas supplies into the intrastate market raised the fear of chronic shortages in non-producing states.

In contrast, residents of Texas, Louisiana, and other gas-producing states complain that the East Coast residents have tried to block offshore drilling, while siphoning off large volumes of gas subsidized at a low price at the expense of the gas producers. Bumper stickers in Texas reflect this sense of frustration: "Turn up the gas and freeze a Yankee."

THE FUTURE

The Natural Gas Policy Act, which passed Congress in October 1978, provided a framework for the future. In broad outline, the major provisions of the Act can be readily summarized. The basic idea was to permit a carefully managed deregulation of the price of newly discovered natural gas. After a decade of small annual escalations, according to a complicated-looking but actually very simple formula, the price of newly discovered gas would be decontrolled at the wellhead. Meanwhile, the wellhead price of newly discovered gas that is consumed in the producing state would temporarily be brought under the control umbrella of interstate gas.

It is easy to see who loses by the move: first, the producers who are now selling gas to the intrastate market. If the newly controlled wellhead price of intrastate gas is less than what the free market price would have been, such producers stand to lose billions of dollars. Second, the consumers lose, at least in the short run. "New" gas sold across state boundaries starts at $1.75 per mcf and climbs with inflation and an allowed real growth rate. "Old" interstate gas prices will initially be unchanged, but as contracts for "old" interstate gas expire, the compromise legislation would allow prices to rise to $.54 per mcf, if the original contract price was less. If the original contract price was higher than $.54 per mcf, the government may determine a "fair and reasonable" price.

Since the legislation means increases in real costs to consumers, it has understandably aroused a good deal of public dismay. But the appeal of the compromise is that it offers some real benefits to many producers and distributors as well as consumers. Producers receive a higher price for interstate gas than heretofore has been legal. Also, the interstate pipeline companies can effectively compete with the intrastate companies for newly discovered gas. And all consumers in non-producing states will have considerably greater assurance of stable gas supplies than in the previous two-tiered market. Finally, Washington gains flexibility to respond to unforeseen events on a truly national basis in the coming years.[38]

Something very important, however, is not changed by the compromise legislation. The residential consumer will still have first claim on the nation's gas supplies, however scarce or abundant they prove to be. Since the *Phillips* case in 1954, the federal government has assigned priorities among users. The priorities are, beginning with the most important, (1) residential users, (2) industries using gas for feedstocks, (3) industries with gas boilers, and (4) electrical utilities.

Another thing not changed is that there still is continued controversy over natural gas. Although the Act represents a considerable political accomplishment in light of the numerous and deep conflicts of interest that the lawmakers faced, its details leave much to be desired. For example, the new law's pricing provisions rely heavily on distinctions between "newly discovered" gas and "old" gas which seem straightforward in principle, but which are certain to be very complicated to determine in practice. Furthermore, contrary to the understanding of the gas producers and pipeline companies, FERC issued a preliminary interpretation that the Act does not allow automatic escalations to price ceilings, as ordinarily occurs when a FERC ruling is involved. Instead, FERC contends that because the Act was not a regulatory ruling, the producers must renegotiate each contract to obtain price increases allowed under the Act.

The recognition of the so-called gas bubble further complicated the picture. A number of factors helped bring about the temporary surplus. During the winters 1976–77 and 1977–78, a number of low-priority users were obliged to switch to oil or coal. Many did not switch back. Other industrial users concerned about the continued availability of gas switched out of gas. And users in general began to conserve because of higher prices. The result was that residential and commercial consumption of natural gas grew slightly from 1973 to 1977; meanwhile, industrial consumption plummeted by almost one third, and electric utility consumption fell by 11 percent. With the passage of the Natural Gas Policy Act of 1978, some of the natural gas supplies from marginal fields and shut-in gas wells previously held for the intrastate market were offered to the interstate market in response to anticipated price increases in that market. The convergence of decreased demand and increased supply created the gas bubble. An important factor in determining how long the bubble lasts will be the willingness of industrial users to switch back to natural gas. Gas-industry observers expect that the bubble, involving in total about 3 tcf, will gradually collapse over the next two to five years.

These complications and controversies in turn suggest that some results of the legislation will be a new wave of litigation and battles over regulation, and, no doubt, considerable efforts by various interests to get legislation rewritten. In other words, the Natural Gas Policy Act of 1978 has hardly ended the religious war over how to price the prince of hydrocarbons.

In general, the Natural Gas Policy Act and any subsequent legislation could have three possible outcomes on the vexing question of future gas supplies. First, higher prices might stimulate exploration to such an extent that the discovery of new reservoirs would support consumption in excess of the current 20 tcf per year, say, 25 tcf per year or more. In that case, new electric generating plants might even be permitted to use natural gas. This seems, on the available evidence, to be the least likely outcome, but it is possible.

Second, and only slightly more likely, is that higher prices could fail to stimulate more than the current 10 tcf of new discoveries per year.

The third and most likely outcome is that annual discoveries will range between 10 tcf and 25 tcf per year. Within that range, a figure closer to 10 tcf means further restrictions on the industrial use of gas as boiler fuel and as feedstock, plus stern enforcement of mandates to convert existing gas-fired industrial and electric utility boilers to coal. A figure closer to the upper end of the range means little change in present use patterns.

It is tempting to say that the most likely outcome is in the middle. But, of course, no one really can say for sure. The essential point is that the nation should not plan on greater quantities of natural gas to stop the rise in oil imports. Indeed, it will be a challenge to find enough new gas reserves to maintain production at current levels.

MEL HORWITCH

With the assistance of Frank Schuller

4

Coal: Constrained Abundance

Coal has been rediscovered. America's most abundant fossil fuel had powered the country's shift from an agrarian to an industrialized society in the late nineteenth and early twentieth centuries. But with the appearance of cheap and convenient oil and natural gas shortly after World War II, coal experienced a dramatic decline in key markets, and soon the local coal delivery truck, the coal-fired boiler used in the factory, and the steam locomotive practically vanished from the American landscape. When President Truman battled United Mine Workers' president John L. Lewis in 1946 during a debilitating coal strike, coal supplied about half of America's total energy needs. When President Carter threatened to use the Taft-Hartley injunction in 1978 during a much longer coal strike, coal accounted for less than 20 percent of total U.S. energy consumption.

The Arab oil embargo and the shocking realization of America's growing dependence on foreign oil really signaled the public's rediscovery of coal. Suddenly the United States was called "the Persian Gulf of coal," and the resource itself was termed the "great black hope." The coal industry was characterized as "a somewhat frumpy middle-aged ballerina rushed out of semiretirement to fill an unanticipated gap in a show that must go on. Suddenly the old girl is back in demand." And so a "new age of coal" was proclaimed. The airline flights that linked the coal areas of Appalachia and the Midwest with the financial centers of the East Coast were soon booked with bull-market deal makers eagerly attempting to acquire coal mines or reserves. At the very least, coal was thought to be the "transition" fuel until effectively inexhaustible energy sources, such as fusion or solar power, became available.[1]

Even as the Carter Administration proposed and issued regulations which tended to retard growth in coal usage, President Carter also made coal's rebirth official in April 1977 when he unveiled his National Energy Plan, which gave coal a key role. Carter specifically mentioned the seemingly huge disparity between America's coal resources and coal's current use. Although coal comprised about 90 percent of total U.S. energy reserves, he said, it provided only 18 percent of total energy production. The President called for over an 80-percent increase in coal production by 1985–from about 680 million tons in 1976 to over 1.2 billion tons.* He further proposed an expanded program of research and development in such areas as mining, burning, liquefying, and gasifying coal. Carter also pushed for the conversion from scarcer fuels to coal "whenever possible." Indeed, he looked to "the expanded use of coal, supplemented by nuclear power and renewable resources, to fill the growing gap created by rising energy demand and relatively stable production of oil and gas." All this, the President proclaimed, could pay "rich dividends." [2]

Indeed, the United States is a Persian Gulf of coal; according to one estimate, it possesses about 27 percent of the earth's coal reserves. But projections of U.S. coal reserves vary widely. The U.S. Geological Survey estimated total identified U.S. coal resources at about 1.7 trillion tons, and postulated hypothetical U.S. coal resources in unmapped or unexplored areas at another 1.8 trillion tons. Projections of recoverable reserves range considerably–from about 150 billion tons to 438 billion tons. But it is safe to assume that the United States has enough coal–if one looks only at the

* This is the energy equivalent of an increase of about 7.9 million barrels of oil per day to about 14.5 million barrels per day, according to the National Energy Plan.

physical resource itself—at any reasonably expected level of production for at least the next hundred years.[3]

But bullish rhetoric and expectations have run into a harsher reality. President Carter's original goal of 1.2 billion tons by 1985 is now extremely doubtful, and reasonable estimates for 1985 coal production range all the way from 800 million to 1.1 billion tons. The Carter Administration's sudden policy shift in early 1979 of encouraging industrial and utility plants to convert from oil to natural gas, not coal, at least in the short term, engendered a feeling of betrayal in the coal industry after the emphasis the President had placed on coal in his National Energy Plan twenty-one months earlier. One coal company executive commented, "They keep saying they're all for us, but every chance they get they stab us in the back." Meanwhile, by early 1979 the coal industry faced a weakening of demand as its key markets grew more slowly than usual and as both it and its customers confronted a bewildering array of uncertain government regulations. For really the first time, even the popular media was reporting that "new fears surround the shift to coal," particularly in the environmental area.

The so-called transition to coal, therefore, may prove difficult, if it is possible at all. True, the United States has an abundance of coal. But coal in turn has an abundance of problems. Its greatest positive attribute, as the General Accounting Office observed, may be simply that "there is a lot of it."

This is not to say that the coal industry will not experience impressive growth in the short term. Even an annual production of about one billion tons by 1985 is still considerably above current production, and would represent a healthy 5 percent annual growth rate.[4] Moreover, the longer-term prospects for coal—fifteen to twenty years—are brighter, with coal's share of its current markets increasing. Even now, coal companies are having some success in managing a difficult operating situation. New participants with strong managerial, technological, and financial resources are entering the industry, and new technologies, particularly in the area of coal utilization, may provide more convenient and less environmentally hazardous ways to take advantage of coal's abundance.

THE SHORT TERM

A ubiquitous and diverse set of obstacles stands in the way of reaching the original Carter short-term goal of 1.2 billion tons by

1985. These constraints are not all of equal importance, but all have a common major characteristic: Each will require considerable time to be overcome, if it can be satisfactorily resolved at all. These obstacles are partially systemic, partially environmental, and partially cultural and sociological. Although they are interrelated, for purposes of clear discussion each barrier will be examined separately.

The System

The coal industry can be conceived as a system that produces, transports, and consumes coal. Consumption is the place to begin, because it triggers activity in the rest of the system and because it is the major short-term bottleneck.

Although coal production climbed during both world wars, it suffered serious declines during the postwar periods. Coal could not compete with petroleum and natural gas in key sectors, and after the Second World War, as shown by Figure 4-1, it totally lost its hold on the retail and, to a somewhat lesser extent, industrial markets. On the other hand, electric utility consumption grew substantially, particularly after 1950, and caused coal demand to achieve record levels in the 1970's.

But even within the electric utility industry, coal's share of the total electric generation market declined between 1955 and 1972. Import controls for oil, which were initially established in 1957, were subsequently relaxed for heavy fuel oil used by boilers, and were finally abandoned completely in early 1973. Stricter clean-air standards were passed in the sixties, which also had the effect of favoring oil over coal. Therefore, oil—not coal—was the attractive fuel for utilities in most areas in the late sixties. In addition, by the end of the sixties coal confronted still another challenge to its utility market: large nuclear power plants. In short, by the early seventies coal's last bastion—its relative position in the utility industry—was threatened by oil and gas, increasing environmental regulation, and nuclear power.

But the coal-consumption prospects appeared to brighten considerably after 1973. Although not directly caused by the oil embargo, coal prices rose after 1973. Still, the subsequent oil price rise increased coal's relative cost advantage. Moreover, there was a growing realization of significant delays in bringing nuclear power plants on line.

Figure 4–1

U.S. CONSUMPTION OF BITUMINOUS COAL AND LIGNITE

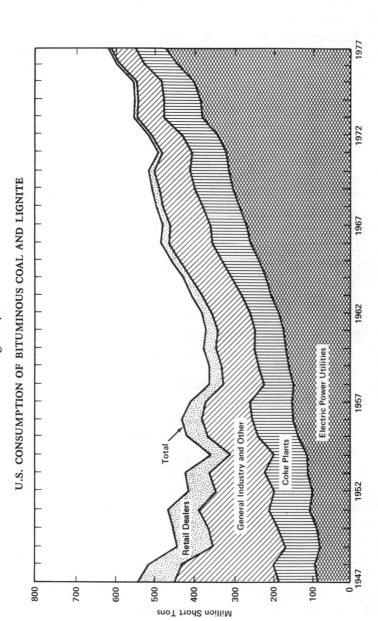

Source: Energy Information Administration, *Annual Report to Congress, Volume III, 1977* (Washington, D.C.: Government Printing Office, May 1978), p. 78.

Federal policy after 1973 also enhanced coal's position in the utility sector, beginning with the Energy Supply and Environmental Coordination Act in 1974. ESECA authorized the Federal Energy Administration to order power plants and other "major fuel-burning installations" to substitute coal for oil or gas as their boiler fuel, a policy termed "coal conversion." But this initial attempt to substitute coal for petroleum was not particularly effective, primarily because of environmental regulations. By mid-1977, of the seventy-four conversion orders issued by the FEA, only fifteen had received approval from the Environmental Protection Agency.[5]

Another government policy—strengthened by the 1978 National Energy Act—discourages the installation of oil and gas boilers for new generating plants. This policy, along with management considerations of fuel price and security of supply, increased the market for coal, especially in the region comprising the states of Arkansas, Kansas, Louisiana, Oklahoma, and Texas. In 1970, 99 percent of all utility fuel in this region was gas, but by 1975 reliance on gas had dropped to 87 percent, with an additional 40 percent reduction scheduled by 1985. However, a sudden shift in policy by the Carter Administration in early 1979—encouraging the short-term use of natural gas instead of coal—confused projections for this region. At this time the Texas Railroad Commission proposed rescission of a 1975 order requiring major gas users to convert to coal by 1985.[6]

The competition from nuclear power has weakened, the cost of oil and gas has increased, and government policy has encouraged conversion to coal. Nevertheless, the utilities' overall ability to change their fuel mix in the short term is limited, and they are also postponing certain coal-fired facilities. Coal's share of the total mix of fuels used to generate electricity is increasing only gradually. The National Electric Reliability Council, which compiles utility data and forecasts, estimated in 1978 that coal's share of the total fuel used by utilities as measured in kilowatt hours would only increase by about one percent between 1978 and 1987—from 48 percent in 1978 to about 49.3 percent in 1987.

Why have utilities made no wholesale move to coal? Their cautious behavior results from the radically different environment that the electric utilities faced after 1973. Rarely does such a stable and prosperous industry confront the kind of upheaval and uncertainty that the electric utilities have experienced since 1973. Until then, the historic annual growth rate for the industry hovered between 7 and 7.5 percent. The growth rate has since plummeted, and the electric utility industry has grown at far below recent historic rates. Recent projections of annual growth rates range between 4 and 5.5 percent for the next decade, and some respected analysts think it could be even lower.

The general causes for the decrease in the growth rate seem clear. One reason is the overall slowdown in the U.S. economy. This has hurt industry, which consumed about 40 percent of the output of electric utilities in 1977. Another obvious reason is that the price of electricity rose considerably after the 1973 oil embargo as utility costs increased. Between 1972 and 1975, the average residential electric price increased by over 13 percent per year, and has continued to rise since then; industrial prices increased even faster.

Price increases, along with government legislation and regulation, have encouraged energy conservation in manufacturing processes, appliances, and buildings—all of which work to reduce growth in the demand for electricity. Other factors tending to constrain the growth rate of utilities include government air-pollution regulations and licensing procedures (which increase costs) and related difficulties in raising capital for new plants.[7] Finally, some people see in electric utilities the embodiment of an older set of values—a commitment to unchecked economic growth—at a time when another set of values has appeared, embodied in the idea of limits to growth. Hence, utilities have also become a target of more general social protests.[8]

The problem for utility managers—and consequently for forecasting future utility coal consumption—is that this radically different and difficult environment might very well continue to change in unpredictable ways. In July 1977, the National Electric Reliability Council warned that the future electric power supply would be in "jeopardy" if planned generating units were significantly delayed and if demand picked up. In fact, this group saw possible power "deficits" in the near future in certain regions.

In any event, long-term planning for the electric utility industry, a task which was once relatively simple, is now a difficult and uncertain responsibility occupying, as it frequently did not in the past, senior management. As *Electrical World* observed, "The [electric] load-growth pattern—or lack of it—since late 1973 has no precedent in utility history. The excursions we have experienced clearly show that traditional trending methods no longer apply, especially for the years in the immediate future." The days of regular, steady growth, comparatively short construction lead times, guaranteed fuel supply, and easily obtainable financing are over. In short, utility electric-load forecasting, as one utility planner noted, is "not easy" anymore.[9]

Even if capacity forecasts could be more certain, there would still be great difficulties in deciding what kind of plant to build. Many estimates indicate that the total electricity-generating costs of nuclear and coal (with desulfurization systems) are close. Thus, the decision to build one or the other is not a straightforward task of selecting the lowest cost option. The coal strike of 1977–78 demonstrated the importance of maintaining a strong interconnected electric transmission system and of relying on a diversity of fuels. The regions that were struck imported large amounts of electricity that were generated by oil and nuclear plants. After the strike, two analysts for the Commonwealth Edison Company observed that "the advantage of having a mix of fuels is obvious: the eggs are not all in one basket. This offers some protection against monopoly pricing, strikes, embargoes, and weather effects."

These uncertainties explain why utilities have remained cautious in adding generating capacity and in deciding on fuel sources, and why, therefore, forecasts for coal use in the mid-1980's can vary so much. It is, above all, demand that will determine the coal industry's growth. As even a rather bullish 1977 report by the National Coal Association observed, "Without a market, a coal mine will not be opened or expanded." After the end of the long 1977–78 coal strike, one investment analyst commented, "It's remarkable that after a 110-day strike the market is as soft as it is. Thank goodness we had such a long strike, otherwise we'd be awash in coal." Moreover, this softness in the coal market increased during the next twelve months.[10]

The growing use of long-term coal delivery contracts gives even greater importance to utility demand uncertainties in determining overall coal production, because such contracts explicitly link the opening of new mines and the expanding of old ones to utility decisions. The practice coincided with the rise of the utility market, the associated increase in power plant size, the growing desire by some utilities to use low sulfur coal from large Western surface mines, and the increasing capital expenses encountered by coal companies and utilities. Without such contracts, coal companies are increasingly reluctant to open major new mines, which further hinders expanded production.[11]

To comprehend other barriers, one must know something about the geographic distribution of coal reserves and its relationship to mining methods. Coal mined in the United States is usually categorized as either Eastern or Western. About 54 percent of coal by weight—but only 30 percent by heat content—is estimated to lie west of the Mississippi, far from the traditional Eastern markets that use it to generate electricity (steam coal) or to make steel (metallurgical coal). Western coals are also generally lower in sulfur content by weight; and sulfur dioxide, of course, is a prime atmospheric pollutant when coal is burned. Consequently, Western coals are generally less polluting, although they also provide less energy per pound. So, on a comparable basis of heat value, the difference in the emission polluting potential between Eastern and Western coals lessens, although it is still significant.

Much of the Eastern coal is extracted by underground mining. Most Western coal can be surface-mined (strip-mined), since the seams often lie less than two hundred feet below the surface. The overburden—the earth above the seam—is blasted and removed by giant draglines or by huge shovels, and the exposed coal is loaded onto trucks, which move it to railroad sidings. According to the 1977 Surface Mining Control and Reclamation Act, the mining companies must then reclaim the land. Moreover, laws in many states already required reclamation.

Surface mining is more efficient than underground mining, which recovers only 50 to 60 percent of a seam's coal. Up to 90 percent may be recovered through surface mining, and the average productivity of surface mines in 1977, measured in tons per worker-day, was slightly less than three times that of underground mines. And by 1977, surface mines produced about 60 percent of total U.S. output.[12]

This geographic distribution of coal profoundly affects another systemic barrier, the transportation network that links the production and consumption of coal. Coal is shipped by various modes of transportation and at times is even transshipped—for example, from trucks to barges. In 1975 about 65 percent was shipped by rail, 11 percent by water, 12 percent by truck, 11 percent used at mine-mouth generating plants, and 1 percent transported by other methods. It is important to remember that the transportation of coal can be a difficult and complex operation, especially when compared to the ease with which oil and gas can be moved through pipelines and in tankers. Although barge can be the least costly method of transporting coal where available, the coal industry will have to rely on the railroad to carry a major share of planned new tonnage because this is the only realistic option in most cases.[13]

This increasing dependence on the railroad will be especially great in the West, which is estimated to be the major growth area in coal production, because its low sulfur coal can be surface-mined as well as because of the expected growth in the use of coal by Western utilities. This makes the railroad the key means of transport, because it is already in place—although not necessarily with sufficient capacity—where there are no navigable waterways. The Federal Energy Regulatory Commission estimated that the railways would carry 62 percent of the coal for new electric generating units between 1977 and 1986.

Can the railroads deliver? In the East, there is already a well-established coal rail network, and many consumers, producers, and railroaders believe that the lines—despite some short-term delays—can handle the additional business. More uncertain is the ability of the railroads to carry the expected growth of coal traffic in the West. The principal rail carriers of Western coal are the Burlington Northern, Chicago and Northwestern, Union Pacific, and the Denver and Rio Grande. To service the large surface mines of the West, these railroads are increasingly using "unit trains," which consist of a chain of about a hundred hopper cars with each carrying a hundred tons of coal. In 1976 the Burlington Northern used about fifty-five such trains per week, and by 1985 is expecting to operate two hundred. Other Western railroads are forecasting similar rates of increase. But the new demand on railroads to carry coal places enormous stress on their capital expenditure requirements for hopper cars, locomotives, physical plant improvements, and maintenance facilities.[14]

Railroads also face other problems. For one thing, Western railroads create their own set of environmental and social hazards. Frequent unit trains can rumble through towns, disrupting whole communities and, in effect, cutting a town in half. Auto traffic is delayed, and general inconvenience engendered. Various studies also project an increased number of train-related accidents. It is no surprise that some Western towns along the routes and environmental groups are resisting the railroads. For example, the Sierra Club has sued to force a more thorough environmental impact study of a proposed 116-mile coal route that will service the huge Wyoming coal reserves. The route, the Sierra Club claims, would allow as many as forty-eight unit trains daily to pass through several small towns. Meanwhile, the cost of bypassing such communities altogether—or of building numerous bridges, overpasses, or underpasses—presents a significant and still largely unknown factor in the economics of coal rail transport.[15]

There is some chance that the railroads might not carry all the increased output of the West, for they face potential competition from slurry pipelines. In a slurry pipeline system, mined coal is successively cleaned, pulverized, mixed with water, pumped through a pipeline, and dewatered for use. In 1978 only one slurry pipeline operated in the United States, a 273-mile line transporting 4.8 million tons annually from Peabody Coal Company mines at Black Mesa, Arizona, to a Nevada utility. But at least five new pipelines have been proposed, including one that would transport 25 million tons of coal annually more than one thousand miles from Wyoming to Arkansas. Slurry pipelines seem to have a cost advantage for long-distance high-tonnage runs of coal, and in any case, their potential development could keep railroad freight rates lower than they would otherwise be.[16]

But the decision to build slurry pipelines is not likely to be decided solely on the basis of cost. Political and environmental considerations are also important. The railroads, determined to maintain a dominant positon, have fought legislation at both the federal and the state level that would give slurry pipelines the right to eminent domain across land owned by the railroads. But pipeline sponsors, through successful court suits and state legislation, are gradually obtaining eminent domain power, without which most proposed pipelines are stymied. It does appear that sufficient water exists for slurry pipeline operations (slurry pipelines, in fact, use less water than mine-mouth generating plants or coal gasification plants), but the water is not yet legally available. And in the West, water rights are a volatile issue. In mid-1978 a coalition of railroad interests, environmentalists, and Western congressmen decisively defeated a coal slurry pipeline bill.[17]

Further, a cloud of uncertainty hangs over any decision having to do with Western coal transportation. There is still some question whether the projected high growth in demand for Western coal will actually develop, at least before the mid-eighties. It must be remembered that both President Carter's energy plan and the 1977 Clean Air Act Amendments call for the "best available control technology" pollution-control equipment for desulfurization—or so-called scrubbers—on all new utility plants. If implemented, this policy would tend to reduce the market for low-sulfur Western coal, and to increase the demand for higher sulfur Eastern coal, because the key economic reason for preferring low-sulfur Western coal— scrubbers not being required—is removed. But the actual impact of the regulation is not clear. Because it is so much less costly to produce surface-mined Western coal, and because smaller scrubbers can be used, less limestone is needed in the scrubbing process, and the amount of solid waste disposal—sludge—is reduced. Therefore, Western coal may still be able to compete for the large Midwestern coal market. In addition, there is still confusion and debate over the precise sulfur dioxide removal requirement. In September 1978 the Environmental Protection Agency proposed a strict 85-percent-removal level with a floor of 75 percent for the lowest sulfur coal. But the Department of Energy, with support from White House staff concerned with inflation, favored a somewhat more lenient approach—85 percent, with a sliding scale for lower sulfur coal. The latter method was more favorable toward Western coal than the EPA's proposed requirement.

Meanwhile, other factors, such as environmental regulation, court litigation, and environmental impact statements, have also slowed down the rush to mine Western coal. Some industry observers claim that the Department of Interior's de facto moratorium on leasing coal reserves on federal lands—which represent at least 60 percent of all Western coal reserves—also is a deterrent; but the Department of Energy indicated in 1978 that even the original Carter goal of 1.2 billion tons could be reached without leasing more coal.[18]

So the development of an efficient coal transportation system finds itself in a Catch-22 situation. On the one hand, if demand substantially increases, the railroads face a severe challenge in transporting the additional coal. On the other hand, there are enough uncertainties about the actual demand for coal to inhibit the railroad's ability to raise the capital required to prepare for the explosive growth, if indeed it does come about.

In short, a critical obstacle to massive coal production is the short-term systemic barriers. Coal demand, particularly from utilities, triggers the rest of the system, and that demand is the crucial factor. Utilities are hedging because of their own set of doubts; meanwhile, the railroads, which in the short term provide the key link between increased production and increased consumption, also hesitate and therefore may not possess the capacity to meet the system's needs. Finally, the producers wait for long-term contracts, especially in the West, before opening new mines. Consequently, all parts of the system are hindered from assembling the required capital, which in turn reinforces the delays in increased coal production and improved coal transportation.

Environmental Barriers

A second major hurdle to massive increases of coal utilization in the near term lies in the resource's physical properties and setting. Coal possesses the troublesome attribute of generating a seemingly endless string of environmental hazards that are ubiquitous and pervasive. As soon as one hazard, such as sulfur dioxide emissions, is identified and solved, it seems that another possible environmental danger associated with coal, such as carbon dioxide emissions, becomes a source of controversy.

As it is, serious environmental problems exist at practically every part of the coal system. During production, underground mining can result in acid drainage, subsidence, and coal workers' pneumoconiosis (black lung disease). Surface mining requires careful reclamation, or the unrestored land will usually remain scarred and unproductive. As already discussed, the transportation of coal creates its own set of environmental effects, including disruptions of communities by unit trains and possible depletion of water by slurry pipelines. Finally, coal consumption generates still another set of serious environmental hazards, including emissions of sulfur dioxide, nitrogen oxide, trace elements (including arsenic, cadmium, mercury, lead, fluorine, and beryllium), and carbon dioxide into the atmosphere, thermal and chemical discharges into water, and the solid-waste-disposal problems of coal ash. In fact, the very process of reducing sulfur dioxide emissions with scrubbers creates a new pollutant, sludge.[19] This entire array of environmental problems creates such a complex—and politically and socially charged—set of issues that it is difficult to envision an easy technological fix to deal with all of coal's environmental effects.

These diverse environmental hazards work in diverse ways to harass large-scale coal utilization in the short term. The hazards have obviously led to greater government regulation, which extends the time needed to open mines and power plants. Such regulations lead to battles between pro-industry and environmental forces in courts, government agencies, and legislatures, which further increases lead times. For example, after four environmentalist organizations obtained an injunction in September 1977 against the issuing of new coal leases in the West on federal lands (where a de facto moratorium had been in effect since 1971), the Department of Interior took five months to secure an initial settlement. Meanwhile, several large coal companies, the National Coal Association, and the American Mining Congress have sued the Department of Interior, asking for major revisions in the 1977 Surface Mining Control and Reclamation Act. One official called the suit "the case of just-about-everyone versus the Department of Interior." [20] Another observed that "the real winners of the new regulation are the lawyers."

The potential environmental impacts of coal also substantially alter the market for key segments of the coal industry. For example, as mentioned earlier, the 1977 Amendments to the Clean Air Act, which limit sulfur dioxide emissions and require desulfurization treatment on all new coal-fired power plants, may substantially reduce the attractiveness of low-sulfur Western coal for Eastern markets.

At least one potential hazard of burning coal (and other fossil fuels as well)—the emission of carbon dioxide into the atmosphere —could have worldwide implications. This ultimately may well be the most difficult environmental problem. In 1977 the National Academy of Sciences warned in a report that a warming of the earth's temperature due to the "greenhouse effect" from increased carbon dioxide emissions might pose a severe, long-term global threat. The Academy went on to say, "The climatic effects of carbon dioxide release may be the primary limiting factor on energy production from fossil fuels over the next few centuries." The warning was soon echoed by other experts. Massive, long-term climate changes, including the melting of the polar icecaps and the shift of prime agriculture zones, have been suggested as major consequences.

There is disagreement as to how much of the increasing carbon dioxide is due to coal burning and other fossil-fuel combustion and to the cutting down of forests. It is certainly true that fossil fuels other than coal produce carbon dioxide. Still, the implication is clear: Coal and other fossil fuels may produce over time a global environmental hazard. Indeed, many critics (including, it should be noted, some advocates of nuclear power) have portrayed coal as the same kind of major danger that nuclear power is alleged to be by its critics. At a recent conference, a pro-nuclear speaker devoted his first twenty minutes to the potential carbon dioxide problems of coal. So, for the first time, it is possible to envision an absolute environmentally imposed limit to the use of coal in many parts of the world.[21]

The carbon dioxide issue aside, there are those, like the members of the Rall Committee on Health and Environmental Effects of Increased Coal Utilization (which submitted its report to the Secretary of Health, Education, and Welfare at the end of 1977, who in turn transmitted it to the President), who suggest that it is environmentally safe to proceed toward the Carter goal of 1.2 billion tons by 1985. But the Rall Committee also says that strong environmental and safety measures must be followed, and it stresses the uncertainties in information and data collection. The committee recommends rigorous compliance with all air, water, and solid-waste environmental regulations, universal adoption of the best available pollution-control technology on new coal-burning plants, rigorous compliance with reclamation standards and mine health and safety standards, and "judicious" siting of coal-fired facilities.

It may very well be that under such strict standards, environmental degradation may not be significant. But in the "real world," strict adherence to the recommendations of the Rall Committee may not be possible for a variety of economic, political, bureaucratic, and technological reasons. In mid-1978, for example, a Massachusetts utility agreed to convert an oil-fired plant to coal only after the federal and state governments exempted the utility from laws that would require scrubbers; it will use low-sulfur coal instead. In addition, the state had to guarantee that it would not impose tougher environmental laws on the plant for ten years.[22]

People Barriers

The great 1977–78 coal strike underscored another significant problem with coal: people. Unlike other widely used energy sources —oil, nuclear and natural gas—coal is more "people-intensive," and because it is, a number of short-term barriers to large-scale coal utilization are raised. Truly stable and planned increases in coal production for particularly underground mines depend on at least three key factors: the coal labor force and its productivity, managerial effectiveness and commitment, and the history and culture of the Eastern coal fields.

Labor force and productivity are usually the only human factors that receive attention. To understand the coal work force, one must remember the decades of mutual distrust between labor and management, the historic stark poverty of the Eastern coal fields, the dangerous and humanly exhausting nature of the work itself, and the miners' feelings of long-term exploitation. No wonder that there have been several severe labor-management wars since the beginning of the twentieth century. Not surprisingly, the coal work force and the union that represented it, the United Mine Workers of America, were often violent in asserting their demands. Before World War II, the UMW's president, John L. Lewis, proudly called his union "the shock troops of American labor." [23]

When Lewis made that claim, the UMW possessed about half a million working members. The postwar decline in production, the rapid mechanization of underground mining (which Lewis encouraged), the move from more labor-intensive underground mining to less labor-intensive surface mining, corrupt and ineffective union leadership after Lewis' reign, the shift of production to the West (which was outside the UMW's Eastern stronghold), and new competing unions—all these took their toll. By 1965, the UMW had approximately only 90,000 working members, which represented about two thirds of the total coal work force. With the rise in coal production in the early seventies, the UMW experienced rapid growth. By 1976, the UMW had well over 160,000 working members, or about 75 percent of the total work force, but the

UMW's control over coal production was still declining. Between 1972 and 1977 the share of coal produced by UMW mines fell from 75 to 52 percent.

Still, even if the UMW's control of coal production is receding—temporarily or permanently—its legacy of conflict, defiance, and distrust of management remains. Moreover, in the coal fields there is a "new breed" of miner, who is young, militant, and frequently well educated. In 1977, a thirty-year-old vice-president of a UMW district in southern West Virginia said of his members, "Now, they're very intelligent. They don't think the company is doing them a favor by hiring them. They think they're doin' the company a favor by workin'." Indeed, coal mining is attracting skilled workers from other occupations.

Coping successfully with the enduring posture of confrontation of the coal work force is critical for stable, growing coal production. Accomplishing as much will not be easy. In 1977, for example, there was an epidemic of wildcat strikes. The president of a UMW local in West Virginia later candidly explained, "Many coal miners have come to see the wildcat strike as the only means available to them to enforce their contractual job rights." [24]

In any case, productivity in underground mines fell drastically after 1969. The decline has been variously attributed to the provisions of the Mine Health and Safety Act of 1969, changes in mining conditions (such as the quality of seams), the introduction of a number of inexperienced miners, the requirements of additional personnel due to union agreements, unscheduled interruptions due to wildcat strikes, and absenteeism.[25] But whatever the reason, the decline in productivity, along with wildcat strikes, frustrated the coal operators during the period leading up to the 1977–78 coal strike. As the strike approached, they saw a union in disarray, with confused leadership and a depleted health-and-retirement fund.

One thing the operators did not see, or at least did not publicly acknowledge, was inadequacy in their own managerial ranks. Although an increasing number of mining engineers are now being trained, coal companies across the country still suffer from a shortage of qualified mining foremen, mine superintendents, and professional general management talent. And firms vary greatly in their interest or commitment to upgrade management staffs, with most coal companies only now beginning to hire professional labor-relations experts and administrators, and to recruit at schools of management. The traditional reluctance to recruit managers from outside the industry stems from the sound belief that mine experience is necessary for managing coal production. But the same perceived requirement tends to spill over into nonproduction areas, which significantly limits the talent pool for new managers and also constrains managerial creativity and innovation in such areas as labor relations.[26]

This does not have to be the case. There are a few underground mines, union and nonunion, which are highly productive, and management policy and behavior are important factors in their success. For example, the productivity of one coal company that operates underground mines in western Kentucky is 50 percent greater than most neighboring mines with similar seams. The firm is nonunion, but it pays basic union-scale wages and provides substantial production incentives, profit sharing, private medical insurance, and a retirement plan; it also pursues a policy of no layoffs and places great emphasis on recruitment. The firm's vice-president explained, "Our money and fringe benefits are better than anybody else's coal mine. Our miners are, perhaps, the richest working people in Kentucky. They have fancy homes. Some of them have boats which they take to nearby lakes. Quite a few are golf fans. They have expensive vacations. And why not? A few of these miners make—in salary, overtime, bonuses and profit sharing—as much as $25,000 a year."

Up to 1977 the firm never experienced a wildcat strike. When striking union members from other mines occasionally picketed the firm's mines, the company made a tacit agreement with its employees—the miners would try to get to work, but they would also avoid any confrontation with the pickets. If they were unable to get to the mine, they would return home. "We appreciated," said the vice-president, "that the decision of a miner to come to work can sometimes create tensions within his family, especially if there is a close relative who is out of work because of his union's strike." Although there are other factors that contribute to the firm's success, its consistent and innovative labor-relations policies and its commitment to implement them are key.[27]

The effects of endemic poverty are also powerful impediments to labor peace and stable production. Coal has been so closely associated with massive social deprivation for so· long that a few boom years cannot be expected to erase basic social scars. Several close observers of Appalachia have seen coal and its after-effects as destroyers of both the human and natural landscape. In addition, certain Eastern coal-mining regions still suffer from poor public services.

But the Eastern coal regions generally have never been entirely isolated from mainstream industrial America. Moreover, prosperity has come to the coal fields. In 1978, the average wage of a union coal miner was between $17,000 and $18,000. The comparable nonunion figure was about $23,000.[28]

In spite of this increasing affluence, however, old and deep-seated social problems remain, as illustrated by the strike of 1977–78. Although the strike itself was "settled," fundamental issues of labor productivity and motivation, managerial thinness, distrust of coal operators, and social insecurity generally still remain. Such human problems present the coal industry with difficult, though not impossible, operating barriers. These problems require dedication, patience, and time, and therefore will impede the increased production of Eastern coal in the short term.

Taken together, the three major types of barriers to the massive short-term utilization of coal—which are systemic, environmental, and human—stand in the way of our relying heavily on coal as an alternative to imported oil.

THE LONG TERM

Although coal may not be a panacea for our immediate energy problems, it will continue to experience steady growth. Moreover, its prospects by the turn of the century are much brighter because of two trends: the growing participation of a diverse and strong set of established companies that are newcomers to the industry, and the associated emergence of new technologies that permit cleaner and more convenient ways to utilize coal.

The New Participants

To understand the long-term importance of the newcomers, it is necessary to review briefly the dramatic changes that have taken place in the coal industry since 1960. Until then, the coal industry consisted primarily of coal-mining companies and a few steel firms and utilities. It was also regional and isolated, and was confined almost entirely to east of the Mississippi. Management was basically home-grown and the labor force was aging. And finally, the industry was experiencing a rapid decline in the railroad, industrial, and residential markets. In short, coal in 1960 was the prototypical "sick" industry.

But during the decade that followed, especially the latter half of it, the earlier decline of coal production and employment within the industry halted and then slowly began to rise, thanks to increased demand from utilities. Also in the late sixties, petroleum firms became major forces in the coal industry as they began to buy operating coal companies, acquired coal reserves, and established totally new coal subsidiaries.

In 1966 Continental Oil Company acquired the mammoth Consolidation Coal Company. At the time Consolidation was the second largest coal producer in the United States, with an annual production of about 49 million tons, or about 9.5 percent of all U.S. production. Two years later Standard Oil of Ohio and Occidental Petroleum Corporation also acquired major coal companies, while Ashland Oil created an affiliate to mine Western coal. Exxon, the largest petroleum firm in the world, also entered the coal industry during the latter half of the sixties, acquiring large federal coal leases in Wyoming and several hundred thousand additional acres in Colorado, Montana, North Dakota, and Illinois. Exxon subsequently established the Monterey Coal Company to operate underground mines in Illinois, and the Carter Mining Company to conduct its Western mining operations. By 1974 at least seventeen of the twenty-five largest petroleum companies had entered the coal business in some fashion, and it was clear that a new group of firms with strong managements and massive financial and technological resources had a large stake in coal.[29]

But this large-scale entry by petroleum firms has become a source of controversy and may place another obstacle in the way of further development of the industry. Some people are afraid that the oil companies will come to dominate the coal industry, with the result that competition within the coal industry or between oil and coal will be limited. Fear of such concentration of market power or of reserve control in the energy field by a few huge oil-based, but increasingly diversified, energy companies has resulted in demands for horizontal divestiture of the petroleum firms. In its typical form, horizontal divestiture would force major oil producers to get out of all but one major energy resource (coal, uranium, or oil/natural gas). Its practical effect would be to force the oil companies out of coal.

Proponents of horizontal divestiture argue that each major energy source can be substituted for another, especially in the production of electricity. Therefore, a firm with interests in more than one energy source might delay the development of a competing fuel. Moreover, the proponents say, there is a great deal of regional variation in the control of coal production or of reserves, and in some areas, especially in the West, control is already highly concentrated.[30]

But the opponents of horizontal divestiture cite a number of studies or opinions from such diverse sources as the Federal Trade Commission, the American Petroleum Institute, the U.S. Department of the Treasury, the National Coal Association, the U.S. General Accounting Office, and the U.S. Department of Justice. These studies show that the coal industry, even with the entry of oil firms, is still quite competitive. Using the most common measure of competition within an industry, concentration of production, we find that the top four firms, after increasing their control of production during the sixties—from about 21 percent in 1960 to about 31 percent in 1970—slipped some during the seventies, to about 25 percent in 1976. As for coal reserves, the top four holders controlled only about 10 percent in 1975. Meanwhile, the National Coal Association reported that in 1976 oil and gas firms controlled about 20 percent of all U.S. coal production and about 13 percent of U.S. reserves. Such concentrations do not appear alarming. According to economic theory, collusion is usually prevented and competition is usually protected when the top four firms control less than 50 percent of production. There is also evidence to support the notion that oil firms spend a disproportionately large amount of money on coal research and development. In 1975, their research investments accounted for about 60 percent of total private funds spent on coal research, at a time when their share of production was 20 percent. In sum, the industry remains quite competitive, and if anything, the entry of oil firms has made the industry stronger.[31]

Nevertheless, the mere threat of horizontal divestiture may restrain coal activity by some oil firms.[32] In any case, competition can be protected by methods short of horizontal divestiture, such as existing antitrust laws, setting limits on the share of reserves any single firm can control, and innovative federal leasing policies. The last can be especially effective. About 40 percent of all U.S. coal reserves (by weight) and at least 60 percent of all Western coal (by weight) lie in government land. Through its huge ownership of the coal reserves in public lands, the government can do a great deal to control regional concentration. And so, alternatives to horizontal divestiture would spare both the industry and the nation much disruption and confusion. The short-term uncertainty would be relieved, and the long-term benefits from developing a strong coal industry would continue.[33]

By the mid-seventies, additional participants and new patterns had emerged which will help shape U.S. coal production for the next twenty-five years. Other types of large companies have joined oil firms in starting or significantly increasing steam-coal production, a pattern which brings in a still larger base of corporate, managerial, and technical resources. Some of the new participants are those whose main business is the extraction of natural resources, principally Amax and Utah International. These two firms are now respectively the seventh and twenty-first largest private holders of coal reserves in the United States.* Also by the mid-seventies, the production of coal from utility-owned captive mines—mines that produce coal primarily for the parent company's own use—increased significantly; as recently as 1970, no utility ranked as one of the top fifteen coal producers in the United States. But by 1976 no less than three utilities—American Electric Power, Pacific Power and Light, and Montana Power's subsidiary, Western Energy— were among the top fifteen. This kind of production represented a significant attempt by certain utilities to cut into the fragmented coal demand-supply network and to integrate vertically.

Although the coal production of steel-company-owned captive mines did not increase substantially, U.S. Steel, one of the two steel companies among the top fifteen coal producers, had decided to diversify and sell steam coal, used to produce electricity, as well as to continue mining metallurgical coal for making its own steel. This meant that still another kind of large firm with huge coal reserves now produces steam coal.[34]

In many ways, the most intriguing new participants are the

* In 1976, Utah International merged with General Electric, the largest merger in U.S. history.

engineering firms and other high-technology companies that have recently moved to enter the coal industry. State-of-the-art technological expertise is now coupled with control of coal. The most dramatic example of the pattern is the formation of the Peabody Holding Company, which in 1977 acquired Peabody Coal and instantly became the largest coal producer in the United States and fourth largest holder of reserves. The holding company was comprised of Newmont Mining Corporation (a diversified international mining company), Williams Company (a diversified fertilizer and chemical manufacturer), Bechtel Corporation, the Boeing Company, Fluor Corporation, and Equitable Life Assurance Society of the United States. Bechtel and Fluor are engineering firms, and both have designed and built gasification and liquefaction plants of various sizes. Bechtel is also trying to develop and promote slurry pipeline systems.[35]

Plans for multiorganization and multisector coal activities also appeared with greater frequency in the middle seventies. These are joint ventures comprised of two or more firms and frequently involving some form of government participation through direct funding or some other type of subsidy. Such enterprises bring together significant managerial, financial, and technological resources, while limiting the risk of each participant, and therefore make possible very large, risky projects in coal. Such projects came into being for large-scale surface mining, particularly in the West, and have been proposed for developing coal gasification or liquefaction plants.[36]

The Wesco and ConPaso projects exemplify such new patterns of organizational relationships. Essentially, both are efforts to mine massive amounts of coal from the Navajo Indian Reservation in the northwest corner of New Mexico, to gasify the coal into high BTU gas,* and then to feed it into nearby gas pipelines. Both projects involve the participation of a gas company, and both plan to use the proven Lurgi gasification process. The projects, however, are delayed indefinitely, due to regulatory decisions, inflation, lack of

---

* BTU stands for British Thermal Unit—the heat required to raise the temperature of one pound of water by 1° F at or near 39.2° F. Gas with an energy value exceeding 900 BTU per cubic foot (cf) is generally labeled high-BTU gas. Low-BTU gas has a heating value of 100–200 BTU/cf and medium-BTU gas, 300–650 BTU/cf.

water access, difficult negotiations with the Navajo Indian tribe, and other problems. But elsewhere similar efforts have been proposed.[37]

New Technologies

Until the sixties, technological innovation in the coal industry focused on mining operations. Although fragmented, it was reasonably successful, involving as it did the participation of mining equipment companies, the U.S. Bureau of Mines, and operating firms. The system permitted a rapid diffusion of several innovations, including the shuttle car in the thirties and forties and the continuous mining machine in the fifties. The traditional pattern of innovation has persisted, although the government, the larger coal companies, and foreign firms are now playing greater roles.

Meanwhile, a new kind of innovative activity has emerged that could greatly improve coal's long-run prospects. On the one hand, increasing attention is being given to making the direct combustion of coal more efficient and cleaner. Such an approach makes sense, because direct combustion of coal accounts for about 85 percent of total coal usage. One of the most promising areas for continued research is fluidized-bed combustion, in which a fossil fuel is burned in a bed of granular particles held in suspension in an air stream. The government has sponsored work in this area for a number of years. This process offers significant potential for increased efficiencies and reduced emissions of sulfur oxides. Studies have estimated that the cost of energy using fluidized-bed combustion is potentially lower than that of energy from burning coal with scrubbers. Still, a number of technical problems remain, although there is hope that this technology will become available in the eighties.[38]

The other new direction of coal research and development is in gasification and liquefaction. Such work is less fragmented than the pattern exhibited in traditional technological innovation in the coal industry. In one sense this policy is attractive because a convenient infrastructure already exists to transport oil or gas. Moreover, energy in either of these two forms is cleaner than can be had by the direct combustion of coal, because the technology to curb pollution from burning oil and gas is more developed and less expensive.

Liquefaction in particular is one weapon against increasing dependence on foreign oil. It is being used for precisely that purpose by another country with plentiful coal reserves, the Republic of South Africa. Although oil-trading ties with its previous major source of foreign oil, Iran, were severed with the overthrow of the Shah, South Africa seemed prepared in early 1979 to survive this shock. It had planned well. It had stored large stockpiles of crude in abandoned coal mines; in any event, coal supplied about 80 percent of the country's total energy needs. Moreover, South Africa has an extensive liquefaction capability. One facility is already operating, and another much larger liquefaction plant will be on line in the early eighties. At that point, oil-from-coal will account for about 35 to 50 percent of South Africa's total petroleum consumption.

In contrast, the American synthetics fuel effort has been less intense and less focused. It was never given the priority it possessed in South Africa. Although some work on coal gasification and liquefaction took place in the United States before 1960, the Department of Interior's Office of Coal Research ignited the beginning of current interest by sponsoring a relatively small research effort during the sixties. In 1963, OCR signed a contract with Consolidation Coal Company to support Consolidation's ongoing Project Gasoline effort. The company's president G. Albert Shoemaker happily told a Wall Street audience in November 1964: "Thus far we have demonstrated that our method is technologically feasible and in an area of such economic attractiveness as to encourage moving into the pilot-plant stage in the near future." After building a small pilot plant in the mid-sixties, Shoemaker was still quite optimistic: "While our projections must be confirmed in the . . . pilot plant, we believe this development offers the potential of creating a major new market for coal."

OCR also sponsored other synthetic fuels (oil and gas made from coal) projects during the sixties, including efforts by the Institute of Gas Technology and FMC Corporation. Optimistic comments on the prospects for a "flourishing" synthetic fuels industry continued to be heard during the middle and late sixties.[39]

Oil and gas companies considered coal a potential feed stock for their refineries, pipelines, and petrochemical plants. The lure of synthetic fuels led them to acquire coal, particularly in the West, where at the time no large market existed for steam coal. In fact, a Texaco official openly said in 1977 that his firm had acquired coal reserves during the sixties because of the optimistic forecasts made for synthetic fuels. Meanwhile, Exxon, in addition to acquiring reserves in the late 1960's, began to build up a coal and synthetic fuels research department.

By the early seventies, however, more discouraging estimates for the cost of getting oil and gas from coal emerged. The "economic attractiveness" mentioned by Shoemaker was a mirage, and the flush of optimism vanished when all participants discovered that liquefied or gasified coal was much more expensive than petroleum or natural gas. A respected engineering study of six gasification processes in 1976 found synthetic gas costs ranging from $3.88 to $6.72 per million BTUs compared to a regulated interstate natural gas price for newly found gas of about $1.40, and an unregulated intrastate gas price that hovered around two dollars per million BTUs. And in 1979, estimates for synthetic crude were over $30 per barrel. Moreover, the government's first attempt to build a liquefaction demonstration plant, the clean-boiler-fuel Coalcon project, was stopped on grounds that still more pilot-plant work was needed, and that other technologies appeared more promising.

Still, the Iranian crisis in 1979 again highlighted the danger of U.S. dependence on foreign oil. In many ways the need for a synthetic fuels capability is more a matter of national security than of comparative fuel economics, especially for liquefaction. The United States needs a credible capability to derive oil from coal. But because no single technology appears dominant, the government, on grounds of security, should sponsor a number of promising technologies at the pilot-plant level. Indeed, the government has shifted away from backing large-scale demonstration facilities toward funding promising technologies. In early 1979 the Department of Energy authorized two gasification demonstration plant designs, and continued to fund at least three joint pilot or demonstration liquefaction facilities.[40] The earlier rush toward synthetic fuels may have been misdirected, but national-interest considerations demand that research and development momentum be maintained in this area.[41]

The dominant force in coal industry R&D and the promotion of innovative activity is the federal government, which has shifted its energy research priorities. In 1963, out of a total federal energy R&D budget of about $330 million, only $11 million was spent on coal, while over $210 million was spent on nuclear fission. But by fiscal 1979 the Department of Energy research budget allocated

$618 million (or 23 percent of the total) for coal and $905 million (or 34 percent) to nuclear fission, with almost half of coal's money going to gasification and liquefaction.

In the private sector, coal-related R&D spending by oil firms rose from about $5.5 million in 1971 to an estimated $42.4 million in 1976, which meant that coal's share of all oil firms' total R&D budgets had increased from 2 percent to 10 percent. Furthermore, coal-related R&D by oil firms is generally less fragmented than the traditional technological innovation process in the coal industry. Oil company research directs itself toward developing whole systems (not just specific components of a process), and the approach usually involves coordinating various technologies and different organizations. And because oil and natural gas companies are interested in gasification and liquefaction, they are playing key roles in the new, integrated kind of R&D now emerging in coal. As a result, of the thirty-six proposed synthetic fuels projects identified by the Bureau of Mines in mid-1976, oil and gas companies sponsored twenty-two.[42]

THE DISAPPOINTMENT AND THE HOPE

Despite its much-touted abundance, coal will not become our major near-term solution to the energy problem. Its use, however, will grow, and it will play an increasingly important role in certain sectors. But coal's potential long-run strength—new, strong participants and new kinds of technological innovation—is gradually emerging. The industry is more vigorous than it has been for decades; it is no longer isolated, and a large, rich, and diverse set of firms now participate. Finally, the government is beginning to pump relatively large amounts of money into the coal industry to encourage the development of new technologies. The strategy for coal is clear: to concentrate on long-term answers, especially through technological innovation, while seeking acceptable ways to utilize coal's steady short-term growth.

Nevertheless, U.S. energy policy must cope with the inevitable disappointment that is even now beginning to develop as it becomes clear that coal cannot be *the* transitional energy source.

This disappointment could engender a pathology that in turn could further cripple coal's prospects in the short term. The industry could become even more conflict-ridden, with the environmentalists pitted against industry, industry against government, and labor against management. Such pathology was exhibited during the great coal strike of 1977–78, which lasted for a little less than four months, and as a coal newsletter in March 1978 observed, it was a struggle "nobody won." The coal strike, in fact, created a new layer of distrust between workers and managers. Fundamental needs in labor-management relations—more mining foremen, skilled labor managers, and specialists, and generally better qualified middle-level managers—went unrecognized. The strike also left the union weakened, and given the way it ended, one cannot be unduly optimistic over long-term labor stability in the coal fields. As one observer stated, "The union suffered from leaderlessness, the operators from arrogance." [43]

At about the same time, however, a more promising way to cope with another of the constraints—the environmental problem—was being pursued. The undertaking, called the National Coal Policy Project, sought to achieve some consensus and cooperation between two long-term antagonists: industry and environmentalists. It was jointly headed by the corporate energy manager of the Dow Chemical Company and a former president of the Sierra Club. The group included task forces on mining, transportation, air pollution, fuel utilization, conservation, and energy pricing, which met periodically throughout 1977.

The project's draft report, which was issued in February 1978 and was widely reported in the press, observed that "in the past, these two groups [industry and the environmentalists] have often met as adversaries across the hearing table, before the courts, or in the halls of Congress in an effort to see that their particular point of view or policy position prevailed. In the National Coal Policy Project, they came together in a cooperative effort to reach consensus, provide guidance for the resolution of national coal policy issues, and to articulate their differences in a useful way." At the end of the first stage, the project's leaders claimed "80 percent agreement." They urged a next phase to implement their recommendations, and in December 1978 they at least announced plans to study a specific series of issues.

The project was attacked by some environmentalists, by some members of industry, and even by some persons within the government, but what was really important was that a potentially significant, new cooperative process was set in motion. Such an effort might begin to replace the worn, debilitating rhetoric of conflict with sensible compromise and with a successful resolution of difficult issues.[44]

Because of its abundance, too much too soon has been expected from coal. The 1977–78 coal strike and the National Energy Project offer two radically different examples of what coal's future might be. On the one hand, the United States can have a coal industry that suffers from internal conflict, labor strife, and environmental contention. This instability itself threatens to harm coal's long-term future. On the other hand, the United States can go down the road marked out by the National Coal Policy Project, finding methods to cope with coal's short-term difficulties and immediate conflicts, while recognizing that coal is not a panacea and miracle transition for America's energy needs. A calm and reasonable approach in the short term, and an overall long-term focus that would include a serious program to utilize coal more cleanly either in direct combustion or as a liquid or gas, will permit more freedom in meeting U.S. energy needs, including choices involving coal.

I. C. BUPP

5

The Nuclear Stalemate

By the end of 1974, the political and economic establishments of the Western industrial countries had agreed on a common response to OPEC and dependence on imported oil: If the 1960's had been a decade of oil, the 1980's would be a decade of nuclear power. President Nixon unveiled Project Independence, which called for atomic energy to provide 30 to 40 percent of America's electricity by the end of the 1980's and even more—up to half—by the beginning of the twenty-first century. French Premier Jacques Chirac expressed the view of most Western leaders when he declared, "For the immediate future, I mean for the coming ten years, nuclear energy is one of the main answers to our energy needs." [1]

Five years have passed since these declarations were so firmly made. For the government and business leaders committed to atomic energy, these have been years of frustration, for the nuclear promise has turned into the nuclear disappointment. A wide range of individuals and groups—in some countries, even political parties—have argued that nuclear power plants are unsafe or uneconomical or both. The critics have attacked the industry for failing to develop feasible, safe, and acceptable methods to dispose of radioactive waste materials, and they have charged that worldwide growth of the nuclear power industry will inexorably result in the spread of the materials and know-how required to produce atomic bombs.

The hazards of waste disposal and nuclear weapons proliferation are only the most recent in a long series of charges leveled against nuclear power. But they have been the most effective. Well before the March 1979 accident at the Three Mile Island plant, near Harrisburg, orders by American electric utilities for atomic-powered generating equipment had all but stopped.[2] Less than half a dozen reactors were purchased between 1975 and 1979. Moreover, since 1975 there have been about twenty cancellations of previous orders and at least twice as many announced deferrals for periods ranging from five to ten years. Abroad the picture is much the same: a rancorous and paralyzing controversy.[3]

On one side are government and business leaders and their allies from the scientific and engineering communities—the nuclear advocates. In most cases the advocates believe that there is no realistic alternative to increased reliance on nuclear power. For them, it is not so much the technology of choice as the technology of necessity. Although many admit the existence of some technical and economic uncertainties, particularly with respect to the long-term disposal of radioactive wastes, they steadfastly maintain that these problems are tractable. This assumption is what separates the nuclear advocates from many of their critics, with the advocates implicitly accepting the motto: For every problem there is at least one feasible solution. They also maintain that few other activities that shape modern industrial society—automobile and air transport, chemical manufacture and transport—could pass the tests of proven safety and proven necessity that many nuclear critics demand for atomic power plants and associated services.

On the other side are the opponents of nuclear power. Some challenge the technology on pragmatic grounds. In their opinion, safe nuclear power plants simply demand more than can be delivered by the limited scientific, engineering, and administrative resources of contemporary society. At the very least, they say, no further growth of nuclear power should occur until a long list of outstanding questions has been answered. Among them are, What are the effects of low-level radiation over long periods? What is the probability of a major accident at an operating nuclear power plant? What would be the consequences of such an accident? How can the dangerous waste products of nuclear power be permanently isolated from the environment? [4]

Some nuclear critics, motivated by images and associations of Hiroshima and nuclear war, go beyond such pragmatic questions and want to put "the atomic genie" back in the bottle as soon as possible and at any cost. No technical or economic facts will easily erase for them the image of the mushroom cloud.

Still other critics attack atomic energy as a product and as a symbol of an entire social and political order to which they are opposed. In Western Europe and Japan those critics have played an important part in the opposition to atomic energy. But in the United States the most influential critics have been those who have challenged the nuclear industry with pragmatic, not ideological, questions.

Pragmatic questions do suggest the possibility of pragmatic answers. But in this case there are wide differences of opinion about how to interpret even basic scientific facts. There are equally wide differences about who should have the burden of proof when alternative interpretations imply alternative conclusions about the safety and economy of nuclear power. To put it another way, the fundamental disagreement is over how to deal with uncertainty. The resolution of differing opinions over how to deal with uncertainty, over how much risk is acceptable or how safe is safe enough—all require judgments in which values play as large a role as scientific facts. [5]

In the absence of broad consensus on how to interpret the facts, the stalemate between the advocates and the critics will persist. What contribution, then, can nuclear power make to the U.S. energy supply in the future? In the first place, it must be remembered that the only application of atomic energy is to produce electricity. In early 1978, the American electric utility industry had the capacity to produce about 515 "gigawatts"(GW) of electricity.* Thus, the entire electric utility industry contributed the equivalent of about 10 million barrels of oil per day of energy in 1977. About 45 GW, or 9 percent of this capacity, was nuclear. However, these nuclear plants were generating about 12 percent of the electricity

---

* One gigawatt is equal to 1 million kilowatts (1,000 megawatts), or the equivalent of 42,000 barrels of oil per day in an oil plant operating full time, or the equivalent of 20,000 barrels of oil per day at the actual 1977 operating rate of the U.S. electric utility industry. [6]

actually produced. The discrepancy between capacity and actual generation exists because electric utilities try to operate nuclear power plants full time. In industry jargon, these plants are "base-loaded," which means that they normally account for a higher percentage of electricity actually generated than of generating capacity.

As of October 1978, there were seventy-one nuclear power plants with operating licenses in the United States, representing about 50 GW of generating capacity. About ninety additional plants, representing about 100 GW of capacity, were either under construction or had the necessary permits. And about forty plants representing about 45 GW had been ordered but had not yet been granted construction permits.[7] Hence, about 145 GW of nuclear generating capacity was "in the pipeline." If all of these plants are built on or near schedule, the United States will have a total of about 195 GW of operating nuclear capacity by the early 1990's—only half of what had been officially projected less than five years earlier. Moreover, in view of the numerous deferrals and outright cancellations of that same period, and the probable effects of the accident at Harrisburg, even this modest growth can by no means be considered a sure thing.

But electricity demand is also growing more slowly than it did for many years. In 1978, most electric utility executives anticipated an average annual growth in electricity demand of about 5.5 percent during the decade ahead, in contrast to the 7 to 7.5 percent that characterized the 1960's.[8] Some even suspected that demand growth in the 1980's might average only half what it had been in the 1960's—that is, 3.5 percent. Demand growth at 3.5 percent, half the historic rate, would still mean that approximately 250 gigawatts of new capacity would be required to avoid an electricity shortage in the 1990's.

In early 1979, more than 100 gigawatts of *non*nuclear generating capacity was also scheduled for completion by the mid-1980's. Hence, even if electric utility companies in the 1980's do not or cannot purchase additional nuclear capacity beyond that in the pipeline now, there still may be no national electricity supply crisis in the 1980's or 1990's. But there are two big ifs: if all the nuclear plants in the pipeline are completed on schedule, and if electricity demand grows only modestly.

But if demand grows at the conventionally accepted rate of 5.5 percent, or if projected coal-fired capacity fails to materialize, then an electricity shortage might well materialize in the early 1990's.[9] To avoid such a costly development, the nuclear advocates insist that the stalemate preventing new orders and threatening existing ones must be ended. Yet even if it were ended, new nuclear plant orders would have no significant impact on the country's energy supplies until the 1990's, since no resolution is likely to shave more than a few years from the twelve to fourteen years currently required to plan, build, and license an atomic power plant. And it is far from clear that the issues that have produced the stalemate can be resolved in the next few years.

The most obvious of these issues is the unresolved question of what to do with the used up, or "spent," fuel from nuclear power plants. Not only does this inflammatory issue stand in the way of new orders, it also threatens the operation of both the plants in the pipeline and those already on line. For it is not even clear that all of the plants that are today producing electricity can continue to do so throughout the 1980's. If some currently operating nuclear plants were forced to shut down because there was no place to put their spent fuel, or because of safety considerations, it would be anyone's guess whether new ones would be permitted to start up. *Moreover, plant shutdowns would mean that the absolute output from nuclear power a decade from now could actually decline from where it is today.*

To understand how the prospects for nuclear power at the end of the 1970's came to diverge so dramatically from those at the beginning of the decade, it is necessary go back some years and review the evolution of the business and the federal government's role in it.

THE EARLY COMMERCIAL TRIUMPH

A group of scientists led by Enrico Fermi operated the first manmade atomic reactor in 1942—in a converted squash court at the University of Chicago.[10] The first power-prototype atomic reactor to be connected to an electricity distribution network in the United States began operation in late 1957, at Shippingport, Pennsylvania. It was a very different machine from the first reactor in Chicago. The first device and others built during World War II as part of the government's Manhattan Project were called piles, for they were little more than piles of graphite blocks into which uranium fuel rods were inserted. The Chicago pile was not designed to produce anything; its purpose was merely to demonstrate the feasibility of starting and controlling a nuclear chain reaction. The graphite slowed down, or "moderated," the neutrons produced by the splitting uranium in order to sustain and to control the process. The later piles of the Manhattan Project also used graphite as a neutron moderator, but were designed to produce plutonium, a new element for use in atomic bombs.

The reactor in Shippingport had a different purpose: to turn the heat of fission into electricity. Its uranium fuel sat in a steel chamber through which ordinary water was circulated at high pressure. An intricate system of pipes, valves, and pumps allowed this circulating water to slow down—to moderate—the neutrons, and to carry heat from the fuel chamber, the reactor core, to an electricity-producing steam turbine. The second large American reactor to produce electricity was based on a similar design. It also used ordinary water as a neutron moderator and as a coolant, but instead of maintaining the water in the reactor core under high pressure, it permitted it to boil off into steam.

Naturally enough, these designs were dubbed pressurized water and boiling water reactors, and together were known as light water reactors. The term *light water* distinguished them from yet another design that used a rare compound of oxygen and a form of hydrogen called deuterium to transfer heat from its core. The descriptive name of this compound was heavy water, and the devices using it were known as heavy water reactors—the design favored by the Canadians.

Pressurized water reactors, boiling water reactors, and heavy water reactors by no means exhaust the technical possibilities for generating electricity from nuclear fission. Since World War II dozens of other systems have been tried around the world. The first reactor to be connected to an electricity network—in England in 1953—used graphite piles as its moderator and carbon dioxide gas to remove heat from its core. During the 1950's, such gas-graphite reactors were the basis of British and French efforts to produce nuclear electricity. The fifteen-year attempt by Britain and France to make the gas-graphite design the basis of the world's nuclear electricity program effectively ended in 1967. A decade later, no manufacturer offered gas-graphite plants for sale. By the early 1970's, the world market for nuclear power plants was dominated by the light water systems that had been developed in the United States as a direct outgrowth of the American Navy's nuclear submarine propulsion program. Only the Canadian heavy water system—nicknamed CANDU—survived as potential competition.

In December 1963, the Jersey Central Power and Light Company announced its purchase of a 515-megawatt light water reactor from General Electric to be built at a site called Oyster Creek. To explain its decision, Jersey Central offered a novel reason: The Oyster Creek nuclear power plant would produce electricity more cheaply than any other generating system.[11] Also unique was that the federal government did not participate. For the first time a reactor would be built to produce electricity in the United States without a direct subsidy from the Atomic Energy Commission. General Electric had offered to build the entire facility for a price that, over the several years of the project, could change only to correct for inflation.

Jersey Central's decision was widely regarded as a milestone in the development of nuclear power technology. It was accepted as proof that the day when reactors would be sold in direct competition with conventional generating plants was very close at hand, if indeed it had not already arrived.

Spokesmen for General Electric maintained that the low cost of the Oyster Creek plant was not unique.[12] At about the same time, GE was also furnishing another utility in upstate New York, Niagara Mohawk, with the major components for its Nine-Mile Point nuclear plant at prices "in line" with those for Oyster Creek. In addition, the company published a price list for nuclear power plants of many sizes. The Westinghouse Electric Company was quick to match GE's price quotations with its own for the pressurized water system. Few informed persons in either industry or government publicly questioned the rosy picture for the future of nuclear power that these events seemed to foretell.

In the months following the purchase of the Oyster Creek plant, four American reactor manufacturers—G.E. and Westinghouse, joined by Babcock and Wilcox and by Combustion Engineering—committed themselves to deliver complete nuclear power generating stations at firm prices, subject to change only for inflation. The reactor manufacturers would assume responsibility for the cost of materials and equipment manufactured by other companies as well as for managing the entire construction project. The manufacturer would turn over the completed plant to the owner, who merely had to start the generating equipment. Such facilities were called turn-key plants, and the Oyster Creek turnkey contract was followed by eight others. These sales were regarded as proof of the reality of commercial electricity generation from light water nuclear reactors.[13]

In 1965, American utilities placed their first orders for nuclear plants for which the manufacturers no longer provided firm price guarantees. The next two years produced a "great bandwagon market" for nuclear plants as utilities ordered nearly fifty systems totaling about 40 gigawatts of electrical generating capacity. The intense competition among the four reactor manufacturers was waged in terms of plant prices and confidential ancillary guarantees on fuel and other factors affecting plant operating costs. In most cases, a utility considering the purchase of a reactor could solicit secret bids from each of the four manufacturers and then bargain among the lowest bidders for the most attractive supplemental guarantees. This remarkable buyer's market was characterized by continuous downward revision of the *estimated* cost of electricity from nuclear plants.

Advocates of nuclear power were ecstatic, many speaking of a "revolution" that had been accomplished. "Nuclear reactors now appear to be the cheapest of all sources of energy," Alvin Weinberg, director of the Oak Ridge National Laboratory, told the National Academy of Science in 1966. By 1968 authoritative forecasts for continued rapid growth of nuclear generating capacity were being made by government and industry. These forecasts, with their astonishingly low cost projections, would have seemed incredible to even the most bullish proponents of nuclear power only five years earlier.[14]

A CHANGE IN GOVERNMENT REACTOR POLICY

In the mid-1960's, just as the great bandwagon market for the light water system was getting going, the federal government's power reactor research-and-development program underwent a major transformation.[15] A close look at the circumstances surrounding the drastic change in government policy and internal organization suggests that it was only indirectly related to the rapidly unfolding commercial prospects for nuclear power; instead, it was the culmination of a ten-year-long squabble between the Atomic Energy Commission and the congressional Joint Committee on

Atomic Energy about the proper scope of the federal government's role in developing and bringing advanced reactor systems to the marketplace.

Lewis Strauss headed the Atomic Energy Commission during much of the Eisenhower Administration. He was a conservative Republican who believed that the chief responsibility for developing the technology to produce electricity from nuclear fission lay with private industry, not the federal government. Several powerful New Deal Democrats on the Joint Committee thought otherwise.[16] Years of increasingly bitter disagreement between Strauss and the Joint Committee Democrats left a residue of antagonism and distrust that affected the government's nuclear power program long after Strauss's departure. In fact, federal power reactor development policy during the mid-1960's was more heavily influenced by the issues of the 1950's than by contemporary events and their implications for the 1970's.

In November 1964, Milton Shaw, a protégé of Admiral Hyman Rickover, was named Director of Reactor Development for the AEC.[17] His main objective was to turn the commission's nuclear power research-and-development activities into the kind of aggressive government-controlled program that Democrats on the Joint Committee had wanted. Shaw was only secondarily concerned with the surprising commercial success of light water systems. The key problem of reactor development, as seen by Shaw and the AEC, was that light water technology wasted uranium. This perceived technical deficiency called for an accelerated government-controlled effort to develop systems that would use uranium more efficiently. A large number of new reactor technologies held this promise. The most sophisticated, and the best according to this criterion, was the "breeder" reactor. But between light water systems and breeder reactors were a number of "second generation" concepts, such as high-temperature gas reactors.

For some time, several different concepts had seemed to the AEC's technical staff to offer roughly equal promise for second-generation nuclear power plants. Prior to Shaw's appointment, the AEC had already taken the position that electric utilities would have to make their buying decisions from the menu of these various second-generation designs. In early 1964, the AEC had solicited proposals for joint government-industry projects to build prototypes of one or more of these new designs. For the rest of the year—

the crucial period preceding the great bandwagon market for light water—the staff of the AEC was preoccupied with sorting out the technical pros and cons of the next reactor technology.

Shaw turned the government program even further away from the remaining development problems of light water systems. He wanted to skip the second-generation technologies and move directly to establish the liquid-metal fast breeder reactor as the highest-priority government research-and-development effort. During 1965, Shaw successfully reoriented the AEC program. His liquid-metal breeder reactor program would attempt to do for this new technology what Admiral Rickover's naval propulsion program had done for light water systems: provide a solid technical base for a prototype construction project. For the first time, the AEC's power reactor research-and-development program would meet the demands articulated for years by the Democratic majority on the Joint Committee on Atomic Energy. It would assume full, not partial, responsibility for shepherding a new reactor technology from engineering concept to full-scale prototype demonstration.[18]

But Shaw's success had an unforeseen and unfortunate consequence of great importance for today's stalemate. The time devoted by AEC officials to bringing about this policy change and their willingness to sacrifice other goals to meet the new commitment to the breeder reactor diverted the government's managerial attention and fiscal resources away from other huge unsolved problems that remained in the development of light water systems.

LIGHT WATER: AN INCOMPLETE SYSTEM

Many business executives are familiar with looking at manufacturing or service operations in terms of systems that require certain inputs to produce some product. Descriptions of nuclear power technology that appeared to fit this approach were common during the mid-1960's, and indeed still are. That technology is often described as a way to produce electricity from steam turbines, in which uranium fission is substituted for the burning of oil, gas, or coal to make steam. According to this beguiling description, nuclear energy is little more than a novel way to boil water. The board of directors of General Electric reportedly viewed the company's decision to enter the nuclear power field as a decision merely to integrate the company's business "backward" into steam-boiler manufacturing. Their customers were subsequently persuaded to buy "nuclear steam supply systems," not "atomic reactors." [19]

The phrase *nuclear steam supply system,* though perhaps technically correct, obscures a crucial point. In nuclear power technology, the pertinent system is much larger and far more complex than simply a new way to make steam to turn a turbine. In fact, the nuclear power system is an interconnected set of subsystems that extend far beyond the atomic generating station itself. The reliable production of electricity from nuclear fission requires the operation of a fuel supply system to provide uranium; a fuel preparation system to "enrich" the uranium and package it in appropriate form; a power-plant operating system to build and maintain the reactor and associated generating equipment; and a spent-fuel treatment and disposal system. Moreover, these several systems are not connected in a simple one-way fashion. There are complex technical and economic linkages among them, and each depends on the other in complicated and often subtle ways. Cheap and reliable electricity from nuclear power requires one of the most demanding creations of modern society: a highly interactive set of extremely sophisticated industrial processes and services spread across many separate plants and requiring many specialized management skills. To believe that nuclear power is merely a new way to boil water is to believe that open-heart surgery is merely a new way to relieve minor chest pains.

In the mid-1960's, the light water reactor development job was far from complete, for only pieces of the massive interdependent enterprise were actually in place. Yet the government's policies, combined with assumptions about the willingness and ability of private industry to complete whatever supposedly minor tasks remained, created the unfortunate illusion that the research-and-development task for light water systems was all but complete. It was as though a fleet of modern jet aircraft, such as Boeing 747s, had been sold to some less-developed country, with the aircraft and engines working perfectly. But to become operational, the fleet obviously would require the support of a sophisticated group of ancillary operations and services, such as airports, air traffic control, and a skilled crew. In short, the aircraft and its engines would be useless without a complex and highly interdependent infrastructure. Of course, such a comparison is only an analogy. Yet it captures an essential reality about the great bandwagon market of the 1960's:

Nearly everyone involved in the initial commercial success of nuclear power helped to create and to sustain the illusion that a difficult task that had barely been started was, instead, almost finished.

In order to sell a nuclear power plant to an electric utility, its manufacturer had to demonstrate that it would produce electricity cheaper than coal or oil. The trouble was that there was no real evidence to sustain the reactor manufacturers' claims. The nuclear power plants that were being sold in the mid-1960's on the *promise* of cheap electricity would not actually begin to operate until the early 1970's. But from the mid-1960's until the mid-1970's, there was little or no effort by reactor manufacturers, by the purchasers, or by the government itself to distinguish fact from expectation on a systematic basis.[20]

In the last half of the 1960's, the buyers were obliged to accept on faith the sellers' claims that nuclear power could produce electricity more cheaply than coal. Each new buyer was, understandably, cited by the successful vendor as proof of the soundness of his (or her) economic claims, and such proof was accepted by other utilities in the United States and abroad, as well as by those companies already committed to light water plants. In this way the rush to nuclear power became a self-sustaining process.

The reactor manufacturers had earlier absorbed large losses from the original nine turnkey projects, but they had expected these losses, which in any event seemed to be more than offset by the marketing and advertising victory that followed. Nevertheless, it soon became apparent that things were not as bright as they might have appeared in the fever of the great bandwagon market. By the end of the 1960's there was already considerable evidence that the 1964–65 cost estimates by government and industry for electricity from light water nuclear power plants had been low.[21] But since the illusion of a completed research-and-development job served so many interests, a consensus quickly developed that the causes of the initial cost overruns were fully understood and were being dealt with.

The nuclear power community had ready explanations and ready solutions. First, economies of scale were seen as a powerful tool for lowering the cost of electricity from nuclear plants. By the late 1960's, light water manufacturers were offering ever larger plants for sale. In 1968, manufacturers were taking orders for plants *six* times larger than the largest then in operation.[22] Second, there was widespread confidence that "learning" effects and design improvements in such key areas as fuel life would help to compensate for the unexpectedly high costs of the plants themselves.

In 1971, M. J. Whitman, an Atomic Energy Commission official, told the Fourth Geneva Conference on the Peaceful Uses of Atomic Energy that "the evolution in the costs of nuclear power . . . would under normal circumstances, be classified as a traumatic, rather than a successful experience." However, "many of the trends which have affected the rise in the investment costs of nuclear plants have had similar effects on alternative methods of generating power." [23] Whitman also asserted that the atomic energy's cost difficulties were a "pre-learning" experience and that future costs would inevitably decline because of learning effects. He concluded that the cost of a nuclear plant delivered a decade after the first commercial sale in the United States would be more than 200 percent higher in real terms, but labeling this a period of "intense learning," he predicted that learning effects would begin to lower costs for plants coming into operation in the late 1970's.

During the first half of the 1970's, these sorts of explanations and remedies constituted an unchallenged conventional wisdom about the economic status and prospects for nuclear power. That wisdom provided the basis for a new surge of nuclear plant orders, and one would not have known from the apparent evidence of the bandwagon market and the claims that went with it that the economic problems of light water systems had barely begun. Nor would one have known that they were directly related to criticism of atomic power plant safety. At the time, most business and government references to such criticism dismissed it as a transient phenomenon related only to local power plant siting decisions. Yet by the early 1970's, the still less than fully developed battle over reactor safety was already having a major effect on the cost of nuclear power and on the ability of industry to complete the job of putting the entire nuclear energy supply system in place.[24]

HOW SAFE IS SAFE?

Since the end of World War II, suspicions and fears had, of course, often been expressed about atomic energy. Many of the same scientists who first saw the great promise of the discovery also stressed its terrible dangers. For most of those who later worked to turn the promise into reality, these dangers, though real enough, were completely manageable. Convinced of this by their own knowledge and experience, they were impatient with outsiders who raised questions. But the outside doubts would not go away; in fact, they grew in both number and intensity as the years passed.

Nevertheless, the nuclear community, and especially its most prominent scientists and engineers, had the credibility to persuade government and business leaders that its view was the technically correct one. By the early 1970's the general tendency among the Western world's business and government establishments was to accept the judgment of the nuclear advocates that doubts about nuclear safety were confined to a comparative handful of noisy and misguided people.

This assessment of the situation was unfortunate. Policymakers overlooked an important political reality: The judgment of the nuclear power community about the acceptability of its technology was being effectively questioned in the United States and in almost all of the Western countries. Moreover, the issue was becoming more, not less, troublesome at the very time of OPEC's price increases.

As opposition to nuclear power developed in the Western democracies, government officials and the affected business executives and nuclear scientists responded in remarkably similar ways.[25] They initially defined the problem as a distortion of certain well-established scientific facts. Given the correct information, the public at large would quickly realize that there was no merit to the questions raised by nuclear critics who were merely exploiting public ignorance. All questions about nuclear safety would vanish as soon as the facts were known. This, of course, has not happened. Why? What keeps the nuclear safety controversy alive?

For the nuclear advocates the answer is obvious: the actions of their opponents. Most advocates believed in 1978, as in 1968, that, as one electric utility executive told us in an interview, "Atomic energy's promise to rescue the world from energy starvation is being killed by the critics, the courts, the bureaucracy, the press, and the politicians." Thus, most persons in the nuclear business insist that the persistence of the critics and their political and legal victories have little or nothing to do with the scientific merit of the allegations made and the safety standards demanded.

Most nuclear critics naturally disagree. The basic problem, in their opinion, is technical. The nuclear critics steadfastly maintain that neither the government nor the nuclear advocates have satisfactorily answered the questions they have asked abut reactor safety and reliability. Indeed, the first thing that impresses an observer of the dispute between the two sides is the breadth and depth of the apparent disagreement among their respective experts. For each of the scientific or technical questions that the nuclear critics have asked about the risks of nuclear power, it was easier in 1978 than in 1968 to find seemingly qualified technicians and scientists prepared to give contradictory answers. In fact, since the early 1970's it has been virtually impossible to make any substantive statement about reactor safety that would not be challenged by either nuclear advocates or nuclear critics as inaccurate or misleading. It does seem, however, that some critics, notably the Union of Concerned Scientists, have shifted the burden of proof to the nuclear advocates on certain key technical issues.[26] But at the same time one must be sympathetic to the frustrations of responsible, technically competent nuclear advocates. One Harvard physicist, Professor Harvey Brooks, speaks for them: "What drives me up a wall is that conservative expressions of scientific caution are seized on by nuclear critics as admissions that nuclear power is unsafe. Unfortunately, extremism on the part of some critics stimulates equal extremism on the part of a small minority of the proponents, which then tends to alienate more scientifically cautious persons and put them, in the eyes of the public, into the critics' camp." [27]

Brooks's comment captures a good deal of what has been going on in the nuclear safety controversy. What it misses is the poisonous legacy of the early years of nuclear power promotion by the industry, the Joint Committee on Atomic Energy, and the Atomic Energy Commission.

A major reason for the persistence of reactor safety criticism is the way the introduction of light water technology was managed by American industry and government.[28] First, the developers of nuclear power had too narrow a view of their task. In many instances they did indeed behave as though they were building and selling a Boeing 747, and leaving the development and construction of airports, radar, and pilot skills to some later date.

A cavalier attitude toward the "outside world" was a second crucial error. During the years of early public visibility—the great bandwagon market—the government agencies and the business interests with the most to gain from successful innovation largely monopolized the technical information about it. Their impatience with questions from outside the club surely contributed to their critics' sense that they were hiding something. By impugning the competence or even the rights of outsiders to question their judgments, government and industry advocates of nuclear power helped to create the impression that much of the "truth" about atomic energy's dangers was being distorted. Government documents have recently been released by the Union of Concerned Scientists that seem to support the contention that much concealment and distortion did in fact occur.[29]

The nuclear advocates' case has also been damaged by the heavily negative connotations attached to nuclear technology. Talk of "unprecedentedly deadly materials," "invisible radiation killers," and lone psychotics or terrorists fashioning atomic bombs may all seem melodramatic and irrelevant to a nuclear engineer. But it is powerful stuff, and the fears it arouses resist the rational calculus of economic costs and benefits more strongly than most officials have supposed.

The nuclear critics have also been helped by the growing demands of many in Western society for more participation in matters that affect them. In the past, it was usually easy for proponents of technological change to establish the desirability of their proposals and implement them. That those who directly benefited were generally different from those who paid the price could usually be overlooked. Those who challenged "progress" could be dismissed as irrational. Much of this has changed during the past generation. An increasingly representative society guarantees the nuclear critics a base of political power, which means that resolution of the nuclear safety controversy will require more than a consensus of established scientific and engineering judgment.

But in 1979 even this relatively narrow consensus did not exist. Indeed, disagreement among apparent experts had spread into yet another technical area: the economics of nuclear power. A tangle of contradictory expert opinions similar to that which for years had characterized the nuclear safety imbroglio had by 1977 overtaken the question of whether nuclear power was, or could ever be expected to be, a relatively cheap source of electricity. It was virtually impossible to make any substantive statement about the economic performance of nuclear power without offending one of the parties in an arcane and increasingly rancorous dispute.

The controversy touched every important factor on which the economic performance of a nuclear power plant depended: the respective investment costs of coal and nuclear generating stations; the proper way to deal with inflation in allocating this investment cost to each kilowatt hour of electricity produced by the plants during their assumed lifetimes; respective fuel and fuel-related costs; and the appropriate way to discount future cash flows in allocating the fixed and variable costs of nuclear and coal-fired power plants.[30]

After weeks of expert testimony on the subject in early 1978, the Public Service Commission of Wisconsin concluded that "there is a wide range of views in this record concerning the relative economics of nuclear and coal-fired generation. These views range from nuclear power's being much less costly than coal to coal's being much less costly than nuclear, *and include the view that it is impossible to tell* [emphasis added]." Several months later, the staff of another state public service commission—New York's—summarized yet another lengthy review with a terse conclusion: "There is no credible bottom line comparison of the total generating costs of nuclear and fossil facilities which can be extracted from this record." [31]

In my opinion, no credible bottom line comparison can be extracted from any existing data. In short, almost six years after OPEC quadrupled the price of fossil fuels—and almost fifteen years after nuclear power supposedly first gained a competitive edge over coal—it is still plausible to assert that atomic energy is or is not competitive by a choice of assumptions that suits one's interest.

Whether an analyst supports or opposes nuclear power, he or she adopts assumptions and cites evidence about relative capital and fuel costs, power plant capacity, and other factors that maintain one or the other position. Consciously or unconsciously, those arguing whether nuclear power is cheaper than coal are simultaneously arguing the larger issue.[32] This means that we should expect no early resolution of the traditional question about the economics of nuclear power. In the coming years there is little chance of an unbiased scientific consensus on whether nuclear power is cheaper than coal.

This is scant comfort to executives of electric utilities and other energy planners confronting investment decisions so large in absolute terms that different answers imply billions of dollars in potential costs and/or forgone benefits in choosing between nuclear and coal-fired power plants. It is also particularly frustrating to advocates of nuclear power because they have believed for so long that the critics would be silenced as soon as the general public came to share the conviction that atomic energy was the cheapest way to generate electricity and satisfy growing demand. Instead, during the mid-1970's electric utilities applying for nuclear power plant licenses have been obliged to prove that their plans make economic sense. And more recently, some have also been challenged to show that nuclear power is cheaper than conservation. The genuine uncertainty about this extraordinarily complex issue and the dedication of critics intent on proving that nuclear plants are both uneconomic and unsafe make the economic calculation an increasingly time-consuming and thankless task.

An economic recession in 1974 and 1975 may have provided a welcome respite for the battle-weary American utility industry. But as the economy recovered in 1976, it appeared that very few utility industry executives were eager to reenter the fray. Sharply reduced projections for future electricity-demand growth were given as justification for decisions to postpone or even cancel previously announced plans to build both nuclear and coal-fired generating equipment.

Whatever the real reasons for deferrals and cancellations, the Carter Administration inherited a stalemate on nuclear power when it entered office in 1977. The nuclear critics had lost in several attempts at the polls to impose statewide bans on new nuclear power plant purchases. But at the same time the nuclear power industry and its friends in government had not been able to reduce the critics' ability to exploit the power plant licensing process to contribute to delays and cost increases.

Already by mid-1977 there was a national de facto moratorium on the purchase of nuclear generating equipment in the United States. As a consequence, the 1979 outlook for nuclear power in the United States is much different from what it seemed to be in the wake of the 1973–74 OPEC actions. Actually, since the early 1970's there have been successive downward revisions in projections of future installed capacity in all of the Western oil-importing countries (Figure 5–1, page 126).

In business and government circles, some optimism persists that the de facto moratorium on orders can be ended in the early 1980's

Figure 5–1

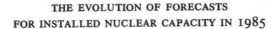

THE EVOLUTION OF FORECASTS
FOR INSTALLED NUCLEAR CAPACITY IN 1985

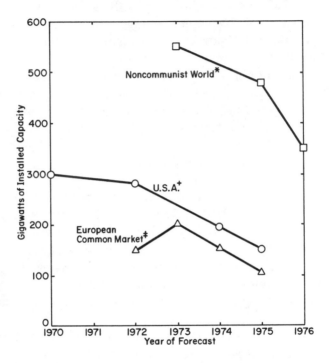

Sources: * For worldwide capacity, reports of the Organization for Economic Cooperation and Development.

† For the United States, constructed by author from reports of the Atomic Energy Commission, the Energy Research and Development Agency, and the Department of Energy.

‡ For the European communities, constructed by author from reports from the planning offices of the European Economic Community.

to supplement the relatively meager 145 GW of new capacity that is in the licensing and/or construction pipeline. The chance of this happening, however, seems small. With the possible exception of a handful of new orders, it is much more likely that the American de facto moratorium on new nuclear power plant purchases will continue at least through the 1980's. It is also likely that a significant fraction of the plants now in the licensing/construction pipeline will be indefinitely delayed, and that many will eventually be cancelled. In the United States the realistic prospects for incremental energy from nuclear power are far more slender than one might infer from official optimism about an imminent end to the de facto moratorium.

The principal reasons are reactor safety—and the problem of what to do with spent reactor fuel. The latter constitutes a physical bottleneck, both for the continued operation of the 50 GW of existing nuclear generating capacity in the United States and for the addition of new capacity.

WASTE: A BOTTLENECK IN THE SYSTEM

Periodically nuclear power plants must be refueled, and spent fuel assemblies must be removed and replaced with fresh ones. In the 1960's and early 1970's, government and industry planners assumed that the spent fuel would be reprocessed to recover still-useful fissionable material from useless and dangerous waste materials as soon as enough reactors were in operation to support the large-scale facilities required for economical reprocessing. Both the government and the nuclear industry postponed important technical decisions about the treatment and disposal of the waste materials pending the start of large-scale reprocessing.[33]

Meanwhile, the nuclear critics gained a new ally—people, both in and out of government, worried about the proliferation of nuclear weapons. International treaties and government secrecy notwithstanding, for thirty years the major impediment to such proliferation had been the scarcity of nuclear explosive material. Even in the 1970's its manufacture remained so difficult and so expensive that it was beyond the reach of all but a few governments. But arms-control specialists began to argue that widespread reprocessing of light water reactor fuel would produce a great deal of material that could be used to make atomic bombs. The plutonium that would be recovered during reprocessing for use as a supplemental reactor fuel could also, though with some difficulty, be used to make nuclear explosives. Moreover, compared to the amounts of plutonium that would have to be routinely handled and accounted for in an economically viable reprocessing facility—on the order of tens of thousands of kilograms per year—only tiny amounts, say 20 kilograms or so, would theoretically be necessary for an atomic bomb.[34]

Consequently, in October 1976, the Ford Administration warned the American nuclear industry that fuel reprocessing might become unacceptable. Six months later, the Carter Administration transformed the warning into an outright prohibition of indefinite duration.[35] For the American nuclear industry, the new government policy also transformed what had been regarded as a relatively minor technical problem, the choice of specific methods to dispose of radioactive waste materials, into an acute operational problem— what to do with spent fuel.

Because the industry has been forbidden to reprocess the spent fuel, it has been piling up in storage areas, specially designed pools of water at nuclear power plant sites. The reactor storage pools are generally large enough to contain the entire reactor core if it must be removed for any reason, as well as one to three years' discharge of spent fuel. One year's normal discharge usually corresponds to about one third of the total fuel in the reactor. In 1978 most of the operators of nuclear power plants expanded the capacity of their fuel storage areas by installing redesigned holding racks. But "re-racking" is only a temporary help. The fuel storage pools at most operating American nuclear power plants could be filled by 1985, and at several by 1983.[36] As these plants exhaust storage capacity, their owners will be forced to transfer spent fuel to other locations or to build and license additional storage capacity—or to shut the plants down altogether.

One expedient would be to transport spent fuel to one or more centrally located common storage pools. Such "away from reactor" pools (AFRs) would be comparatively cheap to build and could safely store spent fuel for many years. A dozen or so such facilities, each about the size of an average industrial warehouse, would accommodate all the spent fuel that is likely to be produced for the rest of the century. But AFRs pose a considerable not-in-my-backyard political problem.

In September 1977, Illinois Attorney General William Scott appeared before a congressional committee considering the nuclear waste problem. His testimony captured the distaste for local "nuclear dumping grounds" that had been building up in many states. Scott said that General Electric's announced plan to increase spent fuel storage capacity at a company-owned facility near Morris, Illinois, was unacceptable to that state's citizens. "Illinois will not passively allow itself to become the nation's dumping group for high-level nuclear waste—when a proper review is complete, General Electric will not be permitted to expand its facility." [37]

By early 1978 at least seven state legislatures had imposed various prohibitions on the construction or expansion of local nuclear waste storage facilities or on the transport of radioactive waste into the state. One of the strongest was a 1976 California statute that actually prohibited future construction of nuclear power plants until the state "finds that there has been developed, and that the United States through its authorized agency has approved, and there exists a demonstrated technology or means for the disposal of high-level nuclear waste."

In this political environment, reliance on AFRs can easily be dismissed as prevarication and deferral of the real radioactive waste disposal problem.[38] A national nuclear waste management and disposal program that consisted solely of shipping spent fuel assemblies from one "temporary" storage pool to another would *look like* confusion and indecision. Of most concern to those opposed to AFRs is that such facilities might become de facto substitutes for a permanent solution to the waste disposal problem. Nuclear critics want to avoid a situation in which large-scale, supposedly interim spent fuel storage facilities allow the continued operation of nuclear power plants in the absence of real progress toward safe, ultimate disposal of radioactive waste.[39]

Many officials within the Carter Administration share the nuclear critics' reservations about heavy reliance on AFRs. They believe that the 1977 presidential prohibition on reprocessing contained an implicit obligation to "do something intelligent," as one official put it, about the spent fuel piling up as a consequence. Many federal officials also believe that the technology and know-how are already in hand to proceed with a safe waste disposal demonstration program that would be better technically as well as politically than merely shuffling assemblies of spent fuel from one storage pool to another—provided some real options are carefully retained for response to experience and new information.

In early 1978 senior officials in the Department of Energy turned their attention to a long-planned facility with the catchy acronym WIPP (Waste Isolation Pilot Plant). The facility had been conceived some years earlier as part of the government program for ultimate disposal of certain kinds of radioactive wastes from atomic weapons production. The idea was that suitably packaged wastes from weapons manufacture would be placed in holes in the floor of a conventionally mined salt cavern several hundred meters beneath the surface of a site in remote southeastern New Mexico. The President created a cabinet-level interagency review group to oversee the program's details.

The technical adequacy of WIPP was sharply attacked, however, by scientists in New Mexico, and the ensuing controversy obliged the Administration to promise the state's governor the right to veto a decision to begin construction.[40]

Perhaps most troubling for the work of the review committee was the growing likelihood that underground salt formations would not be unanimously accepted by the scientific community as the uniquely preferred medium for disposing of nuclear power wastes. In fact, a substantial body of technical opinion held that it was not even very desirable.[41]

Contrary to the rhetoric of the nuclear industry, and contrary to past government policy statements, some troubling gaps in scientific knowledge about long-term radioactive waste disposal exist. These gaps mean that an immediate decision to build, without further experimentation, a full-scale, permanent waste repository in any specific geologic medium is not advisable.[42] It may be that several years of additional study are not needed before any decisions can be safely made about radioactive waste disposal. Yet the history of the light water program does suggest the need for considerable caution.

The experience of building and licensing nuclear power plants before a community-wide consensus existed on how safe is safe enough has proved very costly for the utilities. It has proved even more damaging for the federal government's nuclear research-and-development and regulatory agencies, whose credibility and even legitimacy were demolished in the eyes of many citizens. Repeatedly during the 1960's and early 1970's, government officials had acted less like umpires than partisans, defending the nuclear advocates against the nuclear critics. Today, many influential persons fear that the government will manage radioactive waste in the same way it managed the light water system, pressing forward prematurely and thus generating serious health and safety problems.

Reversal of the Carter Administration's prohibition on reprocessing would help to alleviate the spent fuel pile-up problem. But reprocessing would leave open the question of how to isolate radioactive wastes from the environment. Some scientists and industry spokesmen argue that introduction of breeder reactors would considerably simplify this problem. The reason is that breeder reactors would "burn" the plutonium that is a by-product of light water reactor operation, and hence, theoretically, eliminate the need to isolate this particularly long-lasting radioactive element from the environment. Like most technical questions having to do with the safety of nuclear power, the degree to which consumption of plutonium in breeders would lower the long-term hazards of radioactive waste management is hotly disputed by apparently qualified experts.[43]

Moreover, many nuclear critics claim that even if breeders did simplify the radioactive waste isolation problem, their introduction would create a new hazard: enhanced possibilities for the illicit manufacture of nuclear explosives from plutonium.

The debate on these issues rages in the absence of much relevant evidence. Contrary to a widespread impression, even the world's most technically advanced breeder-reactor development program (in France) is decades from making any significant addition to that country's nuclear power supply. One small pilot plant, called Phoenix, is in operation. A second, much larger plant has been authorized, and initial site preparation is underway. This plant, Super Phoenix, will not operate until at least 1984 or 1985. But even Super Phoenix is not a prototype for commercial breeder reactors. Yet another, still larger plant will be needed to demonstrate commercial (that is, economical) power production. A French government decision to proceed with a third breeder reactor is several years away. In the United States, the Carter Administration has been forced by Congress to allow construction of the Clinch River breeder reactor to begin. This plant is intermediate in size, between Phoenix and Super Phoenix. It will not begin to operate until at least 1986.

Events beyond the early 1990's are, of course, anyone's guess, but the history of the light-water-reactor development effort cautions against expecting too much too soon from a new and highly complex technology. Certainly for the indefinite future there would seem to be little or no realistic possibility that breeder reactors could have any practical effect on the waste disposal problem.

Meanwhile, because distrust of the government is so high on nuclear issues, it will be enormously difficult to find a program for waste disposal whose acceptance extends beyond the nuclear industry's own scientists, engineers, and executives. But without a consensus on waste policy, there can be no end to the nuclear power stalemate. As a practical matter, the nuclear advocates who urge haste in developing and implementing a spent-fuel management and waste disposal program have little means of producing that outcome. The nuclear critics who prefer less haste, however, have numerous mechanisms by which they can express their discontent and slow down the program's progress. These include outright veto in the case of state legislatures and time-consuming court challenges in the case of environmental groups.

If the stalemate is to be ended, decisions on radioactive waste disposal will have to be supported by a body of scientific evidence and a range of informed technical judgment that will withstand the most searching independent scrutiny. It seems clear that a condition for any relief of the spent-fuel bottleneck is massive and rapid improvement in the vigor and technical quality of the federal government's radioactive waste disposal program.

As a first step, government officials will probably have to be candid and admit that uncertainties and gaps do exist in the scientific understanding of the long-term environmental effects of disposal in specific geologic media. In light of past official assurances that no technical problems existed in radioactive waste disposal, such an admission may be painful. But it is absolutely crucial in order to reestablish confidence in the government's ability to manage a safe program. Few persons outside the nuclear industry will accept a supposedly comprehensive new waste-disposal program if it is criticized as technically deficient by apparently qualified scientists.

But candor is only a first step. It must be accompanied by real changes in programs. One highly desirable change would be to abandon WIPP as the first stage of the new government program and salt as the primary candidate medium for a full-scale repository. Instead, the government should propose to develop a number of small or intermediate-scale repositories simultaneously in several different geologic media. This strategy would build on the agreement among a wide range of informed scientists that safe and secure radioactive waste disposal in conventionally mined repositories *is probably feasible*. Even many prominent nuclear critics agree with this proposition. The scientific problem is to carry out the experiments in the field and to analyze the data to identify the specific continental or submarine media.[44]

The government may find very broad scientific support for a program that attacks this problem in a systematic and deliberate way. Explicit support could even come from respected scientists who have been nuclear critics, so long as the program includes provisions for abandoning any candidate repository at any stage in its development if new scientific data show that it is unsuitable.

Of course, such a strategy would be criticized by those nuclear advocates who believe that the only real waste-disposal problem is government incompetence and foot dragging and public misunderstanding and emotionalism. But more perceptive nuclear advocates might well support it. A waste-disposal strategy that accepts the delays that this program would impose is much less likely to encounter delays further down the road during the licensing process. On the other side, even the nuclear critics who are trying to use the waste issue as a way to shut the industry down have an interest in finding a scientifically sound method to dispose of the wastes that already exist.

But a strategy of this sort cannot be implemented overnight. Political as well as technical uncertainties ensure that the first repositories cannot be available before 1990 and may not be available before 1995. The prospect of such long lead times raises at least three practical questions about what to do with atomic power in the meantime:

1. Should the federal government try to prevent the non-safety-related shutdown of *operating reactors* in the early and mid-1980's by building AFRs or by encouraging the expansion of on-site spent-fuel storage capacity?

2. Should the federal government or state governments allow the *plants now under construction* to begin operation in the 1980's?

3. Should the federal government or state governments allow electric utilities to purchase *new nuclear plants* before the first repository is available?

The answer to the first question is pretty clearly yes. There does not appear to be any safety problem associated with indefinite interim spent-fuel storage. Moreover, while these plants supply little more than 10 percent of the country's total electricity demand, their regional importance is far greater. In the early 1980's, New England, the upper Midwest (especially around Chicago), and the southern Atlantic states will be receiving more than one third of their electricity from nuclear power, even if no more nuclear plants go into operation. Although we can not predict the precise effects of a widespread shutdown of operating reactors between 1981 and 1985, it is a good bet that a shutdown of this sort would cause a major regional, if not national, economic catastrophe.[45]

A vigorous government effort to find a scientifically acceptable solution to the ultimate-disposal problem is the key to eliminating the spent-fuel bottleneck that threatens reactor shutdowns in the 1980's. It should make it possible to work out a compromise policy on AFRs to provide the additional spent-fuel storage capacity that will be needed until ultimate-disposal repositories are available in the early or middle 1990's.

The second question—allowing plants under construction to go into operation—was difficult to answer even before the Three Mile Island accident. It has now become a very complex one indeed. The electricity that these plants will produce in the 1980's may or may not be needed. Furthermore, it is reasonable to be cautious about creating new sources of waste before a broadly accepted solution for dealing with them is in hand.

But several tens of billions of dollars had already been invested in nuclear power plants under construction at the time of the Three Mile Island accident. If their owners are unable to complete them and bring them into operation, this will have very serious financial consequences. Moreover, the 100 GW of capacity that these plants represent promise the equivalent of about two million barrels of oil a day in new energy supplies. Because of this and because of the large amounts of money involved, the question of what to do with partially built nuclear power plants will be one of the most acrimonious energy policy questions of the early 1980's. For any one person, the answer will require a complex judgment about reactor safety, the prospects for waste disposal, and expectations about future growth in electricity demand.

Our own belief is that both federal and state governments should proceed with a great deal of caution in licensing any new nuclear power plants for operation. The burden is now plainly on the nuclear advocates to provide answers to specific technical criticisms of standards and practices for reactor design operation and inspection. In addition, the burden is on the Department of Energy to implement a waste disposal program that has the support of pragmatic nuclear critics as well as nuclear advocates.

For all of these reasons, the third question, about new reactor orders, is effectively moot. The waste problem had made extension of the plant order moratorium into the middle or late 1980's highly probable before Three Mile Island. That accident makes such an extension a virtual certainty.

As recently as early 1978, Sir Brian Flowers, one of the European pioneers of atomic energy, declared: "Nuclear power is the only energy source we can rely upon at present for *massive contributions to our energy needs up to the end of the century,* and if necessary beyond." [46]

In the United States there is simply no reasonable possibility for "massive contributions" from nuclear power for at least the rest of the twentieth century. In fact, unless government and industry leaders start now to work with the nuclear critics, many existing plants will run out of spent-fuel storage within four years.[47] The federal government will then face a very difficult choice: shutting down the plants or riding roughshod over the nuclear critics. The time available to avoid the choice is short. *In any case, nuclear power offers no solution to the problem of America's growing dependence on imported oil for the rest of this century.*

If nuclear and other conventional energy sources cannot substantially increase their contribution to U.S. energy supplies, the nation must look to the unconventional alternatives: conservation and solar power.

DANIEL YERGIN

6

Conservation: The Key Energy Source

There is a source of energy that produces no radioactive waste, nothing in the way of petrodollars, and very little pollution. Moreover, the source can provide the energy that conventional sources may not be able to furnish. Unhappily, however, it does not receive the emphasis and attention it deserves.

The source might be called energy efficiency, for Americans like to think of themselves as an efficient people. But the energy source is generally known by the more prosaic term *conservation*. To be semantically accurate, the source should be called conservation energy, to remind us of the reality—that conservation is no less an energy alternative than oil, gas, coal, or nuclear. Indeed, in the near term, conservation could do more than any of the conventional sources to help the country deal with the energy problem it has.

If the United States were to make a serious commitment to conservation, it might well consume 30 to 40 percent less energy than it now does, and still enjoy the same or an even higher standard of living. That saving would not hinge on a major technological breakthrough, and it would require only modest adjustments in the way people live. Moreover, the cost of conservation energy is very competitive with other energy sources. The possible energy savings would be the equivalent of the elimination of all imported oil—and then some.

How could this come about? There is very great flexibility in how much energy is required, and how much is actually used, for this or that purpose. To give a simple example, toast can be made on a barbecue, in a broiler—or in a toaster. The end product is the same—toast—but the methods employed to make it vary greatly in energy consumed and waste produced. The making of toast illustrates the central point of this chapter—that much less energy than is now consumed can be used to achieve the same end.[1]

The barriers to the potential savings through conservation are very great, but they are rarely technological. Although some of the barriers are economic, they are in most cases institutional, political, and social. Overcoming them requires a government policy that champions conservation, that gives it a chance equal in the marketplace to that enjoyed by conventional sources of energy.

But why should the country place greater emphasis on conservation?

We have already discussed the substantial difficulties and very considerable social costs that attend increasing dependence on foreign oil, the limited productive capacity of domestic oil and gas, and the high uncertainties of increased reliance on coal and nuclear. Conservation may well be the cheapest, safest, most productive energy alternative readily available in large amounts. By comparison, conservation is a quality energy source. It does not threaten to undermine the international monetary system, nor does it emit carbon dioxide into the atmosphere, nor does it generate problems comparable to nuclear waste. And contrary to the conventional wisdom, conservation can stimulate innovation, employment, and economic growth.[2] Since the United States uses a third of all the oil used in the world every day, major reduction in U.S. demand would have a major impact on the international energy markets.

A firm commitment to conservation is also required for American foreign policy to become credible on energy and nuclear proliferation issues. How can the United States ask the Europeans and the Japanese to give up the fast breeder reactor and reprocessing if it does not hold out some alternative—that is, less pressure on the international oil market? How can the United States ask them to come to the aid of the dollar when American oil imports, a major cause of the dollar's weakness, continue to grow? In fact, foreigners often seem to pay much greater attention than do Americans to the relation between U.S. energy demand and these kinds of issues.

THREE TYPES OF CONSERVATION

One can pick up a piece of coal, hold it in one's hand, and say, "This is coal." Conservation is far harder to grasp and comprehend. While it involves a host of different things—heat pumps, insulation, new engines—it also involves changes in methods, and even more important, an ongoing commitment to promote and implement it. To clarify matters, we can identify three categories of energy conservation, although the boundaries among them are fuzzy. The first two are not desirable. The third is.

The first category of conservation is out-and-out *curtailment*. When supplies are suddenly interrupted, energy saving is forced as factories are closed and working days lost. This is what happened when interstate natural gas ran short in 1976 and 1977, and during the coal strike of 1977 and 1978. The country can expect more curtailment in the future if sensible actions are not taken now.

A second category is *overhaul*, dramatically changing the way Americans live and work. An extreme example would be the outlawing of further suburbanization, forcing people to move into the urban center and live in tall buildings not equipped with places to park a car. Very few people would willingly accept that kind of energy conservation program.

To many people, energy conservation means only curtailment or overhaul—something repressive and most un-American, involving cutting back, rationing, and unemployment. They see it as the product of an anti-growth crusade led by the granola-chomping children of the affluent. Unhappily, some of the language of the Carter Administration—such as the President's insistence on "sacrifice"—has strengthened this unpleasant and misleading imagery in the mind of the public.

But there is a third way to think about conservation: as a form of *adjustment,* entailing such things as insulating the house, making automobiles, industrial processes, and home appliances more efficient, and capturing waste heat. This can be called *productive conservation,* which encourages changes in capital stock and daily behavior that promote energy savings in a manner that is economically and socially nondisruptive. Its aim is to use less energy than has been the habit to accomplish some task—whether it be to heat a home or to make a widget—in order to prevent disruption later. Conservation, therefore, is not a theological or ideological issue. It should be pursued not as an end in itself, but as a means toward greater social and economic welfare, as a way to promote the well-being of the citizenry. As two prominent analysts, Lee Schipper and Joel Darmstadter, have expressed it, "The most impelling factor in encouraging conservation action is the cost of not conserving." [3]

The Obstacles

One major roadblock to productive conservation is its very character—it is a highly fragmented subject that certainly lacks glamour. There is in this country a natural-enough desire for a technological fix, preferably one big one, that will solve all energy problems—another Manhattan Project, another man-in-space program. Indeed, some survey data indicate that many Americans simply assume that high technology will step in to "save" them just in time.[4] It is certainly easier for the government to organize itself to do one big thing, but, alas, that is not what productive conservation is about. It involves 50,000 or 50 million things, big, medium, and little, and not in the one centralized place where the energy is produced, but in the decentralized milieu where it is consumed. Conservation is prosaic, even boring.

At an energy conference in Los Angeles, the oil people told the heroic story of tapping Alaska's North Slope; the nuclear people talked about advanced nuclear technology; the coal people discussed synthetic fuels and slurry pipelines. The engineer from Los Angeles' Department of Water and Power then stood up to explain how it got the residents of Los Angeles to go along with an effort that substantially reduced consumption of electricity: "We advised citizens to do such things as shut their curtains at night." Here was a conference devoted to the great energy crisis, the moral equivalent of war, and here was a man who was saying that the solution is for people to do such bold things as close their curtains at night.

This is why, when an aide to one of the most powerful senators on energy issues was asked why the distinguished legislator has never given a speech on conservation, the aide replied, "It would either be filled with platitudes or so specific that everybody's nose would fall into his Rice Krispies." [5]

The lack of drama has restricted not only policy and public attention, but also scientific interest in conservation-oriented research. As one physicist has observed, "Little glamour has adhered to research in heat transfer at modest temperatures, relative to heat transfer at the temperature of a nose cone at reentry into the atmosphere; or the hysteresis losses of rubber tires, relative to the neutron losses in nuclear reactors." [6]

A second obstacle to conservation is the way the energy debate has been shaped in this country. When the crisis broke in 1973, the country turned to the experts, the people who had spent their working lives trying to increase energy supplies through oil, gas, coal, and nuclear. Not surprisingly, these people forcefully advocated a rapid further build-up of conventional sources, and their voices were powerful. After all, they belonged to organizations set up to accomplish such tasks, and the organizations had often been admirably effective at doing this, and in the process contributed much to the general welfare. Even if economic self-interest had not been involved, points of view, convictions, and experience acquired over many decades of working to provide more energy would naturally have caused these people to emphasize energy production rather than conservation.

Their voices were all the more powerful because the other side was so weak. What serious "expert" would advocate dampening demand? What economic entity had an interest—economic or otherwise—in promoting such a view, lobbying Congress, spending millions to advertise the case? * Thus the energy suppliers pretty much shaped the terms of the debate, and established what was important and what was not.

The debate has remained one-sided, although certain statements advocating conservation have received some attention. The Carter Administration, in its first year, gave an imprimatur of seriousness

* As Roger Sant, formerly assistant administrator for conservation in the Federal Energy Administration, observed: "Outside of perhaps the insulation manufacturers, there is no organized conservation industry in this country. So we have nothing to compare to the energy producers in terms of marketing, distribution, and lobbying. The oil companies and utilities are busy talking up how much they need to produce. But no one's out there wholesaling conservation by the ton and barrel." Remarks to the Conference Board, mimeographed release FEA, September 30, 1975. The conservation industry has grown considerably since 1975, but it is still a fledgling.

to the case for conservation, though it has since retreated. Public opinion has retreated even more. In sum, the effort to reshape the national energy debate to the point where conservation will be regarded as a serious energy source has been slow, halting, and wholly inadequate. In 1978, for example, a large California utility with ambitious plans for expanding coal and nuclear capacity told the California Public Utilities Commission that it had not yet made a single direct cost comparison between investment in traditional plants and investment in alternative and conservation energy sources—and was unequipped to do so.[7]

A third obstacle to productive conservation may be summarized by the maxim, Let the market do its work. Proponents of this view say that the present levels of conservation activity are a rational response to present energy prices. In other words, price will determine the extent of conservation, and therefore the only way to promote further conservation is through swift deregulation—and much higher energy prices. As should be clear, we are convinced that real prices of oil and gas should move moderately but consistently upward. But that is not enough. Even if prices did begin to rise substantially tomorrow, conservation would still be seriously hampered by political and social barriers. Moreover, those who believe exclusively in price as the answer are unable to answer effectively the question of what—and whose—price. Excessive faith in the market tends to obscure the difficulties and requirements of the needed transition away from the world of imported oil.[8]

The question of price is one of the main reasons why the entire energy debate has been so bitter. There is considerable reason to consider pricing energy closer to its replacement costs, rather than at average cost. But that would mean sharply rising prices. Any significant change in energy prices raises real distributional issues—some people will be hurt and others stand to gain. To say that these issues are secondary is to ignore the way interest groups and regions assert their claims in the American political process. Therefore, an effective conservation strategy has to look for methods to mediate the distributional claims and has to provide incentives and cushions to reduce the hardship of higher prices on consumers.

The fourth barrier is a fundamental misconception about the relation of energy use and economic growth. Many believe in what might be called the iron law of the energy–GNP link. Some economists, as well as a number of decision-makers in business, labor, and government, believe that there is a direct, even inevitable, one-to-one correlation between economic growth and consumption of energy, and accordingly, that encouraging conservation could easily plunge the nation into serious economic straits. The *Energy Report* from the Chase Manhattan Bank went so far as to say that "there is no documented evidence that indicates the long-lasting, consistent relationship between energy use and GNP will change in the future. There is no sound, proven basis for believing a billion dollars of GNP can be generated with less energy in the future." And an internal communication of one of the seven major oil companies said, "There is no empirical evidence to indicate that the coupling of energy to economic growth can be uncoupled." The basic idea behind the iron link was pungently expressed by the head of the Texas Railroad Commission: "This country did not conserve its way to greatness. It produced itself to greatness." [9]

But the iron link has, in fact, yet to be convincingly demonstrated. It remains unproved, as can be seen by looking at the historical record and by comparing the United States with other advanced industrial countries.

There has been wide and erratic variation in the relationship between energy and GNP in the United States. Table 6–1 expresses the ratio of energy to gross national product—that is, the amount of energy growth for every unit of GNP growth. As can be seen in the table, the ratio varies from .63 to 1.45.

Table 6–1

RATIO OF ENERGY GROWTH TO ECONOMIC GROWTH
IN THE UNITED STATES

	Unit of Energy for Every Unit of GNP	Economic Growth Rates (percent)
1950–55	0.63	4.3
1955–60	1.10	2.2
1960–65	0.81	4.8
1965–70	1.45	3.2
1970–75	0.65	2.1

Source: Herman Franssen, *Energy—An Uncertain Future: An Analysis of U.S. and World Energy Projections Through 1990* (Washington, D.C.: Government Printing Office, 1978), p. 17.

Going back to the late nineteenth century, the variations in the relationship between energy and GNP are even greater. As a study by the Conference Board has summarized, "Energy use and economic growth are certainly not independent of one another, but the link between them is more elastic than is commonly assumed." [10]

A similar insight is obtained by comparing the American experience with that of other advanced industrial countries (Table 6–2), which suggests that the realm of possibility for energy savings in the United States is rather broad.[11] Thus, West Germany consumed less than three quarters as much energy for each dollar of gross product as the United States, and France only half. Of course, such comparisons cannot be taken as perfectly equivalent, because of obvious differences involving such factors as exchange rates, political culture, government policies, geography, import dependence, and industrial structure. The broad comparisons do indeed mask a great deal of variation among sectors within countries.[12] Even so, the comparative data seem to make clear that substantial but nondisruptive energy savings are possible in the United States; for while

Table 6–2

COMPARATIVE ENERGY/OUTPUT RELATIONSHIPS, 1976

Country	Gross Domestic Product per Capita (dollars)	Energy Consumption per Capita (tons of oil equivalent)	Energy/GDP Ratio	
			Tons Oil Equivalent per $ Million GDP	Index (U.S. = 100)
United States	5,960	8.3	1,390	100
France	4,740	3.5	750	54
West Germany	4,350	4.5	1,020	73
Sweden	5,460	6.1	1,190	82

Sources: Data derived by Joy Dunkerley, Resources for the Future, from OECD, *Energy Balances,* various issues, and OECD, *National Accounts.* Foreign currencies are converted into dollars at purchasing power parity rates of exchange. One million tons of oil roughly equals 20,000 barrels of oil a day. Data from 1976 are distorted by the varying effects of recession on different countries. Data from 1972 have greater comparability in that all four countries were sharing in an economic boom. In that year, the indexes measured against 100, for the United States, were France, 53; West Germany, 71; Sweden, 77.

the differences among these societies are important, they are not as important as the fundamental similarities—a common set of problems, a common technology for heavy industry, a common architecture, and common modes of transport.[13]

More people are coming to recognize that the iron link is not ferrous at all, but on the contrary, elastic. In releasing a report in 1978 projecting that energy consumption would increase more slowly than economic growth in the years ahead, the manager for energy economics at Shell U.S.A. simply announced, "We have found that we could decouple the two." [14]

When the Lights Go Out

The case of Los Angeles after the embargo provides a dramatic example of how flexible energy use can be. The Department of Water and Power (DWP) provides the city with electricity, half of which is generated from oil. Air-quality control restrictions require the DWP to use low sulphur oil, thus limiting possible sources. Like most utilities, the DWP had enjoyed very rapid growth—a yearly average of 7 percent between 1950 and 1969, and 5 percent in the early 1970's.[15]

In November 1973, shortly after the Arab oil embargo went into effect, the DWP realized that 11 million barrels of already contracted North African low-sulphur oil (more than half its annual consumption of oil) would not be delivered. Early in December, newspapers ran stories with panic headlines like "What to Do When the Lights Go Out."

Facing a substantial shortfall in electricity production, anxious city officials discussed ways to reduce consumption. They talked about limiting the work week, instituting rolling blackouts in various neighborhoods, and hiking prices massively. But they feared that major loss of jobs would result from reducing the work week, that the rolling blackouts would be a nightmare to administer, and that massive price hikes would arouse a storm of protest.

In the middle of December, an ad hoc committee, representing a broad coalition of civic, business, and labor leaders, came up with an alternative—to set mandatory targets for reductions for all customers—but to leave it to the customers themselves to implement the specific cuts. And so, in mid-December, the city council adopted a two-phased Emergency Energy Curtailment Plan, the purpose of which was to "significantly reduce the consumption of electricity over an extended period of time, thereby extending the available fuel required for the production of electricity, while reducing the hardships on the city and the general public to the greatest possible extent."

Under Phase I, to go into effect immediately, customers were to cut back on their use, compared to the same billing period of the previous year. There was a stiff penalty for noncompliance: a 50 percent surcharge on the entire bill. The aim was to reduce the city's total electricity consumption by 12 percent. Phase II, to go into effect at a later date, set higher targets. The penalty for noncompliance with Phase II was to be even more severe—a cutoff of service. But the city never needed to institute Phase II, because Phase I was so successful; moreover, penalties for Phase I were never even applied (although, of course, neither officials nor consumers knew at the beginning that this would be the case).

The response to the targets of Phase I, to everyone's surprise, went far beyond the targets themselves.

	Target	Actual Reduction
Residential	10%	18%
Industrial	10	11
Commercial	20	28

The total drop was 18 percent, against the target of 12 percent. Much of the adjustment in commercial establishments, which accounted for 50 percent of electricity usage prior to the cutback, was done mostly through better control of lighting and air-conditioning.

In May 1974, two months after the Arab embargo was lifted, the program was suspended, but its impact could still be felt a year later; in May of 1975, total electricity sales were 8 percent lower than the 1973 level. In addition, there had been a far greater reduction in DWP consumption than in that of the three other largest electric utilities in California, none of which had adopted such a program.

Los Angeles, in the course of a crisis, had stumbled onto a very effective program—a combination of price and regulation that might be called a semi-market approach. The program brought dramatic savings with a minimum of sacrifice and disruption, and required virtually no investment. Two factors made for the program's success. First, a broadly based consensus emerged among civic leaders that this plan was the fairest and most effective reponse to a pending crisis. Second, while the program set targets, it still possessed a great deal of flexibility and left it to consumers to figure out how and where to make their own cuts, to make their own decisions about "essential" and "nonessential" uses. What this Los Angeles experience tells us, fundamentally, is that in many circumstances, great flexibility exists for energy use.

Let us now look at the possibilities for conservation in the United States. The discussion is organized by consuming sectors, each of which poses different problems. (See Table 6–3 for a breakdown as to sectors.)

Table 6–3

U.S. ENERGY CONSUMPTION BY SECTOR, 1978

Sector	Consumption (mbd) (oil equivalent, mbd)	Percent of Total Consumption
Residential/ Commercial	14.0	38
Industrial	13.3	36
Transportation	9.7	26
TOTAL	37.0	100

Transportation focuses on the automobile, where the existence of only a few manufacturers makes regulation an effective way to communicate the need for energy saving. Industry is characterized by a larger number of decision-makers, generally informed about costs and alternatives and likely to respond to clear economic signals. Buildings, on the other hand, involve millions of poorly informed decision-makers within a highly decentralized milieu where market imperfections loom large and where the capital stock has a very long life.

TRANSPORTATION

Conservation for the most part involves a multitude of decision-makers. One major exception does stand out—transportation, and in particular, the automobile. What is involved here is a relatively standardized product used for a standardized activity—moving people about. Only a few producers are engaged in its manufacture, and they are the key decision-makers. The automobile is the one and only consumer product that has a disproportionate and indeed massive effect on energy consumption. The transport sector uses 26 percent of the energy used in the United States—half the oil. The private automobile, in turn, consumes over half of the transport sector's energy. Indeed, the American car alone consumes a ninth of all the oil used in the world every day. But the car is amenable to technological fixes—and rather swift ones, since the auto stock turns over more rapidly than most other forms of capital. Because about half of the American automobile population is replaced within five years,[16] an improvement in mileage efficiency, which is a form of conservation, can make itself felt quickly on the international energy market.

The Auto Way of Life

American society as it has evolved since World War II has become enormously dependent on the automobile, both as a means of transport and as a source of economic activity. Most suburbs, whose development has depended on the private auto, would be stranded without the car. In addition, one out of every six workers in the United States is involved in the manufacturing, distribution, servicing, or commercial use of vehicles.[17] So the auto system is not one to be carelessly pressured and pushed.

One obvious goal after the October 1973 crisis was to seek to reduce dependence on the automobile. And one way to do that is through car pooling and van pooling, but this has not proved particularly successful.[18] Another way to reduce dependence is to provide more and better public mass transit. But mass transit today is still fumbling, both for capital and for approaches that would make it a workable alternative. The nub of the problem is to find a system flexible enough to connect the high-density urban core at one end, where the commuter works, with the low-density suburb at the other, where the commuter lives.* One must also distinguish among different kinds of mass transit. Buses can be a very major energy

* Although, of course, work locations have become progressively more decentralized as well.

saver, but the heavy rail systems, such as BART in San Francisco and the Metro in Washington, D.C., may actually use more energy than they save because of the energy required first for construction and then for getting passengers to the stations along the line.[19]

Pooling and mass transit do deserve considerable commitment, but both lack the convenience desired by the American public. This is a fact of life in American life. Thus, the most effective short-term strategy for conservation must focus not on reducing dependence on the car and on passenger miles driven, but on increasing the efficiency at which those miles are driven.

Some actions easily taken can have significant impact, such as the 55 mile-per-hour speed limit, which has also reduced highway fatalities. Significant saving can also come from proper maintenance. Poorly tuned engines and improperly inflated tires can impose a 4 to 6 percent penalty on fuel usage. Keeping the national auto fleet properly maintained could bring considerable savings, perhaps equivalent to a third of the oil flowing from Alaska.[20] But the trend toward self-service stations will probably lead to worse maintenance. Just the simple requirement that all such stations have functioning, easy-to-use tire-checking equipment could in itself save considerable gasoline.

Stretching a Gallon

By far the most promising area for medium-term conservation lies in increasing the efficiency of the vehicle, that is, increasing the miles obtained from each gallon of gasoline (mpg), a form of conservation that has very little effect on life style.[21] Autos for the 1974 model year, which went on sale just before the embargo, averaged 14 miles per gallon. The average for domestically produced models was even lower—12.8 mpg. In the same year, it was quite possible to purchase comfortable European or Japanese imports that got 20 to 30 miles to the gallon. Not only was mpg low in American-made vehicles, but it had been declining since the middle 1960's. The auto industry blamed the decline on government safety and air-pollution regulations instituted in the late 1960's. The pollution-control devices did impose a penalty—about 5 percent in 1975—but far more important was weight, the single most important factor in determining fuel economy, which had been increasing rapidly. For example, between 1968 and 1973, Chevrolet's full-sized V-8 four-door sedan gained 500 pounds, an increase typical for all the auto makers. The main reason for this was the development of larger and heavier engines for higher performance, the growing array of accessories and options, and other design changes that responded to marketing imperatives. Some accessories, like air-conditioning, imposed double penalties, adding to weight and consuming energy directly.[22]

Bigger cars meant more value added, higher prices, and a larger profit margin for manufacturers. As a prominent auto executive pointed out in an interview with the author, "One of our top people in the 1950's was known for having said, 'Small cars mean small engines, small windows, small doors, and small profits.'" Also, competition from imports kept profit margins low on small cars.

Thus, before 1973, fuel efficiency was given a very low priority, and it was difficult for anyone inside Detroit management circles to argue convincingly that gas economy would find any response in the market. Moreover, the manufacturers themselves tended to discount the dangers of any energy squeeze: World oil reserves were increasing and the real price of gasoline declining. Detroit was also suspicious of what it thought might be oil-company special pleading.[23] "The industry was unprepared for the magnitude and the suddenness of the crisis," an executive observed. "In a way, the embargo was a boon because it raised the question of vehicle efficiency from an issue for debate within management to one of the highest priority for action."

As things turned out, the embargo helped those in management who wanted to give mileage greater attention, but who had consistently lost out when intracompany decisions were made. Once the oil crisis broke, auto-engine efficiency became a public policy issue. This was a reversal, for in the past, government policy had fostered increased auto use at the expense of mass transit.[24]

Four policy alternatives for reducing gasoline consumption were apparent. One was a substantial increase in gasoline prices through some mix of price deregulation and higher taxes. Price has since been increased only to a limited extent because of intense political opposition. But the existing evidence suggests that higher gasoline prices do not cause people to reduce miles driven very much. In those Western European countries where the price of gas rose sharply after 1973, consumption would drop for a few months and then return to roughly old levels. Apparently, once people have decided to buy a car, they will use it. But it does seem that higher prices will encourage people to buy a more efficient car the next time. Moreover, higher gas prices will strengthen the signal to the manufacturers that fuel economy has to remain a matter of high priority. Hence, the real price of gasoline should move up gently and consistently.[25] Today, both consumers and manufacturers are confused. At the end of 1978, the real price of gasoline was no higher than in the halcyon energy days of 1960, the year that OPEC was founded.[26]

A second method to promote efficiency that is common in Western Europe is a graduated tax on horsepower or weight.[27] But a tax of this kind can inhibit engineering innovations that could make even large cars much more efficient. A third approach was a flat minimum mileage standard for every car, but this would have placed American manufactured cars at a disadvantage compared to imports.

Standards and Regulations

In any case, the issue of increased fuel economy was considered too important to be left to the industry and the marketplace. The federal policy finally adopted was the establishment of fleet-average mileage standards.* It is attractive because it sets a target and allows the manufacturer the flexibility to experiment with ways to meet it. For all cars manufactured after 1977, the Energy Policy and Conservation Act of 1975 established the following fuel economy standards:

Year	MPG
1978	18
1979	19
1980	20
1985	27.5

Thus, between 1973 and 1985, the efficiency is supposed to double. The Act also provided that standards be set subsequently

* *Fleet average* means that the target applies to the average mileage of a manufacturer's entire production. Thus, individual models might fall below the target.

for trucks and vans, and imposed substantial penalties for failure to meet the targets.

Much of the auto industry vigorously opposed the standards, which puzzled many people, including the then-president of the United Auto Workers, Leonard Woodcock, who told a congressional committee in 1975, "Unfortunately, the industry comes down here and always in their public postures takes a very hard line, which, frankly, they do not pursue when we have conversations with them. I do not know what it gets them, because it puts them in a position where their word is, very frankly, doubted." In fact, as GM was publicly protesting the standards, the company was already taking weight out of its model (down-sizing) as the quickest way to improve efficiency. Ford did not oppose the standards, already having concluded that they would constitute the least disruptive form of regulation, which was inevitable.[28]

The issue of regulation, in fact, was at stake for the manufacturers. The automobile industry is currently in the process of becoming a regulated industry, which it does not like. The move began in the middle 1960's with the passing of auto pollution and safety regulations. But the fuel efficiency standards, which could affect profits and market share, were considered more threatening. The manufacturers' bind is understandable, for they are asked to deal at one and the same time with pollution and safety, which can add weight and so reduce miles per gallon, and with the demand to increase efficiency. All this, plus the consumer who wants high performance and an agreeable sticker price.[29]

So it is no wonder GM's president was exasperated when he said during the 1975 hearings on fuel economy, "We do not want any handouts, we do not want any taxes, and we do not want any regulations . . . we do not like that sort of thing." In 1978 Henry Ford II also attacked the move to regulations, saying, "It is no secret that the automobile industry has been the prime target of those who have presumed to tell the American people, through single-minded and hastily drawn regulations, what is best for them." Yet in the same statement Ford approved of the three main forms of regulation: "I am not at all reluctant to say that some automotive regulations have been needed . . . Some obviously desirable goals such as reduced emissions of pollutants and increased passenger protection in the event of accidents could not have been achieved as readily without uniform, across-the-board government mandates. In retrospect, I think it is fair to say also that the law requiring greater fuel economy in motor vehicle usage has moved us faster toward energy conservation goals than competitive, free-market forces would have done." [30]

Despite their complaints, the auto makers have hastened to meet the fuel economy standards, and have succeeded. The cumulative fuel savings that will result from the targets set up to 1985 could be as high as 20 billion barrels over the years 1975–2000—twice the reserves on the North Slope of Alaska.

But these potential savings need to be qualified; for recently gasoline consumption has begun to rise quickly again. One reason may be that actual driving results are lower than those Enrivnemental Protection Agency tests indicate. Also, the sales of light trucks, still in the process of being regulated as to efficiency, have been rising rapidly, and now comprise a quarter of all new vehicles sold. Also, the number of longer trips has increased. [31]

Moreover, there are potential constraints to further improvements in efficiency. The major gains so far have been effected by weight reduction (the average U.S. car will probably have lost about 700 pounds between 1974 and 1981), and by the importation and commercialization of European and Japanese technology. Those easy options will have been fully exploited by the early 1980's. Substantial technological innovation is needed in materials, engine, and design; and this kind of innovation, as opposed to styling, has not been a major priority for the industry or its suppliers. Massive capital investment is needed over a decade for the four U.S. automotive companies, which will increase vehicle costs.*

But the constraints should not be exaggerated. The mileage standards are not really as onerous as the manufacturers sometimes suggest. Volkswagen's diesel Rabbit already gets 40 miles to the gallon. Fifty miles to the gallon in larger cars is not out of the question. And to their surprise, American manufacturers have discov-

---

* Altogether, the auto industry may require $80 billion in capital investment through 1985; but this investment will be aimed at meeting not only fuel economy standards, but also pollution and safety requirements, as well as for general modernization. It has been estimated that the investment required for fuel efficiency will be $5 billion to $10 billion. Some think the figure will be much higher. [32]

ered an unexpected bonus. Down-sizing and improved fuel economy are making American-produced cars competitive on the world market for the first time in decades, and opening up significant export possibilities.

Further efficiencies should definitely be encouraged by government policy that is flexible. So approached, riskier and more far-reaching innovations will not be shunted aside in favor of safer incremental change, which could easily entrench technologies, and so create barriers to further fuel economy.[33]

MANUFACTURING INDUSTRY

Industry is characterized by constant self-awareness. Ever greater effort goes into computing and comparing, in order to better allocate resources, balance processes, and improve product. In other words, industry has a bottom line, and profits are its final test.

Consequently, because energy in the 1950's and 1960's was very cheap, an effort to save it was hardly a priority concern for most U.S. firms.[34] But the 1973–74 dramatic increase in energy prices provided an incentive—though not always a powerful one—for American business to seek energy-saving innovations as part of a broader effort to reduce energy costs.

The process of industrial energy conservation can be classified in three major broad and somewhat overlapping categories: (1) *Improved housekeeping,* which means such things as furnace maintenance, adjustment of lighting, fixing of leaky steam traps. Often, surprisingly large savings can be realized here with little or no investment. (2) *Recovery of waste,* which frequently involves familiar technology. One of the most important aspects is the recovery of waste heat, a major task for industrial retrofit. Another is the cogeneration of electricity and steam. Still another is the reclamation of waste products. To recycle aluminum, for example, requires only 7 percent as much energy as does getting aluminum from ore.[35] (3) *Technological innovation,* which requires major redesign of processes and products and considerable investment in capital stock that embodies the more efficient technologies.

Since the early 1970's, industry has been struggling to integrate energy saving into its bottom line. That there has been progress can hardly be doubted, and the pages of a publication like *Energy User News* contain claims of substantial energy savings in many firms, as seen in Table 6–4.

Table 6–4 [36]

ENERGY SAVINGS IN VARIOUS U.S. FIRMS

Firm	Reduction	Time Frame	Comment
Burger King	17%	1974–77	50% by housekeeping
Lockheed (Los Angeles area factory complex)	59%	1972–77	Almost no investment
Tenneco	17%	1972–77	(1) 50% by housekeeping (2) Recycling waste heat
Colgate-Palmolive	18%	1973–76	Mostly housekeeping
Exxon (U.S. refineries)	21%	1972–77	80% with little or no capital investment (11.3 million barrels a year)
Western Electric (Kansas City plant)	38%	1972–77	Almost no investment

When explored in terms of pay back on investment, the savings can be dramatic. American Can at a big New Jersey facility reduced energy consumption by 55 percent with an investment of $73,000, and the annual savings amounted to $700,000. The Parker Company, a large manufacturer of automotive parts, put $50,000 into saving energy and ended up saving $1.2 million a year on energy costs. The comment of that firm's senior industrial engineer is well worth noting: "Everyone is looking for the innovative approach to energy conservation. That's not where it is. There's nothing sensational about how we saved energy. It's been a whole new ethic for conservation." [37]

A surprising amount can indeed be accomplished by simple, good housekeeping. But there are limits to even the best housekeeping, and over the longer term, especially in energy-intensive industries, greater efficiencies will require investment in new equipment and plants. The evidence on industrial energy conservation is contradictory, and the time span since the embargo relatively short. Still, it has become quite common recently to claim that energy saving in the industrial sector is proceeding rapidly and sufficiently, that is, to suggest that conservation energy is being widely tapped. One may be misled by the voluntary reporting to the federal government by trade associations, which tends to be particularly optimistic about savings and which ignores the effects of economic downturn and changes in product mix and production levels.

At this point, it would seem fair to say the following: Energy conservation is proceeding more rapidly in the industrial sector than in any other part of the economy, and the process has been accelerating since 1973. The energy–GNP ratio in the industrial sector averaged an annual decline of 2.8 percent between 1973 and 1977. Yet, even if this figure is accepted at face value, one might well question how rapid and sufficient it is. In addition, important qualifications are needed. The usefulness of recent data has been sharply constrained by the recession in the mid-1970's.[38] For instance, a substantial decline in demand for aluminum led to the temporary shutting down of older, less efficient aluminum smelters, thus leading to an apparent, but not real, improvement. Moreover, the mix of U.S. industrial output changed, so that the share produced by energy-intensive industries declined. In fact, an examination of the postembargo record of eight high-energy-consuming industries shows that in five out of eight, the energy-output ratio actually *rose* between 1974 and 1976.[39] On the basis of what is known today, therefore, it would seem sensible to conclude that, although industry has posted the best record so far of any sector, it remains far below its potential. Despite the relative flexibility of American industry when compared to that of other countries, conservation has been much slower and more uncertain than it need be. Why?

The Barriers

A number of important barriers retard conservation in industry. To begin with, effective conservation is not something that happens automatically. It requires an organizational response, usually at three levels. There must be a strong and persistent interest and commitment by senior management. There must be an effective energy unit in the company—that is, engineers and managers with the authority of the senior levels behind them, who can implement change at the plant and in the office. And finally, conservation must be built into the fabric of operations, so that energy use is constantly monitored and so that conservation becomes part of the employee's work habits and an element in the manager's annual bonus. "One of our biggest problems is getting people at the plant level to take energy conservation seriously," Goodyear's corporate energy planner has observed. "Yes, even after all this time." The importance of commitment and persistence at all three levels cannot be exaggerated. Indeed, backsliding is often easier than maintenance. "Like all housekeeping, the benefits soon slip away without constant attention," noted an executive of Armco Steel, the nation's seventeenth largest industrial energy user.[40]

How to get and maintain that commitment—and implement it? In many firms, energy is not a significant cost, and so energy saving, whether by housekeeping or by investment, remains a low-priority concern. In such circumstances, the benefit to society of energy saving exceeds the benefit to the individual firm. And current energy policies widen the cost gap between society and the firm by posting prices that do not take into account replacement costs or the externalities that result from dependence on imported oil. But, of course, there are firms for which energy is a large part of costs— for Allied Chemical, 10 percent of sales; for Armco Steel, 15 percent of sales. Even if they want to save energy, some of these firms may not have the capital available to do so.[41]

Another most important barrier is the very high rate of return demanded by many firms for the type of investment in which energy savings is usually categorized. The high "hurdle rates" frequently seem to cluster around 30 percent after taxes. With accelerated depreciation and an investment tax credit of 10 percent, this corresponds to a two-year payback.[42] Companies base the high "hurdle rates" on the judgment that conservation measures do not have the strategic impact of a new product or additional capacity, and that they are easily postponed—this, despite the very low risk involved in energy-saving investment. Then, too, incomplete data and inadequate information about the appropriate technology can engender doubts about making an investment in conservation.

It can also be very difficult to read and adjust to the environment. The president of Inland Steel complained, with some justification, about the lack of "consistent economic, energy, and other regulatory signals which can be used as a basis for orderly decision-making." [43]

The confusions are multiple. The first, obviously, concerns price. Managers are told that the value of energy is increasing; yet the real price of key industrial fuels in the United States, corrected for inflation, has actually declined since 1974, which is not exactly a signal for increased effort to save energy.* There is even more uncertainty about the future: Will the real price increase or decrease? The public dialogue sometimes seems to have been captured by analysts who say, in effect, that the energy problem is a thing of the past, that 1973–74 was only an aberration, and thus that conservation need not become a high-priority matter.

The uncertainty extends to other aspects of the future energy environment. What will be the preferred sources of energy? How secure will domestic and foreign supplies be? And great confusion exists about the thrust and impact of government policies, especially those involving incentives, regulations, and allocations. Many smaller firms doubt whether conservation will pay off financially, while others are afraid that if they trim energy use now, they could end up being penalized (compared to companies with more "energy fat" on them) by a rationing system imposed in some future crisis.

For larger companies, the simultaneous push to conservation and coal conversion has created a real quandary. The two compete for investment dollars, and a shift from oil and natural gas to coal can actually increase the number of BTUs required per unit of output.[44] A leading energy consultant explains why management postpones making investments in conservation, even when the payback time is only two or three years: "They're waiting because of a fuzzy picture on fuel costs and they're waiting for signals from the government." [45]

Cogeneration: Industry's North Slope

One aspect of industrial conservation stands out, for American industry has within its grasp an Alaskan oil strike, a major new source of energy waiting to be developed. This type of conservation

---

* The real price of No. 6 industrial fuel oil decreased 15 percent between August 1975 and August 1978.

energy deserves special attention because of its potential scale and because its availability largely depends upon requisite decisions being made in the political process. The source is sometimes called *combined heat and power,* a term awkward enough to be designated CHP, and perhaps for that reason better known in the United States as cogeneration. But the meaning is simple enough—the combined production of electricity and heat (the latter for either process or space-heating purposes).

Today there are two independent energy delivery systems. One is composed of utilities in which electricity is centrally produced, be it by coal, nuclear fisson, oil, gas, or hydropower. As the electricity is produced, the power plant sends steam up a stack and into the air —or into lakes and rivers—as waste. In the second system, companies generate their own steam for use in the industrial process; in fact, almost half of all energy used by industry is consumed just to produce steam.[46] What cogeneration means is the integration of the two systems.

Cogeneration can take two forms. In the first form, steam (or hot water) from a power station is delivered by pipes to homes and offices to provide heat and hot water. Such systems, called district heating, are quite common in both Eastern and Western Europe, with about a thousand in the nine countries of the European community. District heating schemes, however, are economical only when urban density is high and subscription to the system general. Indeed, for most American cities, the cost of putting in the pipes and other parts of the system would probably be prohibitive. Steam can also be transmitted from power plants to specific consumers nearby. An Exxon refinery in New Jersey, for example, buys its steam from a power station a mile away, just as Harvard buys steam from a generating plant on the Charles River. Consolidated Edison pipes steam to over 2000 office buildings and apartments in Manhattan. Other firms and power plants are exploring such symbiotic relationships.

The second type of cogeneration, with by far the most significant potential, comes from the combined production of electricity and steam at industrial sites. Here the firm produces not just steam but also electricity as a by-product of generating the steam. The process, called topping cycle, can be explained simply. Energy is used to produce combustion at temperatures up to 3600° F in order to get steam that need not be any hotter than 400° F. Obviously, a great deal of energy is wasted in the process. In cogeneration, the high temperature is used to make gas vapor or very high pressure steam, which drives a turbine or a rotating shaft, which in turn generates electricity. As in a power plant, the waste from this process is steam, except that in this case the steam is not waste, for it then goes on to be used for industrial processes. The energy is thus *cascaded* from uses that require high temperatures to those that require lower temperatures. This is called high-quality and low-quality energy uses.[47]

The advantages of cogeneration are substantial—about half as much fuel is used to produce electricity and steam as would be needed to produce the two separately.* And it appears that the return on investment for many industrial firms (and for other establishments, such as hospitals and shopping centers) is quite good. Furthermore, cogeneration gives companies an important hedge against almost inevitable increases in energy prices and against brownouts and other interruptions of supplies, whether caused by oil producers, coal strikes, or bad weather. For utilities, cogeneration can reduce the need to build new nuclear or coal-fired power stations, at a time when marginal costs are higher than average costs—and at a time when the political obstacles to such new capacity are difficult to surmount in any event.

As with so much else in conservation, cogeneration does not require a major technological breakthrough, but rather regaining a path that was abandoned. Around the turn of the century, many industrial establishments in the United States produced their own electricity as well as steam, but they eventually gave it up.[48] For one thing, the utility regulatory system scared them away from cogeneration. For another, the declining real price of electricity (which for industrial consumers was cut in half between 1940 and 1950), combined with lower unit prices for larger consumers, made cogeneration even more unappealing.[49] Companies chose to stay in the businesses they were in and to leave electricity to the utilities. In 1950, 15 percent of the nation's entire electricity supply was generated by industry; by 1973, only 5 percent, a figure that dif-

* At a typical electric utility generating plant, up to two thirds of the fuel's potential energy is lost as discharged waste heat. Meanwhile, industrial waste heat has been estimated at 20 percent of total national energy consumption.

fered markedly from other countries. Twenty-seven percent of all West Germany's electric power is produced by industrial firms, half by cogeneration. British industry produces 20 percent of its own electrical needs.[50]

A number of analyses suggest substantial energy—over twenty percent of total industrial energy use—could be saved in the United States through cogeneration investments that are economically sound.[51]

In other words, industry is sitting upon an easily recoverable, relatively cheap new source of energy. Is it quickly being exploited? Not especially. Why? Because we have here a near-perfect example of obstacles being not technical, but almost entirely institutional and organizational.

There are two key obstacles to industrial cogeneration. The first is represented in the point where industrial cogenerator, utility, and utility regulator meet. The cogenerator cannot unplug himself from the utility, because he will sometimes have to buy extra electricity to meet his needs. At other times, he will have extra electricity which he will want to export, that is, send into the utility grid. So his questions are three. What will the stand-by electricity cost? Will he be able to feed his electricity to the utility? And if so, at what price?

As things now stand, the whole system discourages cogeneration. The cheap rates for bigger users and the absence of marginal pricing still reduce the incentive for industrial firms to cogenerate, while utilities do not want electricity produced by cogenerators. As a regulated monopoly, the utility industry has an allowable profit based upon a return on capital, and an industrial firm's generating capacity cannot figure in the utility's capital base against which the utility rates are calculated. Meanwhile, the industrial cogenerator, even if he does get the utility to accept his electricity, may find himself in a regulatory thicket; for as soon as he sends electricity across the street, he may be deemed a public utility, subject to a number of federal and state regulating bodies, all of which only complicate his life with more forms, more hearings, more lawyers, and more uncertainty.

Fortunately, the environment is changing. The 1978 National Energy Act has eased the danger of industrial firms' being categorized as utilities, and also presses utilities to become more receptive. Moreover, while some utilities remain hostile, others, already short of capital and afraid of having insufficient generating capacity in the 1980's, are showing a willingness to accept electricity into their systems from wherever they can get it.* Slowly, the regulatory system is adapting to the needs of conservation in general, and electricity rates are in the process of being revised so that they encourage, rather than discourage, cogeneration. This is a very difficult, complicated, and time-consuming problem. It is at one and the same time highly technical, highly political, and highly important.[52]

The second key obstacle is that many companies fail to see the great advantages that cogeneration might bring them—they want to be in the business they are in, not in the electricity business. They also demand a much higher rate of return (sometimes twice as high as normal investment, as high as 30 percent after taxes) for such an ancillary investment as cogeneration, whereas utilities are satisfied with a much lower rate.[53]

Cogeneration is only one of the major ways to save energy in the industrial sector. Altogether, it may be economically possible to cut industrial energy use by more than a third through cogeneration and conservation efforts. The total capital investment would be some forty or so billion dollars less than that required for investment in conventional energy sources.[54] Such an assessment is admittedly rough. After all, energy is used in so many different ways in industry that a quantification of potential savings is very difficult to achieve. Since one cannot really predict means and timing of implementation, the extent of overall potential savings cannot be precisely predicted, but about its potential one can be very optimistic.[55]

The Missing Policies

Public policy has a most important role to play in overcoming the many barriers already discussed to industrial conservation.

At present, price signals are very confusing and need clarification. Prices must be allowed to move moderately upward to reflect the real value that energy has today. Hence, those supporting a system of price regulation should face the fact that current costs strongly retard energy conservation in industry—and increase the

---

* Perhaps ownership of the cogenerating facilities could be vested in the utilities.

prospect of much higher prices later. It is obvious that senior management will not take energy-saving investments more seriously until they are convinced that their firms will be paying more for energy in constant dollars in two years than today. On the other hand, higher prices are not enough. Even moderate price increases will cause some discomfort, raising distributive and equity questions, and any price that even begins to approach the true cost of imported oil would be highly disruptive and politically impossible.

Thus some increase in price must be matched with other policy-induced mechanisms that will accelerate conservation and give it an equitable chance to compete with conventional soures of supply. In particular, these mechanisms need to reduce the high hurdle rate, and thereby increase the rate of energy-saving investment and so speed up the turnover of machinery and other capital stock. Between 1947 and 1973, the ratio of energy to output actually did decline in all U.S. manufacturing. The major reason? More efficient technologies embodied in new plants and equipment. As Myers and Nakamura point out, "Turnover of capital stock and expansion of the size of stock have been the principal means by which reductions in energy–output ratios have been achieved." But this can be a very slow process. "An obvious implication for government policy is that tax or other policies that promote investment will speed energy conservation." A most important point follows from this observation: While low economic growth reduces absolute energy consumption in the short term, it most certainly will slow energy conservation over the longer term by retarding investment in more efficient plants.[56]

The 1978 National Energy Act provides a 10 percent tax credit for conservation investment. But given the subsidies and external costs of other energy sources, as well as the high hurdle rates, 10 percent seems much too low. Significantly greater tax credits, up to 40 percent, plus accelerated depreciation and energy-conservation loans, are required. These are especially important in a period of lagging economic confidence, high uncertainty, and consequent low investment.

Other measures should also be implemented. Currently, the government has a very weak voluntary program in which companies report on energy saving. The data, however, is, as already suggested, obscure, difficult to analyze, and often misleading. The reporting system should be made more meaningful, and the targets for savings strengthened.[57] Self-congratulation would then be replaced by more careful attention. Finally, information and education efforts should be intensified, especially for smaller firms.

Dow's War on BTUs

There is nothing automatic about the integration of energy consciousness and energy efficiency into the bottom line. But once the considerable barriers to energy conservation are surmounted, the savings can be very considerable. So the story of Dow Chemical indicates.

The case is worth serious attention, for Dow is one of the three largest consumers of energy in the United States, ranking with U.S. Steel and Alcoa. Thus Dow had good reason to be very sensitive to the sudden price changes in one of its key inputs, natural gas. It was also able to translate sensitivity into action. In the late 1960's and early 1970's, Dow correctly interpreted market signals and long-term trends ahead of most other major American corporations, and it managed over ten years to reduce its energy consumption (per pound of product) by 40 percent—putting it at the forefront of major chemical companies.[58]

Two features of Dow's situation made it special. First, the firm has traditionally been oriented to energy saving. The hobby of H. H. Dow, the company's founder, was power—in the literal, not political, sense. "He loved to generate power in ingenious ways," said J. E. Mitchell, Dow's director of corporate planning. "Right from the beginning, H. H. Dow spent money for efficiency in power generation that was not justified." Dow's interests were shared by the plant manager in Midland, Michigan, Merle Newkirk, who is still known around Dow's headquarters for having been "a nut on power efficiency." He, too, paid little attention to the economics of investment in power, but rather drove for more and more energy saving. Thus, Dow introduced cogeneration into its plants in the 1920's. A continuing tradition of intense interest in power efficiency was established as well.

The second feature had to do with the development of the company. In the 1940's and 1950's, the Texas Division, operating in the Gulf States, underwent an enormous growth on the basis of cheap natural gas—which at first was available simply for the cost of gathering it. The original Midland Division, based in Michigan, continued to use coal, which meant its costs were higher than those of the Texas Division. "We were under the gun to lower costs," recalled Mitchell. "It was a noose around our necks."

The cost crunch became critical during the 1954 recession, when there was a marked downturn in Midland's contribution to corporate profits. The general manager of the Midland Division launched a program to try to compensate for the division's reliance on coal. He instituted an increasingly elaborate system to keep track of the amount of steam and electricity used in the various manufacturing steps, so providing, on a monthly basis, data on cost of energy per pound of product. "From day one," Mitchell recalled, "Midland knew it was fighting an uphill battle on energy costs." The situation improved, but the division recognized that it could never be competitive with the Texas Division, which was expanding in an almost explosive way in the 1950's and early 1960's.

In 1967, reports began coming to headquarters about developing shortages of natural gas, and a study was done in Midland, which forecast pressure on natural gas supplies as prices of interstate and intrastate natural gas began to diverge. In 1968, the director of U.S. operations was persuaded that all energy prices would increase. He coined the phrase "the war on BTUs"—and ordered it waged throughout the company.

But the Texas Division was reluctant to accept such projections. "People had gone down there in the 1940's and 1950's when gas was cheap," Mitchell said. "The people in Texas said, 'It can't happen here.'"

It was not until 1970 that the senior management of the Texas Division finally came around. What helped convince them was that the price of natural gas went up 10 to 15 percent in 1970 and 1971. But even corporate management in Midland was not fully convinced of how serious the problem was until one intrastate natural gas contract increased $5 million in one year.

In 1972, Mitchell presented to the firm's executive committee a prescient forecast of the energy future and what it could mean for Dow. He predicted increasing dependence on imports, tightening supply, and significantly rising prices. The years 1973–80 he called "the Arab era," the result of which would be "a wild, scrambling worldwide fight for hydrocarbons." He added, "There can be no real security of supply during a scrambling readjustment." As for Dow's course, "The only short-term solution is intensive conservation effort." The executive committee agreed, ensuring that there would be full support at the top for an intensified war on BTUs.

The company thus had its institutional traditions, the experience of the Midland Division in how to save energy, and now it also had top management support behind conservation. Throughout the company, targets were set for the amount of energy that each pound of product should require, and daily corporate reporting of theoretical and actual balances of energy was instituted for all operations. The outcome was taken into account in annual job-performance reviews and in the annual merit-raise procedures. Key personnel are now kept constantly informed of energy prices and trends. This procedure, in the words of a conservation specialist in the Texas Division, has "sensitized all levels of supervision to the need for conservation." Thus, Dow's efforts went from the simplest kind of housekeeping (turning off motors when not needed) to retrofitting (putting in heat exchangers) to designing entirely new plants that yield more product with less energy.

Several lessons emerge from the Dow experience. There is a strong need for accurate measurements of energy consumption in order to establish targets, evaluate results, and assess managerial responsibility. Commitment must come from the top to make clear that conservation is a bottom-line concern and not a public-relations ploy. "Because top management gave it such strong support," said Gerald Decker, corporate energy manager from 1967 to 1978, "everybody gave it top priority. That was the secret." Moreover, as another Dow executive expressed it, "Most things in energy conservation are not based on new knowledge, but rather the applying of knowledge we already have in a different environment." The simpler and cheaper things are done first, and then progressively more capital investment is required. "There are no gimmicks left for us," said Mitchell. "From here on, the only way to save BTU's is by reengineering, by building new plants, by spending more capital. The easy part is over."

But Dow's "easy part" in itself—still an exception for American industry—makes a very dramatic point: With relatively little capital investment, Dow was able to increase the productivity of its energy inputs by 40 percent, and make a substantial contribution to company profits in the process. Dow had stronger incentives than most other firms prior to 1973, but still the lesson is clear.

BUILDINGS

Between 36 and 40 percent of U.S. energy consumption is used to heat, air-condition, light, and provide hot water for homes, commercial structures, and factories. The residential sector alone uses 20 percent of all the energy used in the United States. The use in the individual home breaks down like this:

Space heating	53%
Hot water	14%
Cooling	5%
Air-conditioning	7%
Other	21%

During the 1950's and 1960's, efficient energy usage was increasingly neglected in the construction of new buildings and homes. In New York City, office buildings put up between 1945 and 1950 used half as much energy per square foot as those built between 1960 to 1965. The difference? The older buildings use natural light and have windows that open, whereas the newer buildings are sealed and depend on mechanical systems for lighting, heat, and air-conditioning. The same trend is evident in private housing, where convenience and fashion also promoted an increase in the use of energy-intensive household appliances.[59]

Cleverer Buildings

Since 1973, energy has ceased to be neglected. "Architects are now in a period of major reassessment in which the entire selection of materials and assemblies is being examined to determine whether they can perform to satisfy the new energy conservation demands," Richard G. Stein, one of the nation's most energy-conscious architects, has observed. "Many of the materials that would normally be slowly phased out will now be abruptly rejected. . . . The hope for the future lies in the fundamental reversal in our present commitment to the sealed building, with its massive plant for manufacturing the air and delivering it at predetermined temperatures and velocities and its large lighting apparatus that substitutes a universal switch for selectivity." [60]

This same reassessment is being made by builders and buyers. How extensively? It is not yet clear.

But there are certainly some significant examples of change. The modern, tall, glass-faced, sealed office building of the 1960's and 1970's is an energy-intensive creature, annually using between 150,000 and 250,000 BTUs per square foot—even up to 400,000 BTUs per square foot. But a new IBM facility at Southfield, Michigan, will use only 51,000 BTUs per square foot, a fifth of what the building might have used had it been built in the early 1970's. While the drop in consumption is dramatic, there is nothing dramatic about the methods—double glazing for the windows, more insulation, and lower lighting levels.[61]

The General Services Administration has established an energy target of 55,000 BTUs per square foot per year for its new office buildings. "It is important to realize," said Fred Dubin, the consulting engineer who helped the GSA (and a number of companies) establish that target, "that the order of magnitude of these savings in the new and existing buildings can be done with readily available off-the-shelf hardware, equipment, and systems—with thoughtful, discriminating, innovative design."

Buildings, as some architects express it, are becoming even more "clever" in their use of energy. Requiring no conventional heating plant at all, the twenty-story headquarters of Ontario Hydro in Toronto (a city that is definitely cold) uses only 65,000 BTUs per square foot. The warmth is provided entirely by capturing the waste heat given off by lighting, office equipment—and employees.[62]

Similar changes are likely to follow through design changes in new residential construction. A test house constructed by the Oak Ridge National Laboratory will require only 20 percent as much electricity for heating, cooling, and water heating as would a conventional house.[63]

The continuing development of what has been described as "energy-conscious design" will promote increasing energy efficiency in new construction. For some, such design comes as a reaction to obvious waste, over-engineering and over-lighting, to what is seen as an excessive artificiality in modern building and an insensitivity to the natural environment.[64] For more people, however, emphasis on energy-conscious design is much more the result of a perception of rising costs and the uncertain supply of energy.

The trend is being reinforced slowly by changing building codes and loan requirements, which increasingly stress energy efficiency. As regulations are stiffened, so building stock will become more energy efficient. The Department of Housing and Urban Development is updating its minimum property standards for residential and commercial buildings to tighten energy standards, without which no federal financing will be available. Also, some local governments are developing their own standards. Seattle, for instance, is investigating different approaches to make conservation in buildings an "equal option *with* generation." [65]

Another important factor is a movement away from a "first-cost mentality" to life-cycle costing. Traditionally, the purchase price of a house, rather than that price plus operating cost, has been the chief concern of builder, buyer, and financing institution. Two thirds of all the new single-family homes in 1971, for instance, were built for speculative sale. The builder was therefore interested in keeping the selling price down, and worried much less about longer term energy costs.[66] Obviously, a shift is occurring to looking at life-cycle costs in some form, which has been encouraged not only by the interest of the buyer, but also by lenders concerned about rising utility bills affecting mortgage payments and under-insulated homes losing value. Indeed, the United States League of Savings Associations has pointed out that energy charges as a percentage of operating costs for private homes rose from 18 percent in 1973 to 28 percent in 1975.[67]

Retrofit

The trend toward energy-conscious design is promising, but there is a catch, a large one. Unlike the auto stock, the building population turns over very slowly. In 1972, a record year for new housing starts, new homes accounted for less than 3.5 percent of the total stock.[68] Moreover, major constraints will slow down the diffusion of new designs. These constraints range from building and health codes to availability of materials, to trade-union practices, to architecture and engineering education. So new building designs can have a substantial impact only in the long term.

In order for energy saving to be promoted in existing buildings, an aggressive retrofit campaign must be mounted. *Retrofit* is a space-age term, describing the upgrading of a complex system through the insertion of improved components. In the case of Dow, it has meant the addition of new equipment to existing manufacturing processes. In buildings, it generally means changes in equipment and structure to improve thermal and lighting efficiency.

The evidence so far indicates that a program of retrofit brings savings that astonish those who embark on the strategy. IBM, for instance, was already aware of rising energy prices before the embargo, and so launched a conservation campaign in early 1973. The initial goal was to reduce energy use by 10 percent in thirty-four major locations in the United States. The savings have far exceeded the goal: by the end of 1977, consumption was 39 percent lower than the 1973 preconservation levels. If the company had not embarked on this campaign, its energy costs for the period 1974–77 would have been $90 million higher. That realization, observed a senior IBM executive, "provides a powerful incentive to save." What really impressed the executive was that two thirds of the savings was achieved with little or no capital investment. "In the past," he said, "abundant and low-cost energy supplies did not make conservation a key management concern, and so initial energy conservation measures yielded large savings without capital investment . . . The methods of achieving these initial savings are not very technical or profound. They amount to turning off lights, changing temperatures, shutting down equipment when not needed, fine tuning building systems, and other similar techniques." [69] The 3M Company has had a similar experience, saving considerable energy simply by plugging air leaks, which requires virtually no investment.

Its electricity supplies threatened by the coal strike of 1977 and 1978, Ashland Chemical, a division of Ashland Oil, found that in a matter of weeks it could cut back on electricity use by 25 percent in the buildings that comprise its headquarters complex with little investment and "little inconvenience." The lessons cited by the official in charge of the program provide an insight into the barriers to conservation: "Our management group was generally surprised at the extent of our success, and the amount of discretionary items that use electricity. Unfortunately, it might well take an imminent crisis like the coal shortage to motivate the type of action that we initiated. The key elements here are to develop a sense of participation on the part of all the employees because all the employees can contribute in some manner to energy savings . . . A crisis situation gives one the opportunity to effect change of this nature. Once the change has been developed, it is considerably easier to maintain the change. . . . In addition to the strong emphasis on communication and participation on the part of the employees, another lesson that developed from this experience is that a little additional planning during the design and construction of a physical facility would give the management of that facility considerable more flexibility and opportunity for savings on electrical use." [70]

Sixty-Seven Percent Less

The possibilities in the residential sector parallel those in the commercial-industrial sector. The American housing stock is characterized by extreme diversity. Of the 80 million or so all-year residences in the United States, 50 million are detached single-family houses. More than a third of all residences were built before 1940, when there were few or no standards for insulation. In the late 1950's, the spread of air-conditioning and electric heating provided an impetus for thermal insulation. Only then did insulation manufacture become a major industry, and double glazing and storm windows gained consideration in new housing. Even so, the economics at the time did not exactly create a clamor for significant thermal protection. By some estimates, 30 percent of the residences in the country may be completely uninsulated. Altogether, two thirds probably need additional thermal insulation. [71]

What savings are possible? The Federal Energy Administration found, with a test house in the Washington, D.C., area, that adding standard insulation devices decreased the total annual energy requirement of the house by 25 percent. Additional insulation increased the energy savings to 35 percent. Standard Oil of California conducted a three-year demonstration study of homes in Portland, Spokane, and Seattle. With an investment of $981, the fuel consumption of the Portland house was reduced by 50 percent, with a rate of return on the investment of about 25 percent. The study concluded that 50 percent energy savings are possible as economically attractive investments in a substantial part of the nation's housing stock. These results are consistent with a number of other tests around the country. [72]

The Washington Natural Gas Company, serving the Puget Sound

area, went into the energy conservation business after the embargo, selling "conservation kits" for attic insulation. By "kit," the company meant that it not only provided but also installed the insulation, guaranteed it, and financed it. The cost was about $200. The reduction in heating energy, 22 percent. The company advertised the kit with the message that it would cost consumers more not to buy it than to buy it. By November 1977, the utility had sold 14,000 such kits, and it estimated that its advertising and promotion created a demand that led to an additional 42,000 jobs for other contractors. The utility is now selling a more elaborate kit, which includes attic insulation, pilotless natural gas furnace, and automatic day-night thermostat. This, the company estimates, has reduced energy used for heating by an average of 36 percent.

Such programs also help the utilities cope with sometimes urgent problems of supply. Washington Natural Gas estimates that the savings in the 56,000 homes insulated as a direct or indirect result of its program has freed gas for 16,500 new houses without requiring any new supply. "What all of that meant," said the utility's president, "was that we had been sitting atop a new gas field for years, and didn't recognize it." [73] This was a most valuable discovery at a time of rising prices and supply uncertainty for natural gas. Similarly, some electric utilities hope that retrofit can reduce the need to invest in new generating capacity, and help avoid some of the difficult choices between coal and nuclear.[74]

The most important evidence to date for the underrated value of retrofit comes from Twin Rivers, New Jersey, a community of 3,000 well-constructed residences a few miles from Princeton. For five years, a group from the Center for Environmental Studies at Princeton University intensively studied actual energy consumption in this community. The results are extraordinarily rich for understanding energy use in the real world. The researchers found that a 67-percent reduction in annual energy consumption for space heating was possible with a relatively simple package—interior window insulations, basement and attic insulation, and plugging of air leaks. The costs were hardly exorbitant; at the present price of natural gas, they would bring a 10-percent return on investment, which is better than the bank. The conclusions point to the great potential of systematic conservation: "Among the ways of conserving household energy, there are no spectacular technical fixes. There is only a catalog of small fixes, many of them drab and unimpressive in isolation. It is therefore easy to dismiss conservation of household energy as an incremental business and to seek bigger solutions elsewhere. But the catalog is fat, and many of its entries are cheap. With patience, groups of small and even tiny fixes can be put together into large assemblies that overall can produce impressive results . . . It does not appear to be impossible, in fact, that under present technology and economic conditions, space heat in houses could be a minor rather than a major consumer of fuel." [75]

What would be good for the homeowner and for the utility would also be good for the nation. If simple insulation packages—such as six-inch ceiling insulation, storm windows and doors, caulking, weather stripping—were installed in 20 million poorly insulated homes, it has been estimated that energy use in the residential sector could be cut by a quarter.[76]

Where Is Policy?

Yet, only grudgingly encouraged by government, progress toward a meaningful retrofit program has been disappointingly slow. This failure of public policy has been one of the biggest lapses in the nation's lapse-ridden effort to cope with the world of imported oil.

For the first three years after the crisis, senior government policy-makers in the energy area suffered from an excessive faith in the efficiency of the market. Prices, they thought, would stimulate just the right amount of retrofitting and just the right changes in construction practices for new buildings. A small example: In March 1975, an official in the Department of Housing and Urban Development requested permission to change minimum property standards to increase energy efficiency by promoting "maximum reduction in cost with a minimum, if any, increase in construction cost." HUD higher-ups rejected the request, saying, "Although most of the items submitted are energy saving, it is more effective to have the market dictate the additional thermal requirements." [77]

Those who advocate exclusive reliance on price forget about the considerable imperfections in the very decentralized housing market with its millions of decision-makers. The homeowner is typically ill-informed about conservation, how to analyze energy use and calculate savings, whom to go to for advice and installation, how to finance, what to put in. Another problem is mobility. By 1970, only 54 percent of all household heads were living in the same houses as 1965. If you think you are going to move in a couple of years, why invest? [78] Moreover, many whose houses most need retrofit are the people least able to afford it. What those who argue for exclusive reliance on the free market forget is that there are real people with real problems, for whom high energy costs create genuine hardships. One such real person is Florence Leyland, an elderly woman who owns a three-bedroom house in Waltham, Massachusetts. In one twelve-month period, Mrs. Leyland had to spend $550 of her $3,223.20 total income, all of which comes from Social Security, on heating. [79]

Some encouragements for retrofitting were included in the Energy Policy and Conservation Act of 1975 and the Energy Conservation and Production Act of 1976: A modest program of weatherization assistance for the elderly and the handicapped was established; targets are now being set for improved efficiencies of appliances; and demonstration projects and energy audit procedures are also being established.

The 1978 National Energy Act provides for a 15 percent tax credit on investments in residential conservation, but not to exceed $300. Although better than what has happened so far, the proposals are still rather modest, and still do not do enough to give conservation the chance it needs and deserves.

One must ask how effective even the expanded Carter package will be in overcoming the obstacles described by the Governor's Energy Advisory Council in Texas: "In the absence of government leadership, typical consumers will experience a long period of high expenditure for energy until they come to demand energy efficiency as a priority. The period can be shortened and its impacts lessened by the adoption of appropriate government policies."

It is nothing short of ridiculous that now, almost six years after the embargo, the United States does not yet have a broad-ranging national program of incentives to encourage retrofit. The speed with which retrofit can deliver substantial savings argues for a much more stimulative public policy, with tax credits up to 50 percent. Such a policy would signal the importance of retrofit and would encourage homeowners, entrepreneurs, and manufacturers. It would make retrofit economically attractive for some homeowners, and not only attractive but possible for others. Standards and regulations also need to be pushed. So do demonstration projects. Residential homes comprise the most decentralized sector of energy consumption, and therefore public education and information is particularly important. An Energy Extension Service, modeled on the Agricultural Extension Service, now operates on an experimental basis in ten states, and is a program that should be expanded and supported throughout the nation.[80]

But how to encourage the development of an efficient system that will actually deliver conservation to the individual homeowner? The National Association of Home Builders Research Foundation has projected that the retrofitting of 40 million single-family homes might require the establishment of 6,000 retrofit businesses, each generating at least $400,000 of business a year (at least 500 homes a year) and many thousands of jobs. Some system of licensing, training, and control will be required. But it must also be recognized that as independent operators, retrofit contractors will encounter the considerable skepticism that homeowners feel toward the home-improvement industry and the "aluminum siding boys." [81]

One way to speed up retrofit is to give utilities a stake in it, so that they deliver conservation along with energy.[82] How would they do this? They would do energy audits for the house, recommend retrofit measures, subcontract to independent businesses to do the work, but guarantee the work themselves, thus maintaining quality control and reassuring homeowners. In this way, automatic day-night thermostats and furnace adjustments can be combined with structural changes. To help finance the work, the utility would "loan" the money to the customer, who would pay it back as part of the monthly bill.

Thus, with a stake in conservation, utilities will not see energy saving as a threat to their well-being. But Congress imposed a ban in the 1978 National Energy Act on any more utilities' moving into the retrofit business, a self-defeating measure that will make it much harder to deliver conservation to the homeowner. The ban will cost the nation many hundreds of millions of barrels of oil equivalent a year. The state of California, which has encouraged utilities to go into retrofit, has better perceived the situation. "We find these utilities can serve usefully in getting insulation into residences," the chairman of the state's Energy Resources Conservation and Development Commission stated. "They are the only institutions in our society that get into these homes once a month with some kind of billing or a meter reader. They have a sense of what is out there; what energy consumption is. For that reason, it is important to take advantage of this contact with the consumer so we can target where we are going and systematically cover the marketplace. Many of the concerns that have been expressed about consumer protection and the design of the program can be handled adequately by an intelligent public utility and commission or other State agency." [83]

Retrofit, of course, can mean things beyond insulation and storm windows. The heat pump is a device that became commercially available in the 1950's, but is still far from being widely used. Its working principle is simple: In the summer, the pump removes heat from the interior of a building and discharges it outside. In the winter it does the opposite—extracting heat from the outside and pumping it into the building. It can deliver the desired interior comfort much more efficiently than conventional electric heating, for the heat pump can produce up to three times as much output in thermal energy as it receives in electrical energy input. [84]

Standard setting and efficiency labeling have now begun to be applied to home appliances. If accelerated, the effort could also lead to substantial savings, without affecting people's standard of living. Because energy use by appliances has been increasing much faster than energy use by heating systems, the savings could be very important. Almost a third of residential energy use—6 percent of the national total—is now consumed by major home appliances. [85]

One respected group of researchers at the Oak Ridge National Laboratory has concluded: "A judicious combination of government regulations (appliance efficiency standards, thermal standards for construction of new residences), incentives for weatherization of existing houses, and research and development to produce new technologies can yield a future in which residential energy use in the year 2000 is at roughly the present level. Such a combined program can also provide large economic benefits to households." [86]

Behavior

Increased efficiency in existing and new buildings will result not just from one set of decisions or from one overall fix, but from an interplay of factors—prices, incentives, regulations, research and development, changing techniques and methods of operation, and changing human attitudes. That last should not be underestimated.

The consequences of behavior are crucial. Reducing the thermostat from a twenty-four-hour setting of 74 degrees to 68 degrees during the day and 60 degrees at night can reduce heating loads by as much as 20 percent. Setting air-conditioners at 78 degrees instead of 72 degrees can reduce energy requirements by 15 percent. As the Princeton Twin Rivers project found, some residents use twice as much energy to heat and cool their townhouses as do other residents in identical structures. Retrofit must therefore be combined with an ongoing, consistent, clear, and non-threatening public education campaign. Homeowners otherwise may assume that their one-time decision to weatherize allows them thereafter to forget about energy —which it certainly does not.[87]

CONCLUSION

Conservation is a blanket term that covers several sectors, many activities, and many different kinds of decision-making. It is therefore difficult for public policy to cope with. But let us try to analyze the overall potential in three ways—by projection, by inventory, and by observation.

Many projections about future energy demand have been bound by conventional wisdom, folklore, the inability to incorporate political forces, and the habit of looking at things from the supply side.[88] One of the most significant efforts to break habit's shackles is the recent report of a panel on energy futures assembled by the National Academy of Science. The panel looked at four different plausible scenarios for energy demand in the United States. The scenarios were constructed out of careful engineering analyses of trends in various sectors, combined with econometric and input-output analyses. The results were extraordinary—that in the year 2010 "very similar conditions of habitat, transportation, and other amenities could be provided" in the United States using twice the energy consumed today, *or almost 20 percent less than used today.* And this is with continuing economic and population growth. The fundamental conclusion is "that there is much more flexibility toward reducing energy demand than has been assumed in the past." [89]

Another way to assess the overall potential is through inventory, which can be done by physical modeling. Here, analysts take accessible, often off-the-shelf technologies and compute the cumulative effect on energy consumption of their substitution for conventional technologies. Two scientists followed this procedure in a study for the American Physical Society based on the year 1973.[90] They calculated the savings that would arise from such steps as

- Installing heat pumps
- Increasing refrigerator efficiency by 30 percent
- Reducing heat losses from buildings by 50 percent through better insulation, improved windows, and reduced infiltration
- Implementing cogeneration for half of direct heat applications in industry
- Using organic waste in urban refuse for fuel
- Improving automobile efficiency by 150 percent over 1973 levels

The results are very impressive:

(in millions of barrels per day)	
Total U.S. energy consumption (1973)	36
Potential savings	−15
Hypothetical consumption	21

This hypothetical consumption is 40 percent less than the actual consumption in 1973. *In other words, in 1973, the same U.S. living standard could theoretically have been delivered with 40 percent less energy.* These savings, in BTUs, are almost as much as all the oil— not just imported oil—used that year.[91]

The results fly in the face of conventional wisdom. Yet the changes required are considerably less daunting than developing the breeder reactor or making a wholesale conversion back to coal.[92] Sometimes one concludes that the real challenge of energy conservation is not to do it, but rather to believe that it can be done.*

---

* The physical-modeling approach has been carried much farther in a major new study, *A Low Energy Strategy for the United Kingdom,* by Gerald Leach, Christopher Lewis, Ariane Van Buren, Frederic Romig, and Gerald Foley (London: Science Reviews, 1979). Rather than starting with elasticities and some large model of demand, it takes as its starting point the investigation of energy use, and the multiplicity of changes and adaptations that might reduce the amount of delivered energy required to provide a service. Energy demand is broken down into some 400 separate categories determined by end-use, fuels, and appliances. Projections of demand are then laboriously built up, "brick by brick." The conclusion? That gross national product could

There is yet a third way to conceptualize the real potential for energy conservation, and that is by observation, using such information and understanding as is gained in this chapter. One notes the difference between cars that get 13.5 miles to a gallon and cars that get 33.5 miles to the gallon; between industrial plants that do organize themselves to reduce energy per unit of output by 30 percent and those that do not; between homes that are retrofitted that use 40 percent less energy than homes that are not. Here is evidence for a wide band of flexibility in energy use with current technology.

Miserly Public Policy

But the reader should not be deceived. Nothing will happen automatically, and the obstacles to conservation are manifold. To overcome them in a politically acceptable and nondisruptive way requires adroitness. The movement toward greater energy efficiency, toward greater tapping of conservation energy, will be governed by a complex interaction between government and society. A public policy is required that shapes strong coherent signals, all of which point in the same direction.

It is disheartening to compare the role of public policy in the United States with that of other Western countries, especially when one remembers that the United States is the dominant energy consumer on the world scene. Canada, for instance, has appropri-

---

triple in the United Kingdom in the fifty years 1975–2025, but that a series of simple, known technical fixes could keep energy demand pretty much where it was in 1975. The work shares the same spirit of energy pragmatism that, we hope, is central to our own volume, for *A Low Energy Strategy* makes the following points that also apply to the United States: that there is "immense scope for energy conservation"; that "a low energy future" need *not* be "bleak and repressive"; and that major social change is not required. "Britain—and by implication other countries—can move into a prosperous low-energy future with no more than moderate change. All that is necessary is to apply with a commitment little more vigorous than is being shown today by government, industry, and other agencies some of the technical advances in energy use which have been made, and are still being made, in response to the oil price increase of 1973–74." But, as this chapter has argued, the barriers can be considerable to that somewhat more vigorous commitment. The computer time for the *Low Energy Strategy* study, it might be noted, was provided by British Petroleum, and is to that oil company's credit.

ated $1.4 billion to subsidize housing insulation. The program was instituted after comparing the costs of developing new Canadian hydrocarbon resources with the costs of retrofit. The expenditure, given that Canada's population is a tenth of ours, would be the equivalent of $14 billion in the United States. No such commitment has been made here.[93]

The French government, convinced that an "energy transformation" was at hand, embarked on a major energy conservation program, perhaps the most ambitious in any major industrial country. While some of its elements would not be suitable to American society, the program does indicate how a democratic society can make conservation a high priority without being high-handed. One of the most significant lessons to emerge from the program is the absolute need for coherent signals. As Jean Syrota, the director of the French Energy Conservation Agency, expressed matters: "You have to put things together. You have to do regulations, financial incentives—not only price—and publicity all at the same time. These three means of action must be coordinated. . . . It is indispensable to be helped strongly by politics, by the government, especially by the government. . . . If the government's actions are not positive, then they are negative, harmful. If government does not show interest, then all sectors of society imagine that it is not important." [94]

Yet there is something ironic about the French program, for France's energy consumption per capita is only 40 percent of America's. Also, what it does has far less impact on the international energy system, since it uses only about 10 percent of the oil that the United States does.

What conclusions emerge from this investigation into three very different energy-consuming sectors of American society?

First, there is a great deal of flexibility in how and how much energy is used for various activities and to deliver various services and amenities. In the past, it has not mattered greatly how much energy was consumed for this or that purpose. There was little economic rationale in attending to these questions. It often was cheaper to leave windows broken in a factory than to repair them. But the world has changed. It is no longer economically, politically, or environmentally sound to ignore possibilities for much greater efficiency in energy use.

Second, the best way to conceptualize conservation is as an alternative energy source. As such, we can compare it to other sources in terms of payback, ease of recovery, disruption, and environmental effects. Which is cheaper—a barrel of new production in some distant and hostile terrain, with the risk of a dramatic increase in price, or a barrel saved by insulation? Which is safer—continued reliance on imported oil, or the heat pump? Real choices about direction exist. In general, conservation appears to be the energy source that calls for the greatest emphasis in the short and middle term, since it is often the cheapest, most accessible, and least disruptive. The United States can, in effect, quickly produce millions of barrels per day of conservation energy. Obviously, conservation is no final answer in itself, as some are quick to remind us. Only recently, the energy newsletter of Citibank said, "It cannot be stressed too strongly that this country cannot conserve its way out of the energy problem," and instead argued that increased domestic production could provide an alternative to imported oil. But those who so argue are more and more hard-pressed to demonstrate that domestic oil, gas, coal, or fission provides any final, or indeed even any better, answers than does conservation. Conservation certainly buys the United States time, and given the difficulties that attend the other sources, provides more immediate relief than do high-capital, high-technology alternatives.[95]

Third, conservation energy is not so simple to recover as it might seem. Unfortunately, it is a diffuse source, and it has no clear constituency in the way that oil, gas, coal, and nuclear do. Public policy must be its champion. With such a commitment, many different strategies will be needed. Public policy must create a hospitable environment for the expeditious exploitation of this source. If we had decades, then the market alone, working through gradual rise in prices, would be sufficient. But the decades are not there. For conservation to make the kind of contribution it should in the relevant time span, there must be found that adroit mixture of signals —of price, regulation, incentives, and information. Only in that way can conservation actions become as economically attractive to individual decision-makers as they are to the society at large.

The American system is particularly responsive to incentives, and that is where public policy has been particularly loath to intrude. Up to now, the failure of public policy has been its inability to assess the true prices and true risks of conventional alternatives, and its consequent inability to measure against them the costs of incentives that will promote conservation.

What are the "principles" of a meaningful energy conservation policy?

1. The pricing system should give clear and consistent signals about energy. This means steadily rising real prices, combined with measures to reduce the resulting hardship, but with the goal that prices should eventually reflect replacement costs.

2. Incentives should give conservation a fair chance against imported oil.

3. A permanent information and education campaign should be instituted about the problem of imported oil and the possibilities of conservation.[96]

Each of the three energy-consuming sectors need somewhat different strategies. All should build on existing programs.

1. Automobile manufacturing involves very few decision-makers. Regulatory policies applied with flexibility are most effective. Higher gasoline efficiency should be pursued vigorously and stringently in post-1985 standards. Buses and car-pools should be stimulated, and perhaps free public transportation should be experimented with in a few cities. Gas prices should show real increases.

2. For industry, while subsidies are useful for demonstration projects, the preferred method to encourage conservation should be substantial investment tax credits, accelerated depreciation, and loans (the last is especially important for smaller businesses). To such signals businessmen habitually respond. Firmer conservation targets should be set for major energy-consuming industries, but familiar and flexible methods should be used to encourage them.

3. Buildings involve the greatest number of decision-makers. Regulations perhaps geared to performance, not specifications, are extremely important for encouraging energy-conscious design in new construction. They can be encouraged through federal loan agencies and savings and loan associations, which can also encourage retrofit through their lending policies. This suggestion is an example of the principle that existing and familiar institutions should be used to deliver conservation to the public. Another example would be the giving to utilities of a major stake in retrofit, that is, for generating conservation energy. For the homeowner, direct subsidies, extensive tax credits, and perhaps exemption from

property taxes for conservation improvements—these are required to speed and spread retrofit. Certainly the marginal costs of imported oil justify such assistance, thus making conservation as valuable to the individual homeowner as it is to the society, and also making the signal visible, not merely a haze on the horizon.[97]

The United States can use 30 or 40 percent less energy than it does, with virtually no penalty for the way Americans live—save that billions of dollars will be spared, save that the environment will be less strained, the air less polluted, the dollar under less pressure, save that the growing and alarming dependence on OPEC oil will be reduced, and Western society will be less likely to suffer internal and international tension. These are benefits Americans should be only too happy to accept.

MODESTO A. MAIDIQUE

7

Solar America

During most of this century, solar energy seemed to interest only dreamers, tinkerers, and radicals. But because of the oil embargo, the sun has become a serious alternative source of energy. The issue has now become how much solar energy, what kind—and when. According to the organizer of International Sun Day, "Forty percent of our energy could come from solar energy by the year two thousand if we make some dramatic moves *now*." The editor of *World Oil* disagrees, saying that the source will have the impact over the next quarter-century of "a mosquito bite on an elephant's fanny."

The two estimates lay out the range of the debate on solar's near- and middle-term potential contribution to America's energy needs. We believe that given reasonable incentives, solar could provide between a fifth and a quarter of the nation's energy requirements by the turn of the century. Without the incentives, it could end up amounting to little more than that mosquito bite.[1]

New technology is not required to realize solar's potential, for the kind of relatively low level technology needed for a 20 percent contribution is already here, or very close to being here.[2] What does stand in the way are a series of economic and institutional barriers, which must be overcome in the early 1980's if solar energy, like conservation, is to have a fair chance in the marketplace against conventional sources. Moreover, the considerable near-term and middle-term potential of solar points toward a more general conclusion. It is not unrealistic to envision a Solar America, a society that relies not on exhaustible hydrocarbons, but on renewable sources of energy.

A more responsive public policy, however, is required to overcome the obstacles to near-term solar energy. Certainly, government support for solar has increased dramatically in the last few years, but the support has been weighted much more toward high-technology research-and-development (R&D) projects, which may or may not be able to make a contribution in the twenty-first century. Meanwhile, the near-term, more certain Small Solar opportunities are not yet receiving support commensurate with their likely potential.

This distinction must be understood if the potential for solar is to be comprehended. Solar energy, in fact, is not a single new energy technology, but a blanket term that covers a diverse set of renewable energy technologies, some new and some quite old. The energy form includes proposals for colossal microwave satellites and huge ocean platforms, as well as for wood, Colonial America's principal fuel.

The collection of so many different sources under the term *solar* has confused the thinking about the issue. In an effort to simplify the matter, the Department of Energy has divided solar into eight different individual categories which can be organized into three major groups: [3]

1. *Thermal (heating and cooling) applications*
   Heating and cooling of buildings—including hot water heating
   Agricultural and industrial process heating
2. *Fuels from biomass*
   Plant matter, including wood and waste
3. *Solar electric*
   Solar thermal electric—such as the "power tower"
   Photovoltaics—solar cells
   Wind—windmills
   Ocean thermal electric
   Hydropower—hydroelectric dams

Each of the individual categories can be further elaborated. For instance, biomass includes not only wood, but also technologies to improve yields of sea algae farms, research on new vegetable oils, and gasification of manure. The unifying conception here is that solar energy is the energy that arrived on the earth from the sun "recently"—during the last hundred years or so.[4]

Solar requires new ways of thinking about energy. Procuring energy has conventionally meant buying fuel—oil, gas, and coal. Most forms of biomass also require buying fuel, but other types of solar energy will mean buying equipment. The fuel—sunshine—is free. This distinction makes economic comparisons difficult, for frequent purchases of fuel must be measured against a one-time investment in equipment.

Would the coming of solar energy bring, as some have suggested, fundamental changes to American society? Technological change always does imply some social change, and that will surely be the case with solar energy. But the "revolutionary" impact of solar has been exaggerated. Certain established companies would be affected much more than society as a whole. Firms such as General Electric and Westinghouse are presently the prime suppliers of generating equipment to the nation's electric utilities. A shift to certain large-scale centralized technologies, such as the power tower, for the generation of electricity would transfer some of the generating-equipment business to leading aerospace manufacturers, such as Boeing and Martin-Marietta, which have played a pioneering role in solar electric technologies.

A more significant change could occur from the widespread use of on-site solar technologies, such as solar heating. The term "on-site" derives from the limited area within which the energy produced by a system is consumed—a house, a factory, or at most, a small cluster of structures.[5] The spread of on-site technologies could have significant impact on the traditional distributors of energy—the utilities—which could find themselves partially bypassed, leading to a leveling off or actual decline in demand for their services. The managerial responsibility for generating energy would then, at least in part, move out of the hands of the utility executive into those of the energy consumer himself. Still, a shift of this sort hardly means the disappearance of the utilities or a revolutionary transformation of American society as some "small is beautiful" advocates have argued. In short, on-site solar energy technologies are unlikely to trigger a fundamental decentralization of American society.[6]

A reasonable assessment of solar prospects requires analyses of very diverse forms of energy. We have chosen to concentrate on four of the most important technologies. The first, solar heating, could have a major impact on U.S. energy consumption in the next ten years. So could the second, biomass, especially wood and waste.[7] The third, the "power tower," is the leading recipient of research for the central solar generation of electricity.[8] The fourth technology, photovoltaics, converts light to electricity using the semiconductor, the basis of the transistor and integrated circuit industry. Each of the four solar options differs in its level of technological advancement, its market potential, and the economic and institutional barriers it faces.

SOLAR HEATING: A NEW GROWTH INDUSTRY

Solar heating is the most technically mature of all the on-site solar technologies.[9] It is hardly a new technology at all, but rather represents a return to a path abandoned only during the last hundred years or so. Up until that time, most human cultures sought to incorporate the sun itself into the design of its architecture. The surviving baths at Pompeii, for example, demonstrate the use Romans made of solar water heating. Channels carrying water to the baths were built open to the sun and lined with grooved black slate, so that the water became heated as it ran through the grooves on its way to the bathers. The Mesa Verde cliff dwellings in Colorado were constructed with massive rock overlays, which provided shade from the high and hot summer sun, but allowed the rays of the lower winter sun to penetrate. Over the centuries, many very different cultures have designed and built structures that embody the basic principle of solar architecture found at Mesa Verde.[10]

Today, a design that manages to take advantage of the sun with few or no moving parts is called passive solar heating. The structure itself is sited and landscaped so that it becomes in effect a large solar collector. A home faces south. In winter, the sun's heat is captured through large double-glazed windows, and the heat is retained by the walls during the night. "A passive system has only one moving part," one student of the subject has observed, "the earth moving around the sun." [11]

Such architecture, ignored for several decades in the United States, has received increasing attention since the embargo. Energy-conscious design, where conservation and solar energy overlap, can be very effective. In Nacogdoches, Texas, for example, a retired Air Force colonel built a U-shaped passive house. In his previous home, which was nearby, he paid $850 a year for winter heating and summer air-conditioning; in the new house, the annual bill fell to $260. The passive approach also works in the colder climates; in Maine, an 8,000-square-foot, energy-conscious commercial building saved its owners about $400 a month in heating bills.[12]

There is a fundamental proposition underlying effective solar design, such as was employed in these structures. First, the heating needs of the building are minimized by improved insulation, weather-stripping, and similar conservation practices. Only then is an investment made in solar heating. In sum, the maxim is, Conserve first, then heat. In this manner the total investment required—in conservation and solar design—can be minimized, for conservation substantially reduces the cost of the solar design required. This improves the return on the overall investment, for, in general, investment in conservation yields the highest economic return, followed closely by passive solar design, and then by active solar hot water heating and space heating.

Much experimentation and learning—and relearning—is now occurring with passive solar heating. Enough is already known, however, to argue that incorporation of such design into all structures need not add appreciably, if at all, to building costs, especially as the methods become more familiar.[13] Passive design, however, has a drawback. Its major impact will be on new structures, for, as noted in the preceding chapter, the building stock turns over slowly. However, this is compensated by the relatively rapid rate at which passive design is being adopted in new homes and buildings. Additionally, passive retrofits, such as attached greenhouses, can be used in existing buildings anywhere in the country. But the real impact of passive design will be felt toward the late eighties, as the cumulative number of new structures becomes comparable to the present stock.

For most of the next decade, solar heating will make its principal impact in the form of active systems, especially for heating water, which are retrofitted onto existing structures. These are called active systems because they involve mechanical moving parts. Panels about three feet by seven feet are bolted on the roof. Generally made of aluminum, glass, plastic, and copper, the panels catch and concentrate the sun's rays, which in turn heat water, air, or some other

medium that flows in pipes through them. Fans or pumps then cir-
culate the medium through a heat exchanger in a water-filled storage tank. The hot water in the storage tank can be used either
directly or to heat the house by pumping it through a conventional
radiator network.[14] (See Figures 7−1 and 7−2.)

Active solar heating and solar hot-water heating are really variations on the same system, differing only in cost and scale. Three
solar panels may satisfy the hot-water needs of a house, whereas a
hot-water and space heating system in the same house may require
ten such panels. The larger system will cost several times more, but
will also save several times as much fuel. A solar heating system
does not, however, completely eliminate the need for conventional
sources. A typical system supplies one half to two thirds of heating
needs, which means that a backup conventional heating system is
required for periods of sustained cloudiness. And, of course, conventional electric power is needed for lighting and appliances.

Many people still assume that solar energy is something for the
future, awaiting a technological breakthrough. That assumption
represents a great misunderstanding, *for active and passive solar
heating is a here-and-now alternative to imported oil.* The potential
for solar heating is vast, because it is well suited to most new residential and commercial buildings and to about one third of the
nation's 55 million existing dwellings. By the year 2000, active and
passive solar heating could replace 3 million barrels daily of oil
equivalent (mbdoe).[15]

In 1975, one could still have only speculated about the possibil-
ties for a solar heating industry. Today it is a vibrant, growing industry. Sales, including installation, increased tenfold in three years,
from $25 million in 1975 to $260 million in 1977. In 1977, 3,300
space and water heating systems, 63,000 solar hot-water systems,
and perhaps 35,000 swimming pool systems were sold. During the
first half of 1978 there were only small gains from these levels due
to uncertainties over federal tax credits. Strong growth resumed
again in 1979 after Congress passed a 20 to 30 percent tax credit
for solar systems. The overall potential over the next two decades
may be yet five hundred times larger—over one hundred billion
dollars. The development of the industry and the market potential
permits one to be increasingly optimistic about solar energy.[16]

It also has all the classic characteristics of a new industry—frag-
mentation, with no dominant design or leading manufacturer. No

Figure 7–1

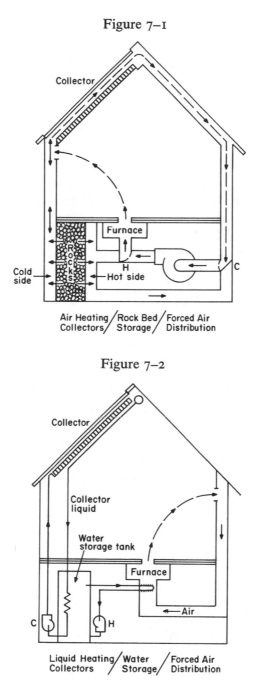

Figure 7–2

Figures 7–1 and 7–2 reprinted from *Survey of the Emerging Solar Industry,* Solar Energy Information Services, 1977, p. 51.

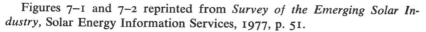

manufacturer has more than about 3 percent of the market, but a great deal of entrepreneurial energy is evident, as hundreds of firms—hungry and hoping for their share of a potentially huge market—compete under such names as Sun Works, Sun Power, Sunsave, Sunspot, Sunburst, and Sunco.[17]

And as happens in new industries, one can expect major changes in the structure with marginal competitors dropping out and leading firms emerging. But what comes out of this will not constitute a high-technology industry. Rather, a given company's strengths will come from its ability to refine its product lines and to work out an effective system of promotion and distribution. As one solar executive observed in an interview, "The key to success in this business is marketing and logistics." This will require effective selection and training of distributors, plus advertising and the development of a quality image. The solar industry might well follow the model provided by the central air-conditioning business, where a handful of firms dominate the manufacturing, but where installation and service are local businesses. The utilities might also play a public information and general contracting role.

For an infant industry to become an established and growing one will require the emergence of dominant firms with widely accepted systems. A potential leader in the industry is Grumman, the only high-technology aerospace firm seeking aggressively to position itself in low-technology solar heating. Grumman decided that the near-term opportunity for solar energy lay in heating and hot water, and as a result, chose to deemphasize competing against the other major aerospace companies for high-technology solar contracts. In short, Grumman has based its solar effort not on technological expertise, but on managerial and marketing skills, and it has mobilized the internal resources that this kind of activity demands and hired outside executives that complement its existing organization. In 1977, Grumman's $1 million advertising budget was larger than the sales of most solar heating firms.[18] Two other major firms, already active in solar heating, are also potential leaders—Honeywell, the dominant firm in temperature controls, and Lennox, which is the leader in heating, ventilation, and air-conditioning products.[19]

But the point of large-scale diffusion is far from being reached, and it could be delayed for many years. This is the paradox of solar heating: While it is already practical and already being commercialized, it is far from clear that solar heating will fulfill its potential as a real alternative to imported oil. There are two barriers, economic and institutional, that must be overcome for solar heating to make a significant contribution. First, possible customers for solar heating want a fast return on their investment—that is, to recoup in a few years the dollars spent on installing a system through dollars saved on fuel costs. The fuel savings need to be substantial, for a hot-water heating system typically costs $1,600 to $2,400 for a single family home, from a few hundred dollars to $5,000 for a passive space heating system, and for an active space and hot-water heating system, $5,000 to $13,000. The Energy Project at the Harvard Business School distributed 4,000 questionnaires to a randomly selected population across the country. Three key points emerged: (1) The idea of solar heating is very attractive. (2) People are poorly informed about how such systems work and how to get one installed. (3) On average, a five-year payback—that is, the investment recovered in fuel savings in five years—is necessary to attract serious consideration. And in order to get 80 percent of the respondents to think about installing a system, a two-year or better payback was required.[20]

Thousands of people are buying solar systems today, but as one Massachusetts state government official explained in 1977, "At this point the solar energy industry is an elitist phenomenon." The marketing manager of a leading solar firm described the 1977 solar consumer as "a man in his late forties or fifties, typically an engineer or an architect earning forty to fifty thousand dollars a year. He buys equipment for status or philosophical reasons. Economics are not a factor." This profile, however, is changing. Solar energy is also being adopted by some middle-income people in their thirties. Moreover, the size and character of the solar market will be expanded by a provision in the Military Construction Act of 1979, which stipulates that if "cost effective," that is, if they can pay for themselves over the next fifteen to twenty-five years, solar heating systems must be incorporated into all new military housing and 25 percent of remaining new military construction.[21] But it is clear that in order to make an important contribution, solar heating must become accepted by a much broader range of people than is currently the case, and basic household economics will be an important factor.

Similar economic considerations also affect decision-making

by businessmen. The Energy Project surveyed the largest employers in Massachusetts, over half of whom had considered solar in the last few years. Some were interested in it as a hedge against higher conventional energy prices, although none had thought of it as a way to protect themselves against supply interruptions. But all had decided not to install solar because a perceived ten- to fifteen-year payback was incompatible with the two- to four-year payback criteria required for a capital investment. All, however, were monitoring the development of the industry.[22]

In another survey we canvassed firms that had installed solar heating. In all but one case, the main reason for so doing was advertising and public relations. As a bank official explained, "It is visible, and we wanted to attract attention on a busy road." [23]

Our surveys support the contention that the rate of return, perceived as too low, constitutes the critical economic obstacle. It need not be. Even by conventional analyses, solar heating can provide reasonable paybacks. But the difficulties of analyzing solar by conventional methods are considerable, and they bias the comparisons against solar. To begin with, analysis must take into account weather, technology, and the cost of the fuel assumed on the comparison. An eight-year payback in Albuquerque, New Mexico, may be a fifteen-year payback in Augusta, Maine. Beyond this, several considerations very much in solar's favor are often overlooked. This can be illustrated with an example—the installation of a solar hot-water system with overnight storage capacity in an oil-heated home in the northern United States. Let us assume that the system costs $2,400, including installation and all accessories, and that it saves $200 a year on home heating-oil costs. Four different methods can be used to evaluate the rate of return.[24]

1. With a simple analysis, $2,400 divided by $200, the payback is twelve years—much higher than the five-year payback required for wide national acceptance. Still, the system provides a good return, for at slightly over 8 percent the interest is better than a savings account and about the same as a high-quality corporate bond.[25]

2. But consider taxes. The $200 saved per year can be regarded as tax-free income, which can be very important if the homeowner is in a high bracket. In a 33 percent bracket, the return is equivalent to a pretax 12 percent return, and in a 50 percent

bracket, 17 percent. Thus, the return is better than the highest yield bonds.

3. A more sophisticated analysis would consider additional factors. Assume that fuel costs rise 2 percent faster than inflation, and that the system's life is twenty years; further assume an initial 20 percent down payment and twenty additional mortgage payments. In other words, visualize a system purchased at *today's* prices—and saving energy tax-free at *tomorrow's* prices. The results—a straight 11 percent, which would amount to a 24 percent pretax return for someone in the 33 percent bracket, and a 40 percent pretax return for someone in the 50 percent bracket.[26]

4. But even that calculation does not go far enough. If one considers the benefits and costs to society at large, it must be noted that solar does not carry the external costs of coal or nuclear. Furthermore, it requires more person-hours per BTU of energy supplied than the conventional energy technologies. Moreover, if an incremental barrel of imported oil costs $35, or three times the 1978 market cost, the real payback shrinks from twelve to four years—which is certainly a most attractive investment. For a person in a 50 percent bracket, it amounts to a two-year pretax simple payback.[27]

One can easily argue about the precise numbers in the analysis, but the exercise should make clear that solar energy is far more "economic" than conventionally assumed. The greatest single barrier, however, is to get the homeowner to see beyond that simple twelve-year payback.

There are two other important economic barriers. One is the cost of borrowing money. An individual will have to pay interest rates up to 20 percent higher to save a kilowatt through solar energy than to add a kilowatt of capacity through utilities. The installation of solar heating may also impose a penalty in the form of increased property tax, a sore subject with homeowners today.[28]

On Sun Day, 1978, President Carter declared in his speech, "Solar energy works. We know it works. The only question now is how to cut costs." [29] The first proposition is true, but costs are not the only question. Powerful institutional barriers also impede the acceptance of solar heating.

One of the most formidable is building codes. Each American locality has its own code—the result of geography, climate, building materials, and local political forces. Of the over ten thousand municipal building codes, only a handful provide for solar energy. When codes do not include solar energy, powerful disincentives can be created. A New Hampshire man, for example, wanted to install solar hot-water heating. He applied for a building permit to make the necessary modifications in the structure of his house, but an official in the town planner's office told him that solar energy was not in the building code's lexicon. So, after going through all the documents, hearings, and other procedures required to obtain variances from building codes, he decided he did not want solar after all. And even when building codes provide for solar, they can be discouraging. Initially, the planning committee for Coral Gables, Florida, rejected solar roof-top collectors outright. It then reversed its decision, but set such strict controls on aesthetics that costs were substantially increased.[30]

Another institutional barrier is lack of skills in installing and maintaining solar systems. The construction industry, like other fragmented low-technology industries, does very little research and development, and is slow to adapt to technological changes, such as the "new" solar heating techniques. Indeed, recent experience suggests that poor installation is a far greater problem than the reliability of the solar equipment itself. Although there is an abundance of solar firms, solar skills are in short supply. A solar engineer or architect with more than three years' experience is a rarity. But the weakest link in the chain is the solar installation technician, often a plumber with no experience in working with solar equipment. A review study of a solar domestic water-heating experiment set up by three New England utilities found that installation problems were the main reason for poor results in the first year of operation. In sum, a general mood of consumer dissatisfaction over poor operating results could have a crippling effect on the willingness of homeowners to take the risk of installing solar equipment.[31]

Another obstacle is legal uncertainty. Access to the sun is not a transferable property right in most of the United States. As of January 1978, only six states have dealt with the issue, and only in New Mexico is access to sunlight legally a property right. Yet such a guarantee is essential to eliminate one of the main risks associated with installing solar energy—loss of access to the sun. The sun's rays strike solar collectors in the United States at any average angle of 35 degrees (plus or minus 20 degrees, depending on location).

Obviously, light must pass through the air space above a neighbor's property before it can be collected. Without legal protection, a rational potential solar buyer will think twice before committing capital to an energy source that depends on a neighbor's whims— or on the growth of his trees, or on the construction of a high-rise apartment house where the neighbor used to live.[32]

The attitude and role of utilities is another major stumbling block. Most solar-heated homes will still need to be connected to electric utilities for a source of backup supply. Thus, the rate structure and attitudes of utilities are very important to the successful implementation of solar heating on any scale. As things now stand, utility rate structures can discourage conversion to solar, affecting institutions as well as individual homeowners. A Massachusetts bank, for example, decided to make a major investment in an advanced design solar space and hot-water system for one of its new branch offices. All went well until the bank found that its application for the low all-electric rate had been turned down by the local utility. The power company claimed that by using solar space and hot-water heating, the bank had forfeited its right to the all-electric rate, even though it employed no conventional fuels—that is, no oil, gas, or coal. Because loss of the right would have nullified any savings that might have been achieved with the solar system, the bank reversed its decision, abandoned plans for an overall solar space heating system, and installed only a smaller system for hot-water heating. With the smaller system, the bank managed to qualify for a lower electric rate, but was still denied all-electric status.[33]

The problem with—and for—the utilities goes deeper. The utilities can see solar heating competing with their own role as producer and converter of energy, and like conservation, solar can be perceived as a threat to the utilities' self-defined growth program. Many utilities, therefore, show indifference or even hostility to solar energy and conservation, and express considerable skepticism about the contribution that solar heating can make during the rest of this century. But according to an executive at Houston Lighting and Power, "When solar becomes significant, the utilities will have to become involved in its implementation. Our role should be in installation and maintenance, and not in selling the equipment. But our participation may be dictated more by legislation rather than by our own choice." [34]

Legislation and regulation have had and will have considerable effect on what utilities will and will not do. One of the few utilities involved in promoting solar energy equipment is in California—not coincidentally, the state that has the largest solar incentives in the country. Other utilities around the country are also beginning to investigate the possibilities of playing an innovative role in the diffusion of solar energy. *But without utility support, or at least cooperation, solar home heating will be greatly retarded.* A survey in Pennsylvania concluded that "if solar energy is ever to achieve substantial adoption rates, utilities will have to play a pivotal marketing role." [35]

It is these economic and institutional barriers, and not technological obstacles, that will determine whether solar heating will make a major contribution in the next two decades. Solar space and hot water heating could provide 3 million barrels daily of oil equivalent by 2000 . . . or not until 2015 . . . or not until 2025. Everything depends on how effectively the barriers are surmounted.

The question, therefore, becomes, How to give solar heating a fair chance in the marketplace against conventional sources? [36] First, economic incentives are necessary. We believe that such incentives should be framed with the traditional innovation pattern in mind—that is, the problem is to get over the sluggish first phase, and then allow the innovation to proceed on its own momentum.[37] Thus, incentives should promote the refinement and diffusion of the products and of associated solar skills during the first phase, and the best method may well be a federal tax incentive that declines over time. A big push in the first phase, followed by eventual abolition, is a strategy based upon market pull, which is what is required in a highly decentralized market difficult to reach through regulation.

Earlier, we pointed out that a simple five-year payback is required to attract over half of the respondents in our survey, a finding consistent with other studies. Consequently, the 20 to 30 percent tax credit provided by the 1978 energy legislation is too low. A 55 to 60 percent tax credit would seem much more appropriate, for a twelve-year payback would thereby be reduced to five years. Ours is not a farfetched proposal, for California already offers a 55 percent credit, which has certainly sped up the spread of solar in that state. Such an incentive would also encourage manufacturers and distributors to commit more resources to the solar business. In general, our recommendation would remove an increasing share of U.S. energy supplies from the threat of rising prices and supply interruptions.[38]

The institutional barriers must also be attended to. Since solar heating equipment is a capital good, the value of tax credits are limited unless accompanied by appropriate financing. For the most part, the homeowner can obtain solar financing only at the standard rates for home-improvement loans, which are typically well above the prime rate. One way to get at the problem is by revising the charter of the Federal Home and Mortgage Insurance Corporation so that it can encourage better financing terms for solar equipment.

Building codes must also be reformed to facilitate action by state legislatures and city councils.[39] Therefore, a model solar heating code is required. The code would need to take into account regional differences, and would also need to include provisions on sun rights and recommendations for property-tax exclusion. Finally, the U.S. Energy Extension Service, now only an experimental program, should be strengthened so that it can provide services to individual and institutional consumers, as well as to the construction industry and solar heating manufacturers. In that way, information and skills about solar systems can swiftly be made available.[40]

WOOD AND WASTE: OLDER THAN ROMAN POOLS

An ideal solar collector has already been designed. Requiring virtually no maintenance, it is economical and nonpolluting; it uses an established technology and it stores energy. It is called a plant.[41]

Indeed, organic matter from plants and animals, or biomass, constitutes another near-term and accessible form of solar energy. Photosynthesis, occurring naturally worldwide, stores more than ten times as much energy annually in plant form than is consumed by all mankind. But very little of this energy is tapped, particularly in the developed countries.[42]

This has not always been the case. Less than a century ago, wood was the United States' principal fuel. As recently as 1900, wood accounted for 25 percent of the country's total energy,[43] but by 1976 it provided less than 1.5 percent, or about .5 million barrels per day. And most of the 1.5 percent—.4 mbd—came from the forest products industry, which burns tree wastes. By comparison, 8 percent of Sweden's and 15 percent of Finland's energy needs are met by wood.[44]

True, the theoretical potential of biomass is very large,[45] but within the continental United States, most of the land suitable for biomass production is legally withdrawn from timber harvesting or is already used to produce food, feed, fiber, or timber. Thus, only about 20 percent—400 million acres—is commercial forest land that can really be considered available for fuel. Even 20 percent would make for the equivalent of 3 million barrels a day. Achieving this contribution, however, could still encounter opposition from environmentalists. Another 1.5 mbd is economically recoverable from municipal solid and liquid wastes and from animal manure; this recovery would also help to solve the problem of what to do with the waste generated in America.[46]

There are several different methods for *bioconversion,* which is the term for the transformation of biomass into usable energy. One is the simple direct burning of solid wood or other plants. The second is the conversion of biomass into a liquid. Brazil, for instance, has a goal of replacing 20 percent of its gasoline with alcohol derived from plant matter, primarily sugar cane, by 1980.[47] A third method is a biological process in which bacteria break down organic waste into methane gas. Considerable research is now going on to develop new sources of biomass, such as quick-growing plants, that would provide greater energy intensities.[48]

But for the remainder of this century, wood and related waste products will continue to be the United States' principal source of domestic biomass. There are three major markets where the potential 3 mbd of additional wood and wood waste could be consumed: the forest products industry, residential wood heating, and industrial firms and utilities that employ coal-fired boilers.

Today the main consumer of wood for energy generation is the forest products industry, which derives 45 percent of its total energy needs from burning bark and mill waste, primarily. But the industry could become totally self-sufficient in about a decade without great difficulty by burning additional quantities of wood and wood waste, thus saving an additional million barrels of oil a day equivalent.[49]

Another major market for wood could be residential wood heating. Sales of wood stoves have been booming since the oil embargo. One survey estimates that the number of installed wood stoves has increased from one million in 1974 to as many as five million in 1978. If the new stoves' wood-burning capacity were used at the same rate as the old capacity, a fivefold increase in home wood consumption—from .1 mbd to .5 mbd—would be realized. Sustaining the present level of sales of about one million stoves a year would mean that capacity would increase at the rate of one quad * every five years.[50]

But there are, of course, short-term limits to supplies of wood and waste. If every American household installed a wood stove, the demand might rise to the oil equivalent of 7 mbd of wood and forest waste. This considerably exceeds the near-term supply of about 3 mbd of wood and forest wastes that may be obtainable annually without resorting to sophisticated forest-management techniques and new tree species.

But will this biomass forest yield be used? Economics will again be a major determinant. The problem can be illustrated by looking at a third category of potential use—industrial firms and utilities that presently employ coal-fired boilers. Wood and dry crop wastes have an energy content of about 16 million BTU per ton. By comparison, coal—wood's old rival and still, for many purposes, its principal competitor—ranges from 16 million BTU per ton for Western coals to 25 million BTU per ton for the good Eastern coals.[51] In other words, wood is roughly competitive with Western coal. However, the economics of any further comparison are influenced by two offsetting factors: Coal is easier to transport and to use than wood and dry crop waste; coal, however, is a notoriously bad pollutant, whereas wood contains, on the average, less than one tenth of the sulfur content of coal.[52]

Yet the constraining effects of both factors can be dealt with. Coal pollution can be greatly reduced by adding scrubbers. And the difficulty in transporting wood can be significantly reduced by compressing it into half-inch-diameter wood pellets that are more convenient to handle than coal. Such being the case, the cost of both processes must be included in any fair economic comparison. Without sulfur scrubbers, the cost of both fuels is comparable ($32 per ton delivered, or $2 per million BTU for wood and Western coal, or $1.30 per million BTU for Eastern coal). But this price for wood

---

* 10^{15} BTU equals one quad, or about half a million barrels of oil daily.

can be obtained only if the consumption point, the pelletizing plant, and the forest source are all within a few hundred miles of each other. At longer distances the economics generally favor coal. But when the costs of sulfur scrubbing are included ($15 per ton), shipping the wood pellets becomes economical over an additional several hundred miles. A full comparison must be more elaborate, but the important point is that, under many circumstances, wood and dry crop wastes can compete with coal.[53]

The pelletizing described above is the key to wood's competitiveness as an industrial boiler fuel. Only if wood is pelletized can it be used as a substitute in a somewhat modified coal-fired system, and pelletizing also simplifies transportation and storage. Burning pelletized wood can be economical. One firm in Tennessee installed a $2.3 million boiler capable of firing 500 tons of pellets or wood chips per day, and has found that the savings in conventional fuel costs provided a payback of less than five years.[54] A utility in Vermont, Burlington Electric, generates one third of its electricity production with wood pellets in a wood-chip burner. (In Hawaii, sugar cane residues are used not only to make process steam and electricity for sugar mills, but also to provide almost 20 percent of total utility power on the island.) [55]

Pelletizing is not only economical, its technology is also simple and familiar. The basic process is at least twenty years old, and the equipment and techniques are very similar to those employed in agricultural feed manufacture. Although only a few firms are now active in pelletizing, interest is picking up. According to an executive of a pelletizing systems manufacturer, "Only a year ago it was difficult to get people to talk to us. Now even the large firms are calling." [56]

Wide diffusion of pelletizing would require substantial supplies of raw biomass, necessitating either the management of present forests for fuel source or the establishment of large "energy plantations" devoted to the growing of energy crops. The latter constitutes one of the best hopes for long-term renewable energy, and it is beginning to attract considerable interest. However, two barriers stand in the way. The first is cost. For energy plantations, biomass would represent the main income, not a supplementary one, as for timberlands. When forest economics are analyzed on the basis of cutting down the entire tree for energy, the cost of the pelletized wood triples to $5 or $6 per million BTU. But research in improved tree cultivation and tree species could reduce this cost by half during the next decade.[57]

The second obstacle is land—a great deal of which would be required for large energy plantations. According to one estimate, to produce the oil equivalent of 10 mbd, a billion acres would be needed, which is all the commercial forest land in the United States. But improved techniques could reduce the amount of land required.[58]

There is much to recommend wood and forest waste as a source of energy. The process of growing trees is familiar, the economics are attractive in many situations, pollution is low, and storage capability is inherent. In the short term, the oil equivalent of 3 mbd of energy may be achieved by the increased use of wood and wood waste in residential heating and in the forest products industry. In the longer term—twenty-five years or more—four times as much energy may be available from this source via a system of energy plantations.[59]

The other major near-term biomass source is municipal and animal wastes, which a biological process can turn into perhaps an additional 1.5 mbd of useful energy. Organic waste can be transformed into a low-BTU gas, containing methane. In less sophisticated forms, the technology has been used in sewage-treatment plants since the turn of the century. One of the best examples of the methane approach is the innovative system designed and managed by Biogas of Colorado, which "harvests" the manure produced by 40,000 head of cattle at a huge feedlot in Lamar, Colorado, and converts the waste into enough methane to provide half the fuel for a 50 megawatt power plant.[60]

Federal energy policy should encourage bioconversion research, especially for small and medium-scale products. A major effort is needed to familiarize foresters, farmers, and businessmen with the benefits and potential of biomass.[61] Bioconversion incentives should also be incorporated into utility regulations.

Today, biomass has strong regional rather than national support and interest. California's Energy Resource Conservation and Development Commission, for example, funded a demonstration of a biomass gas producer from an old Swedish design, even though the federal government backed out.[62] Federal programs should encourage, not discourage, such innovative activity at the regional level.

Meanwhile, large biomass projects, such as energy plantations, may not have an impact in this century. But well-developed, locally based technologies could provide 6 percent of U.S. energy needs by the end of the century, as opposed to less than 2 percent today.

THE POWER TOWER: IN THE IMAGE OF NASA

High in the French Pyrenees sits a sparkling ten-story parabolic mirror that looks from a distance like an oversized diamond nestled in a sloping green valley. It is a 1 megawatt solar furnace—a power tower.

In 1977 this solar furnance, which lies near the town of Font-Romeu, was the only solar thermal electric system in the world that was pumping power into a conventional electric grid. Although the structure was designed primarily for achieving high temperatures for research purposes, the French tapped off some of the heat to prove a point: that solar energy could be used to generate conventional electric power.[63]

The system works by reflecting the sun's rays onto the large parabolic mirror via an array of smaller mirrors perched on an adjoining hillside. The parabolic mirror focuses the incident light onto a small area where a boiler is placed. The steam produced by the boiler is piped down to a small building that contains a 100 kilowatt steam turbine-generator combination. "The French showed it works," one French energy official has observed. "But the Americans will commercialize it. We've worked for a decade with minimal budgets to get where we are. Yet the American solar thermal electric budget—for next year alone—is higher than the total we've spent thus far." [64]

The budget for the U.S. solar thermal electric program is indeed growing rapidly. Despite criticism of the high costs of the program and skepticism about a centralized "technological fix" that has grown out of the nuclear experience, the solar thermal electric program, better known as the power tower, commands nearly a fifth of the entire federal budget for solar energy.[65]

It is one of several large-scale methods that have been proposed for the centralized generation of electricity, all of which fall under the rubric of Big Solar. Some of the others are ocean thermal conversion, microwave space power satellites, and farms of large windmills (that can generate a hundred of more kilowatts of power). All involve complex high-technology systems. But farthest along technologically, the power tower is expected to continue to command a major share of the federal solar energy funds through the 1980's. For these reasons it best illustrates the problems, the costs, and the potential of Big Solar. Although the French official is right about American budgets, commercialization of the American system is still far from certain.

The U.S. power tower is a low-temperature system designed specifically to generate electric power. The giant parabolic mirror of the French tower is replaced by a concrete tower several stories high, which is capped by a steam boiler. Several acres of remote-controlled mirrors, or heliostats, capture sunlight and beam it to the boiler. The remainder of the system is similar to the French design.[66]

The American power tower is presently at the prototype stage. Eighty percent of the cost of the early stages will be borne by the federal government; later, the participating utilities will increase their share to 50 percent. The first commercial station, a 100 megawatt system, is projected for the early 1990's.[67]

The cost of the heliostats, or tracking mirrors, is the dominant element in power-tower economics, for at present, the mirror system accounts for up to two thirds of the overall plant cost. The Department of Energy's goal is to reduce the present system's cost by one tenth—that is, to $70 per square meter. But many in the industry doubt that the material-intensive costs of the heliostats can be reduced beyond $140 per square meter. At that level, a Jet Propulsion Lab analysis projects total plant costs of $2,000 per *peak* kilowatt, or about three times the plant cost of an *average* kilowatt of conventionally generated electric power. And the plant's full output will be available only for a portion of the day—on certain days.[68]

Balanced against this, however, is that the power is available when most needed in warm climates during hours of peak sunshine —and corresponding peak air-conditioning load. This fact, combined with the need for the direct light available on clear, sunny days, explains the choice of southern California and New Mexico as the first power-tower sites. If the power tower is analyzed as a backup system or peak power system for conventional electric power generation, which is the way that it will be initially used, it becomes a more economic proposition. However, even the smaller, less efficient gas turbines presently used for peak power needs are more economical than the *future* cost of power towers, assuming there will be a reduction of heliostat cost to one fifth of present levels, which is far from assured.

In addition to the economic hurdles, several technical obstacles remain. The maximum range at which the focusing mirrors can successfully operate has not been determined, nor is it known how difficult or costly it will be to keep the mirrors properly focused. Estimates for "maximum" size of an individual power-tower plant range from 10 to 100 megawatts, and some suggest that crews of hundreds of cleaning personnel will have to wipe the mirrors continually while scores of technicians adjust the heliostats to maintain the plant's power levels. Others have raised concerns about the blinding effect of hundreds of acres of mirrors on pilots flying over the area. Resolving these issues and proving commercial feasibility could take the rest of the century. Only then could commercialization begin.[69]

Two Polar Alternatives

The power tower and active solar heating both take energy from the sun's rays, yet they dramatize how different are the alternatives —and the requirements—that reside under the name solar energy. The power tower is, in many ways, the antithesis of on-site solar heating. Virtually everything is different: the market, the manufacturers, the time horizons, the economics, the institutional barriers, the technical risks, and the degree of fit with the national research establishment.

Solar heating technology is here *now,* but the customer base is dispersed and numbers in the tens of millions. The economics are marginal in some parts of the country, and the technology faces numerous institutional barriers, such as outdated building codes and utility rate structures. Solar heating would place an increased "managerial" burden on the homeowner and the businessman. Additional equipment would be installed on their property, and energy would be received directly on site rather than via conventional distribution channels.

The power towers would be sold to a concentrated group of customers—the electric utilities—and the principal components would be manufactured by such high-technology firms as Boeing,

Martin-Marietta, and Lockheed. Construction would proceed along the lines of aerospace mission programs. In fact, the Big Solar centralized projects fit nicely into the technical and managerial practices and expertise of the nation's research establishment. Indeed, many of those directing government energy research and development programs—coming out of the Atomic Energy Commission, the National Aeronautics and Space Administration, and the Department of Defense—are accustomed to dealing with major high-technology companies. A Big Solar program is a convenient and familiar way to proceed.

Moreover, the power tower would require no significant changes in the system that delivers energy to the consumer. As before, the utility would make the investment, gather the energy, convert it to electricity, distribute it to its customers, and collect the monthly bill. But it must also be made clear that the technology of the power tower has yet to be fully demonstrated, that the remaining technical risks are moderate to high, that the economics are poor, and that commercialization is probably, at best, fifteen to twenty years away.

The other Big Solar research technologies—ocean thermal conversion, large wind machines, solar satellites—are similar to those associated with the power tower, but here the costs and technical risks are even higher and the time horizons likely to be longer.[70] The microwave solar-power satellite concept, for instance, would require orbiting space platforms with literally square miles of photovoltaic collectors. It could cost upwards of $70 billion per system. Even though the concept is only at the feasibility stage, critics have been quick to point out the military risks of obtaining a sizable amount of our power from orbiting satellites. Ecologists, meanwhile, are concerned about the impact of bombarding several square miles of the earth's surface with microwaves, and the accompanying risks to airline travel.[71] Unfriendly powers can also be expected to react negatively to the potential military value of hundreds of controllable microwave patches. Supporters, however, point out that the satellites could supply round-the-clock power and could convert the sun's rays into usable energy before raw energy is lost in its passage through the atmosphere. In any case, whatever the merits of the various arguments, the microwave satellite solution, if it can be shown to work, is clearly a twenty-first century one.

Table 7–1

ACTUAL AND PROJECTED FUNDING LEVELS FOR SOLAR RESEARCH,
DEVELOPMENT, AND DEMONSTRATION [a]

FISCAL YEARS 1975–1983 [c]

(in millions)

	1975	1976	1977	1978	1979	1980	1981	1982	1983
Thermal Applications									
Heating and Cooling of Buildings	18	27	93	104	129	98	91	36	28
Agricultural and Industrial Process Heating		—	8	10	11	23	19	16	16
Fuels from Plants	5	5	10	20	52	79	68	72	75
Solar Electric									
Solar Thermal	10	15	51	60	70	80	159	237	205
Photovoltaics	8	15	59	57	91	125	151	100	100
Wind Energy	7	12	21	33	51	76	90	90	90
Ocean Thermal	3	4	14	35	52	110	35	68	68
TOTAL [b]	49	78	258	319	455	590	613	618	582

[a] Estimated by solar program officials
[b] Totals may not add, due to rounding
[c] Years 1975 to 1978 actual, 1979 to 1983 projected

Sources: Data compiled from *Government R&D Report*, vol. III, no. 5, September 15, 1977, p. 3; U.S. General Accounting Office, "Federal and State Solar Energy R&D Development and Demonstration Activities," RED–75–376, June 10, 1975; "The Magnitude of the Federal Solar Energy Program and the Effects of Different Levels of Funding," Report of the U.S. Comptroller General, February 2, 1978.

In summary, none of the Big Solar programs can be expected to achieve significant short-term (twentieth century) results. They are twenty-first century technologies. Yet it is expected that in the early 1980's these programs will continue to absorb the dominant share of federal solar energy research funds (see Table 7–1).

Much of the nation is already "wired for electricity," and if the price of conventional fuels rises dramatically as the turn of the century approaches, Big Solar electric will become increasingly attractive and the technical risks may be gradually peeled away. But before the momentum of research pushes investment in these technologies to the hundreds of millions or even billions of dollars, it is important to recognize that they represent uncertain long-term—not short-term—solutions to the nation's energy problem. And large investments in future technologies should not deflect attention from the problems of successfully diffusing existing solar technologies.

PHOTOVOLTAICS: ROOM FOR BREAKTHROUGHS

The logic of photovoltaic conversion is very persuasive. Silicon, the principal raw material used in the manufacture of photovoltaic cells, is the most abundant solid element on earth. And photovoltaic cells alone convert sunlight into one of the most highly prized energy forms—electricity.[72]

Photovoltaic cells are made through the same processes used by the semiconductor industry to manufacture transistors and integrated circuits. Pure silicon is selectively "contaminated" with different gaseous impurities at high temperatures. These impurities define electrical boundaries between the elements of cells that are connected by depositing a thin layer of aluminum on the circuit. The resulting cells are photovoltaic—that is, they will generate an electric voltage when light falls on them. The theory behind the phenomenon was first developed by Albert Einstein in his classic 1905 paper.[73]

The early photovoltaic cells were used to provide electricity for orbiting space satellites, with clusters of them covering the butterflylike wings that opened when the satellite achieved orbit. Cost was not then a problem; reliability, technical performance, and sophistication were the key considerations.

These early cells were manufactured under the same tightly

controlled and costly processes used for making transistors for air-craft electronic systems. They were indeed very expensive. Early systems cost as much as $1 million per peak kilowatt, 1,000 to 2,000 times more than the cost of producing electricity by conventional means.[74]

But what happened to the transistor industry is now happening to photovoltaics. The space program's need for lightweight, reliable electronic amplifiers created a constant, strong stream of federal purchases for what were then often $200 transistors.[75] The rapid fall of semiconductor prices is now legend. A commercial version of the same transistor can today be purchased for $.20. The cost of photovoltaic cells has also been dropping dramatically. The average commercial price for photovoltaic systems in 1977 was slightly over $15,000 per peak kilowatt, but some of the latest quotes for new installations are as low as $3,000 per peak kilowatt.[76]

Even though present price levels reflect a decrease in cost to less than a hundredth of the original satellite cells, they still restrict the current market for photovoltaics to special applications, such as unmanned railroad crossings, remote microwave repeaters, marine battery recharging, and pipeline leak warning systems. But the market for these photovoltaic applications grew by 100 percent from 1976 to 1977, when shipments exceeded $10 million.[77]

The continued reduction of costs could open up a vast market of conventional electric power generation via photovoltaics. At the heart of the cost reduction problem is the high cost of single crystal silicon. As one semiconductor executive explained, "Eighty percent of the cost of a photovoltaic system is in the materials, and eighty percent of the material cost is in the silicon." [78]

There is a wide consensus that in order to penetrate the broader market, the industry must move away from single crystal silicon. A reduction in cost to a tenth of present levels (or $300 to $500 per peak kilowatt) is necessary to make photovoltaics competitive with conventional fuels. And reaching these cost levels will probably require a radical move to very thin cells, up to 100 times thinner than the thickness of today's cells, or to amorphous (uncrystallized) semiconductors—or both. Researchers at RCA and at Energy Conversion Devices have already demonstrated the feasibility of thin-film amorphous cells with moderate efficiencies. Stanford Ovshinsky, the inventor of the thin-film semiconductor, estimates that in a few years electricity generated by amorphous cells could be cheaper than conventionally generated electricity.[79]

So, assuming that the cost barrier is crossed, the potential for photovoltaics is very large. But in what form would the potential be realized? Photovoltaic cells could be used in two distinct ways, with quite different effects on the American energy supply system. The cells could be used on site, like solar heating. But photovoltaics as a generator of electricity would have greater effect than solar heating on the role of utilities, for if on-site power generation by photovoltaics became widespread, the utility industry would have to redefine its business, emphasizing power distribution instead of power generation: Utilities would have to purchase power from on-site installations, mark it up, and distribute it. Unless low-cost means of electrical power storage were developed, the utilities would still supply conventional backup power. A structural change of this sort would require imaginative solutions on the part of regulatory commissions, and would also require the support of the utilities themselves, which would by no means be certain.

The utilities generally would favor another mode of using photovoltaic energy, which would mean building large, centralized photovoltaic "farms" owned and operated by the utilities. Here, it would be business as usual for the utilities, save that they would substitute photovoltaic-fired power plants for nuclear or coal-fired plants.

It is too soon to discern the likely direction. But it should be pointed out that photovoltaic installations do not exhibit dramatic economies of scale, and any such economies are likely to be absorbed by distribution costs. On the other hand, utilities would have the incentive and organization to pursue centralized use of photovoltaics. Thus, it may well be that both centralized and on-site applications would grow rapidly, side by side, if costs came down sufficiently.

One can certainly say that photovoltaics thus far constitutes a success story. With only relatively small volumes of sales, costs today have dropped to three tenths of one percent of original costs. Moreover, advances can come quickly, for as one specialist has put it, photovoltaics are "embedded in what has been one of the most fertile environments for innovation in the twentieth century—the semiconductor industry." [80] As has been the case in that industry, increased volume will bring increased experience, improved processes, and reduced costs. Investments in research will produce fundamentally new production techniques that could lead to even lower costs.

Both applied research and the creation of a market for photovoltaics should be vigorously pursued, but thus far federal efforts in the photovoltaic areas have been relatively conservative. There is widespread speculation in the industry that in 1978 some semiconductor firms were ready to sign contracts at the Department of Energy's 1982 goal of $2,000 per peak kilowatt—given a sufficiently large purchase order. Such firms know from past experience that increased volumes will reduce their costs geometrically. Under the right conditions, a $1,000 price goal may be attainable. If only a part of the $2 billion of federal funds slated for the Clinch River breeder reactor were directed instead to photovoltaic purchases, the $1,000 per-peak-kilowatt price could be achievable very soon—compared to a $5,000 per-peak-kilowatt estimated cost at Clinch River. It is therefore obvious that radical approaches to solving the cost barrier should be stimulated.[81]

Policy for photovoltaic devices should have two objectives: First, to create a market by federal purchases to drive present technology to lower costs; second, to fund R&D and demonstration projects for thin-film amorphous semiconductors and other high-risk proposals. The payoff in either case could be very large.

CONCLUSION

How much of America's energy needs could be derived from solar energy by the year 2000? Estimates vary over a very wide range [82] and comparative analysis is difficult, because different analysts work with different definitions of solar. Some begin and end with solar heating, whereas others include all the solar options except hydropower. Recently, however, many analysts have begun to converge on a definition similar to the one presented at the beginning of this chapter, which includes all the "recent" solar cycles.[83] Even after adjusting predictions to our definition of solar energy, a selection of high-quality projections still ranges from 7 to 23 percent by the year 2000. The variation is demonstrated in the following Table 7–2.

Table 7–2

SOLAR CONTRIBUTION BY YEAR 2000 †

	Solar Energy (oil equivalent mbd)	Total Energy Consumption (oil equivalent mbd)	Solar As a Percent of Total
President's Council on Environmental Quality (CEQ) [84]	12	50	23%
Walter Morrow [85]	14	79	16%
Stanford Research Institute 1 * (Business as Usual)	5	73	7%
Stanford Research Institute 2 * (Low Solar Cost)	10	70	13%
Stanford Research Institute 3 * (High Fuel Cost)	6	45	12%

* The Stanford Research Institute developed three scenarios.[86] The first is a business-as-usual scenario. The two remaining scenarios are solar-emphasis scenarios. In case 2 the emphasis is achieved by lowered solar costs, whereas in scenario 3 the change comes primarily from increased prices of competing fuels.

† Calculations based on quads given in original sources.

The main reasons for the disparities come from different assumptions made about the following five variables, each difficult to predict:

1. Prices of competing fuels
2. Overall levels of domestic energy consumption
3. Rate of federal investment on solar energy
4. Rate of technological advancement of solar technologies
5. Rate at which institutional barriers to solar will be overcome

The factor over which the most direct control can be exerted is the third—the realm of federal policy, about which there is considerable confusion. But a plausible and sensible goal can be identified: to achieve a fifth of the nation's energy from the sun by the year 2000, and to lay a basis for a fuller transition to renewable energy sources in the twenty-first century.[87]

Key public policy decisions are required in the very near future. In order for the rapid diffusion of solar energy to begin, a commitment must be made to proven existing technologies. The only realistic two options for the short-term are wood and wood waste, and on-site solar technologies, such as solar heating, small hydropower, and small wind. The short-term problem is not technology, but accelerating diffusion. As stated before, we recommend self-extinguishing tax incentives to get these technologies through the early sluggish phase of commercialization, combined with measures to overcome the institutional barriers.[88] Incentives would not only stimulate demand, they would encourage manufacturers to make larger commitments and take greater risks.

If the experience of California is any guide, such a program could have a dramatic impact. In 1977 the California legislature passed a 55 percent solar tax credit—the single largest financial incentive in the country to encourage the use of solar energy. Both active and passive systems as well as solar electricity generation systems were included in the initial tax credit. The tax credit was extended in 1978 to wind and process heating systems, and was supplemented by dozens of additional bills dealing with financing, utility involvement, job training, and solar rights. Several municipalities in the state, such as the city of Davis, have also passed ordinances that encourage the use of solar design by setting minimum standards for the thermal performance of buildings. In San Diego county an ordinance will become effective in 1980 that requires all new residential buildings with guaranteed solar access to use solar hot water heating systems. According to the director of the Solar Cal Office, "One reason why we have been able to accomplish so much in California over the last three years has been a political commitment to renewable energy resources . . . Political leadership should be a key element of any commercialization strategy." It's the political climate, not the weather, that will govern the future of solar energy.

California's aggressive solar commercialization program is paying off. In 1978, while solar industry growth stagnated, due to suspense over federal action regarding tax credits, solar industry sales in California continued their rapid rise. By the end of 1978 California had over 30,000 solar installations. Thus far, one hundred manufacturers and two hundred installers with solar experience are doing business in the state. And instead of resisting the spread of solar energy, some California utilities are promoting the use of solar hot water systems.

With about one tenth of the U.S. population, California now has about one quarter of the nation's solar applications. As of the end of 1978 there were 3,000 space heating and cooling installations— about as many as there were in the entire country a year earlier. Based on present growth rates, the state is projecting that in two years (1981) it will have 150,000 space-heated and/or cooled homes, and 200,000 solar water heating installations, a tenfold increase in total solar installations over 1978 levels. Several of the large new housing developments in Davis, each of which includes hundreds of homes, are using solar design to obtain 50 to 75 percent of their heating requirements. By 1985 California plans to have 1.5 million solar installations and 30,000 people employed in the solar industry. If the same degree of market penetration were to occur for the country at large, this would correspond to about 15 million solar heating and cooling installations, and a solar industry employment of over a quarter million people.

Extrapolation of California's projections to the remainder of the country, however, are complicated by the mix of fuels employed, climate, architectural characteristics of the housing and commercial building stock, the rate of new building construction, and—most importantly—incentives. Thus, though solar hot water and space heating is feasible across the entire country, we recognize that the rest of the country will lag behind California for several years. Indeed, one of the main arguments of this chapter is that vigorous action by the federal executive branch and Congress—and their state counterparts—will be required to minimize this lag.[89]

With such a program for the nation at large, it should be possible to obtain about 20 percent contribution by solar—the equivalent of, say, 10 million barrels a day—by the year 2000. On page 214 is a rough guide to the breakdown of solar sources, assuming total U.S. energy consumption of 50 mbd of oil equivalent in the year 2000.[90]

Although our projection resembles other recent estimates, it differs in two key respects. Unlike others, ours does not envision significant contribution from the Big Solar research technologies. Any contributions made by high-technology sources, including photo-

Energy Source	Oil Equivalent (mbd)
1. Solar Space and Hot-Water Heating (Including Active and Passive)	3
2. Other On-Site Technologies *	2
3. Wood and Forest Waste	3
4. Hydropower (Large Installations) [91]	2
TOTAL	10

* See Reference 9.

voltaics, would be a bonus. Also assumed in our projection, and not in most of the others, is a concentrated near-term effort to diffuse on-site solar innovations.

For the longer term, the technological alternatives are less distinct. It is not even clear in some cases, such as in the case of solar satellites, whether the concept is feasible. In other areas, such as in ocean thermal power, technological feasibility is promising, but there is great uncertainty about size, cost, operating efficiency, and timing. A solid ten to fifteen years of research and development will be required of the research technologies before final priorities can be assigned and some possibilities ruled out. Congress should, in the meantime, be prepared to proceed incrementally, to fund at least demonstration versions of the centralized electric technologies (power towers, solar satellites, ocean thermal, large wind machines), and to commit substantial funds to research, development, and commercialization of photovoltaics and energy plantations. By 1990, when technological risks have been satisfactorily reduced (at least for some of the technologies), it should be possible to decide what to commercialize.

Federal investment for solar research and development should rise above the billion-dollar level.[92] By comparison, the United States presently spends almost $2 billion on research in and development of nuclear energy, a technology whose future is still uncertain.[93] Of course, alternatives must be ranked according to their market potential and weighed by their technological risk. High priority should be given to further development of on-site solar, photovoltaics, and energy plantations; medium priority should go to solar thermal electric and large wind machines; and lowest priority should go to solar power satellites and ocean thermal conversion.

Thus, a realistic program to implement solar energy would consist of three phases: First, short-term incentives for a rapid commercialization of existing technologies. Second, a ten-year search to identify the best of the new alternatives. And finally, commercialization of the alternatives identified in the second phase, leading toward a gradual transition to an increasingly Solar America in the twenty-first century, when hydrocarbons would be reserved for premium uses.

Admittedly, what has been described here is a vision—but an eminently practical one. It deserves a fair chance.

ROBERT STOBAUGH
AND DANIEL YERGIN

8

Conclusion: Toward a Balanced Energy Program

The purpose of U.S. energy policy, we have argued, should be the managing of a transition from a world of cheap imported oil to a more balanced system of energy sources. It is clear that domestic oil, gas, coal, and nuclear cannot deliver vastly increased supplies, although it is equally clear that these sources cannot be ignored. America needs all of them. Broadly speaking, however, the nation has only two major alternatives for the rest of this century—to import more oil or to accelerate the development of conservation and solar energy. This is the nature of the choice to be made. Conservation and solar energy, in our view, are much to be preferred.

The character of the choice has been obscured by a long series of bitter battles and confrontations in Congress, regulatory agencies, the courts, and the press. These are likely to continue for some time, for they involve enormous sums of money, competing perceptions of the nation's interests, environmental and other social costs not easily quantifiable, disagreement over the nature of the technical facts, and conflict over basic values.

THE BLEAK OUTLOOK

Some of the battles were going on before the OPEC price revolution, but they were then, by comparison to what happened afterward, only minor skirmishes. Indeed, the OPEC countries precipitated the largest of all domestic energy wars. The OPEC price hike, of course, imposed a heavy cost on all U.S. energy consumers. At the same time, it did increase the value by some $800 billion of the proved U.S. reserves of some 42 billion barrels of crude oil and 250 trillion cubic feet (tcf) of natural gas.[1] That amounts to about $10,000 for each American family. Quite naturally, such a sudden and dramatic increase in value raised the question of how the "windfall" would be distributed. Who was to receive the benefits? Consumers, through price controls that prevented energy prices from rising to world levels? The federal government, through taxation? Producers, through higher prices for domestic oil and gas? And how would the windfall be split among producing and consuming regions?

The distributional questions ignited the most visible fight in the effort to formulate a U.S. energy program. So intense and heated has the debate been that it led participants to make exaggerated statements that have discredited their various positions. Congress, which by and large wanted the consumers to get the windfall, pretended that there was no energy crisis and ignored the problems posed by imported oil. The Carter Administration, which wanted much of the windfall collected in taxes, argued that the supply of natural gas would increase if the price went up to $1.75 per tcf, but that above that price, supply would increase only slightly. Meanwhile, the oil and gas companies, who wanted the windfall for themselves, contended that the decontrol of prices would give them the money and incentives to find the new supplies required to solve U.S. energy problems—a scenario quite at variance with experience, especially in the years since 1973. For the dramatically increased cash flows and higher prices for new oil and gas, although causing a sharp jump in exploration and development activity, have failed to prevent domestic production and reserves from declining.[2]

Finally, two additional groups helped obscure the picture. One consisted of those economists and engineers who used information gathered under one set of conditions, primarily pre-1973, to make projections of what would occur under another, radically different set of conditions, post-1973. The result—highly optimistic projections and estimates of how easily the United States could increase its domestic energy supplies and how difficult it would be for OPEC to sustain its high price. This group of technicians generally tended to support, at least implicitly, the oil companies' position: that is, if

U.S. prices were deregulated, U.S. oil imports would drop dramatically and OPEC prices would decline.[3]

The other group laid heavy blame on what they perceived as the oligopolistic structure of the petroleum industry. They tended to favor the congressional position—that is, to give most of the windfall to the consumers by keeping prices low. Their "solution" to U.S. energy problems has been to break up the oil companies—although there is little reason to think that divestiture would bring forth additional oil from the ground. And there is good reason to think that the oil companies have made a positive contribution to developing the coal industry.

To be sure, our description of the debates and the participants is simplified. There are many nuances and variations, and even some dissenters, within each camp. But the essential outlines are correct. Of course, one could hardly expect that there would not be a major political war, for an $800 billion bounty is well worth the struggle.

Yet the battle has been so intense, the field so littered with charge and countercharge, that the most important point has been obscured. Regardless of how the $800 billion is divided, there is little likelihood of a substantial increase occurring in the domestic production of oil and gas to meet anticipated demand. In fact, the United States will be fortunate if it finds enough new oil and gas to keep production at current levels. Entrepreneurs have searched the continental United States for so long (120 years), and so thoroughly (over two million wells), that it would be foolish to base a national policy on the supposition that the absolute quantity of U.S. production could increase beyond what it is today—10 million barrels daily of oil and natural gas liquids, and the equivalent of nine million barrels daily of natural gas.[4]

True, at the margin, more new oil and gas will be found at higher prices than at lower prices, which will make up for some of the inevitable decline in production from existing wells. It is also true that consumption will be somewhat lower at a higher price than at a lower price. For these reasons, a move toward having a consumers pay higher prices to producers makes good economic sense, although the extra output and reduction in consumption will be less than has often been advertised.[5] The redistribution of income, however, raises very real economic and political problems, which means that changes in price must be gradual.

There are two other conventional domestic sources proffered as solutions to imported oil—coal and nuclear. Of course, coal and nuclear power have obvious benefits—to the users of electricity, to people who find employment (a job in a boom town is better than no job at all), to the companies that mine coal and uranium, and to those who profit from the economic activity associated with these industries. But coal and nuclear are also embroiled in controversy, although of a different kind than oil and gas. The conflicts over coal and nuclear have tended to bring out, rather than hide, the fact that only a modest contribution in increased energy can be counted on with reasonable certainty from these sources. Moreover, the conflict over coal and nuclear has been less about the distribution of huge sums of money than about the distribution of side effects, that is, the externalities, or costs, indirectly borne by the members of society at large rather than paid for directly in cash by the consumers of the energy.[6]

External costs, in the form of environmental and health problems, are rung up at every step in coal's journey from the mine along the railroad to the boiler where it is burned. When an economist or engineer says that the cost of electricity produced from coal is three cents a kilowatt hour, he is ignoring a number of important externalities. In the mining and transportation stage, the costs include acid drainage from mines and the disruption of life in Western communities by noisy trains hauling coal. The costs are even greater when coal is burned, for it releases sulfur dioxide and a host of other pollutants. The costs include smoggy skies, emphysema, and the (unknown) consequences for future generations of increasing the temperature of the atmosphere by producing carbon dioxide.

There is another question. To what extent should external costs be "internalized" as a cost of coal production, transportation, and consumption? Internalization costs money—for the installation of scrubbers by utilities to remove sulfur dioxide, for the building of overpasses above railway tracks, and for the return of strip-mined land to its original contour and condition. But even after a number of such costs are internalized, many external costs still remain, and different people quite naturally place different values on them. A person with a large inheritance might well ascribe a greater value to blue skies than to overall economic activity, while an unemployed person might have the reverse preference, even though both live in the same city. The problem is accentuated when people live in different cities. New Yorkers do not necessarily want to risk increased chances of lung disease so that coal can be burned in Ohio to provide Cleveland with electricity.

Because environmental and health problems have increasingly become issues in the political process, the outlook for coal is considerably less promising than officially projected. Its contribution to U.S. energy supply will fall far short of the Administration's stated goal. In fact, it would be unwise for the nation to plan on an increase between 1976 and 1985 greater than the oil equivalent of 3 million barrels daily, rather than the 6.5 million barrels daily shown in the Administration's National Energy Plan.[7]

Nuclear energy has a set of external costs even more controversial. Bitter disagreements about reactor safety have been a major cause for the many delays around the country in the completion of nuclear power plants. In turn, the delays, along with the increasingly stringent safety requirements, have caused such dramatic cost escalations that a second major controversy has erupted: Does it cost more, the same, or less to generate electricity from nuclear power than from coal? The question is difficult enough to answer when only the direct costs carried by the utility are considered. It is virtually impossible to answer when an effort is made to measure the relative external costs of coal and nuclear.

Furthermore, what makes nuclear power unique among the energy sources are the pervasive external costs it imposes on future generations—generations that will inherit caches of spent fuel that will emit radioactive rays for many centuries.[8] Some of the technical problems of managing this dangerous material might be simplified if it were "reprocessed" to recover uranium or plutonium, both of which are potentially useful. But reprocessing leads to new problems. Plutonium, a material that can also be used to make atomic bombs, would become more readily accessible. Thus, the promotion of reprocessing technology might lead to the proliferation of nuclear weapons.

In the 1960's, the nuclear advocates rushed ahead to attempt one of the most massive technological innovations in history—seeking to convert the world's electric generating system from fossil fuels to nuclear. In so doing, they confused dreams with reality. Utilities made their decisions to install billions of dollars of generating plants on the basis of costs *estimated* by the manufacturers without an adequate base of relevant experience. The proponents also ensured an increasing supply of dedicated enemies by the manner in which they dismissed the opposition, attacking them as revolutionaries, Luddites, and misguided children. In general, the advocates failed to conceive of nuclear power generation as an overall system, one that starts with the mining of uranium and ends with a satisfactory disposal of spent fuel. Meanwhile, nuclear critics fought a war of attrition in fish and game commissions, county commissioners' offices, state regulatory agencies, and the federal courts. Wherever the critics went, they asked one question: "How safe is safe enough?"

Today, the nuclear industry faces widespread shutdowns of operating plants starting as early as 1983 unless the problem of handling spent fuel can be resolved. Space limitations make the current method of storing spent fuel in pools of water at the nuclear power plants unsatisfactory for any substantial period of time. A workable solution will have to satisfy a wide cross section of the public, including at least some of those normally considered to be nuclear opponents. The safety questions highlighted by Harrisburg could, obviously, force shutdowns at any time.

At the most, during the next dozen years, nuclear energy might add a million and a half barrels a day of oil equivalent to the million and a quarter barrels a day it provides now, thus reaching a production level of slightly under 3 million barrels per day of oil equivalent. That level would require the continued operation of all existing nuclear capacity, plus the completion and operation of all new capacity currently under construction or on order. That, we stress, is a very bullish scenario. Indeed, we regard the nuclear condition to be far more uncertain, and believe that the energy generated by nuclear power could actually undergo an absolute decline within ten years. It is, therefore, unwise to rely on nuclear to make any substantial contribution to reducing dependence on imported oil for the rest of this century. It would be equally unwise, however, to foreclose the possibility of a longer term contribution by nuclear energy. It is, after all, pretty clear that our major objective should be to create realistic energy options, not to deny them.

In other words, the prospect for dramatic *increases* in domestic supplies from the four conventional fuels—oil, gas, coal, and nuclear—is bleak. Total U.S. energy consumption today is the equivalent of 37 million barrels a day of oil. Of that, 27 million is derived from the domestic production of the four sources.[9] Over a period of ten years or so, that 27 million could *perhaps* be stretched to 32 million. But during that period, overall consumption would almost surely have risen, so even the increase of 5 million barrels daily would not ease the problem of imports. However, the country certainly cannot afford to ignore measures that *might* marginally increase domestic supplies. Thus we favor

1. The leasing of offshore oil and gas properties, under strict environmental regulations, and in a manner to promote rapid development.

2. Decontrol of *newly found* oil and gas, and of oil obtained by using enhanced recovery methods to recover oil left behind after normal recovery methods have been used.

3. Government assistance for technologies that could provide new supplies—such as coal gasification and liquefaction, and shale oil. The increased instability in the Arab/Persian Gulf argues in particular, on security grounds, for increased capability to liquefy coal.

4. A major attempt by the government to find an acceptable method to dispose of spent fuel from nuclear reactors and to ensure reactor safety.

Unless measures such as these are pursued to a greater extent than they currently are, there might be very little, if any, increase above the 27 million barrels a day of oil equivalent from the four traditional sources. Indeed, there could be an absolute decline, potentially making the United States even more dependent on imported oil than current forecasts indicate. But whether the four domestic sources increase somewhat, remain constant, or decline, the broad choice before the United States is the same—increased dependence on imported oil, or a transition to a more balanced energy system in which conservation and solar play large roles.

THE PROBLEMS OF IMPORTED OIL

The problems flowing from increasing quantities of imported oil are so large that they deserve some reiteration here.[10] There is the obvious risk of a supply interruption because of Middle Eastern politics. The United States can protect itself against that by having an adequate amount of oil-storage capacity, perhaps a volume equal to six months of imports. Such a program is inexpensive compared with the possible economic losses from another embargo.

The greatest single matter of concern posed by increasing U.S. oil imports is their potential link to world oil prices: The higher U.S. oil imports, the higher will be world oil prices.[11] There is no accepted theory about the exact nature of the link. But the role of U.S. imports is crucial, for if the United States does not act to dampen imports, Europe and Japan will find it harder to restrain their own demand. For example, lags in the development of such energy sources as nuclear or solar, or an inattention to conservation in the United States, will contribute to similar conditions in Europe and Japan.

Another difficulty in predicting world oil prices with much certainty stems from the fact that Saudi Arabia dominates OPEC and thereby sets the world oil price. But there is no way to determine exactly on what basis the Saudis set the price. It is reasonable to assume that the most fundamental goal of the present ruling family is survival, that is, to keep themselves in power. There are forces pushing them in different directions. On the one hand, they desire good relations with the United States, and want to avoid creating economic conditions in Western Europe that would lead to left-wing governments.[12] They fear that higher prices will augment the arms-buying potential of their neighbors, such as Iraq. On the other hand, they are not anxious to sell their only real asset—oil—at too low a price. Furthermore, they are becoming increasingly wary of producing substantially greater quantities of oil in order to pay for massive industrial development, especially with Iran as an object lesson. Also, the Saudis are under pressure from the more radical Arabs to raise oil prices and not increase production.

Thus, predicting oil prices is as much art as it is science, requiring sensitivity to political nuance as much as dexterity in mathematical analyses. Our estimates should be viewed accordingly. The reader will recall from Chapter 2 our illustration of how costly imports might be. A plausible set of relationships suggests that U.S. oil imports of 14 million barrels daily might well result in a world oil price 40 percent higher than would U.S. imports of 9 million daily. U.S. imports could have so large an effect because they directly influence the policies and thus imports of the other industrial nations. Thus, in our illustration, a 5-million-barrel difference in

U.S. imports is assumed to be accompanied by a 5-million-barrel difference in the imports of other industrial nations; hence, the total impact on OPEC exports would be 10 million barrels of oil daily. If U.S. imports are at the upper level (14 million), OPEC would be operating near capacity and would perhaps be exporting 40 million barrels daily. The world oil market could be very tight indeed. But if U.S. imports are at the lower level, OPEC might export only 30 million barrels daily and likely would have considerable excess capacity.

If higher oil prices were to result, they in turn would cause national income to be lower than would otherwise be the case. And if history is any guide, an economic slowdown might be accentuated by policies adopted by governments to combat inflation and balance-of-payments deficits. Even without sharply higher prices, the United States and other governments could still feel forced to adopt deflationary policies because of higher oil imports.

In any case, increasing imports from OPEC result in increased tension and suspicion among the nations of the West. In the summer of 1978, the Europeans and Japanese substantially escalated their open criticism of America's failure to adopt policies to slow the growth of imports. Not surprisingly, the foreign leaders were having political difficulties at home implementing conservation measures when their motorists were paying two dollars or more for a gallon of gasoline while Americans were paying only 65 cents. And the foreign leaders were openly afraid that America would "drain the world of oil" or was positioning itself to preempt Saudi oil in a tight market in the 1980's.

We did not attempt to place a monetary cost on these political liabilities. But we did determine a range for the economic costs of imported oil. The range is quite wide, reflecting many uncertainties. We use the bottom part of this range—say, $35 a barrel—as a guide to the cost of imported oil to the nation as a whole. This $35 cost, of course, is about double the world oil price and, because of U.S. price controls, is about triple the price paid for oil by U.S. consumers in early 1979.

A FAIR CHANCE

What to do? We favor reliance on the marketplace. We do not think an ever-more-regulated system is the answer to the problems posed by energy. But if the market is to resolve the problems, its distortions must be corrected so that all energy sources, including conservation and solar, will be able to compete on an equal economic footing. Some might think that we are proposing a transformation of the market. On the contrary, we argue that without a transition to a more balanced energy program, the market system itself in the years ahead will inevitably become increasingly constrained by regulation and disruption. Although both incentives and sanctions have a role to play in the process of equilibration, the emphasis should be placed on incentives. The pursuit of profit has, after all, served American society well in the past, and clearly the carrot makes for better politics and more acceptable change than does the stick.[13]

That last point is crucial, because a politically acceptable program that can make a significant contribution to a solution of the energy problem is better than one that might theoretically solve it altogether but which has no chance of being adopted. Indeed, given the elusive nature of true costs, the uncertainty of using GNP to indicate how well-off the country is, and the difficulty of making accurate predictions as to what effects different policies would have, we do not think it is useful to think in terms of an "economically optimum" energy policy.[14] A vastly improved energy policy, yes. But an economic optimum, no.

Of course, it is not a new idea to intervene in the marketplace to correct for distortions that come under the heading of market imperfections (such as lack of information) and market failures (such as the externalities of a polluting smokestack). Because of continuing imperfections and failures, it is unrealistic to expect an uncorrected "free" market to solve U.S. energy problems.[15] The issue of income distribution makes this even more unrealistic. It is hardly surprising that the government has intervened to ease the discomfort and pain that an uncontrolled energy market causes many citizens. It is predictable that the political process will continue to provide the guidelines by which the market will allocate the costs as well as the benefits of the transition to a new energy balance.

Currently the U.S. price structure for energy is heavily distorted. In addition to the distortion brought about because the price of imported oil fails to represent its true cost, American consumers are receiving enormous price subsidies to use domestic sources of conventional energy. Although we have not made a detailed study of these subsidies, it appears that they are perhaps $50 billion a year or more. Controls keep the prices of domestic oil and gas far below their economic values. Moreover, regulations hold electricity prices at only a fraction of the replacement cost of power plants. This gives consumers large subsidies indirectly to use coal and uranium (as well as that oil and gas used to generate electricity). In addition, coal and nuclear power bear external costs of sufficient importance to cause major political controversies.[16]

The cornerstone of our thinking is that conservation and solar energy should be given a fair chance in the market system to compete with imported oil and the other traditional sources. Of course, a straightforward alternative, and one sometimes suggested by economists, would be to impose a tariff on imported oil and let all other energy prices rise to whatever level market forces take them. We judge, however, that a tariff high enough to offset oil's true costs would not be politically acceptable; neither would the proposal to let the prices of all other domestic energy prices rise to such a high level. A high tariff and free domestic prices also would be especially inflationary because of the automatic price escalations caused by wage and other costs that are tied directly to the consumer price index. Not only does the public view inflation as their number-one problem, but the country's leaders are also becoming increasingly aware and concerned about the deleterious effect of inflation on the nation's social fabric. True, gradual movement should be made toward higher, more realistic energy prices, but a reasonable goal is the level of world oil prices.

Accordingly (and this follows our preference for the carrot rather than the stick), we favor financial incentives to encourage consumers to use conservation and solar energy, not because there is anything virtuous about these energy sources, but because they make good economic sense.[17] Of course, an important variation on this alternative is theoretically possible—that is, to take our principle of the "fair chance" with imported oil and apply it to conventional domestic resources by giving financial incentives to producers of oil, gas, coal, and nuclear power. But we have serious doubts that this program would be politically acceptable. For some time the government has not allowed finders of new oil and gas to obtain even the world prices; to obtain a large premium over world prices would seem impossible, at least for many years to come. One must recognize that many people question whether even world prices, much less a financial incentive in addition to world prices, would be justified by the incremental domestic energy forthcoming—especially given the lack of success in keeping U.S. oil and gas reserves from declining even after the dramatic price jumps that have occurred since 1973, and also given the environmental costs and resultant political barriers associated with coal and nuclear power.

Thus, it appears to us that the only viable program that would politically reduce U.S. dependence on imported oil would be for the government to give financial incentives to encourage conservation and solar energy.[18] It is worth noting, however, that analyses—albeit of a preliminary nature—suggest that conservation and some forms of solar energy are cheaper for an additional ten million barrels per day of oil equivalent than are conventional energy sources.[19]

We do not claim to know exactly the appropriate level of financial incentives. But the fact that the true costs of imported oil are perhaps three times its current U.S. market price suggests an incentive payment or other form of offsetting subsidy of two thirds of the cost of implementing conservation and solar energy. We do not, however, recommend this large an incentive, for we acknowledge readily that our calculations are only crude approximations. Furthermore, for the most part we would think somewhat lower incentives would probably be sufficient to encourage a substantial increase in investment in conservation and solar energy.[20]

Three questions must be answered at this point: What kind of program is needed? How can the program be financed? What would be the likely consequence? We answer each of these questions in turn.

Of the two "new" energy sources, the most immediate priority is conservation, which can reduce America's dependence on imported oil until solar energy can make a substantial contribution. The conservation measures depend on the nature of the sector.

• In transportation, an obvious form of subsidy would be to experiment with free public transportation in some municipalities. But since the automobile is such a central part of American life, we doubt that even free public transportation would reduce gasoline consumption dramatically. The mandatory standards for fuel efficiency seem to be especially attractive because the government has to regulate and monitor only a few companies. Stronger standards, which do not hamper flexibility and experimentation, should be set for the post-1985 period. And a gasoline tax high enough to have a similar effect on consumption is politically unacceptable.

• In the industrial sector, there are many decision-makers, but they generally have better information than most private citizens about the real costs of alternative energy choices. To give them appropriate signals and incentives, we suggest offsetting payments in the form of investment tax credits and accelerated depreciation up to 40 percent of capital costs. This seems to be better than forcing conversion to coal because of the environmental and health problems caused by coal.

• The residential-commercial sector is highly fragmented. While the potential savings from retrofit of existing dwellings can be very great, market imperfections, in particular poor information and limited access to capital, currently present great barriers. Under these circumstances, we suggest tax credits of up to 50 percent of retrofit costs, with rebates for lower income groups. Because it is desirable to use existing organizations where possible, electric utilities should be encouraged to deliver energy conservation. This, of course, means changing regulations to make the conditions attractive enough for them to do so. The new building market can be reached through revised building codes and through the mortgage requirements of loan agencies.

Solar energy presents different problems. Although the technology used in solar heating is widely known, the application of the technology has been very limited. It has long been recognized that the benefits to society of the diffusion of technology exceed the benefits that accrue to the entrepreneur. Individuals and companies are more risk-averse than society as a whole, and information generated by an installation that does not work is often of less use to the innovator who failed than it is to society. Therefore, solar energy deserves and needs greater governmental assistance than conservation. Thus, we suggest that at least until installations are commonplace, solar users receive an offsetting payment equal to about 60 percent of installation costs.[21]

Such payments are especially justified for the smaller solar units for hot water and space heating that would be located on site since they can make an important contribution fairly quickly. It is unfortunate that Small Solar does not have the government support that high-technology programs like the power tower do. The high-technology programs have little potential to help the country in this century. The use of on-site solar installations in existing homes is hampered by some of the same barriers that slow residential conservation—homeowners who are short of capital and information, and averse to risk. Offsetting payments should be accompanied by an attempt to use existing institutions, such as utilities, both for delivery and for financing. For new homes, office buildings, and factories, a large subsidy also would likely prove effective. Again, as with conservation, a considerable long-term education campaign is required. For example, solar heating will certainly prove more cost effective if used in buildings in which a conservation effort has already been undertaken.

In order for conservation and solar incentives to work, certain additional barriers must be broken. At present, for instance, an industrial firm wanting to cogenerate electricity and steam can face a regulatory obstacle in selling electricity to its utility. Solar energy is hampered by lack of standardized building codes, confusion over the right of a person to prevent others from blocking the sun, utility regulations that deny a solar house all-electric rates, and property taxes.

Our recommendations for conservation and solar energy are rough guides. It is impossible to determine precisely how large various subsidies should be, nor do we know what subsidies Congress and the Administration would approve. We do know, however, that they should be a good deal larger than any that have been approved or officially proposed so far.

If the nation is to make the transition to a more balanced energy system, the government must be the champion of conservation and solar for several reasons. First, the conventional energy sources have a host of allies, witting and unwitting, in those analysts who understate the external costs generated by these sources. They not only tend to underestimate the disadvantages of imported oil, but also tend to underestimate the environmental costs of conventional energy. Studies, for example, often equate the health costs of pollution with lost wages plus medical expenses. Presumably, in such a formulation it "costs" society very little for a non-working wife to contract lung cancer from emissions if she dies quickly, so that large medical bills are avoided. Second, the conservation and solar energy industries do not have as many companies and workers involved in them as do other energy sources. Imported oil, for example, has a powerful constituency among those who produce, refine, and distribute it. The international oil majors take in more money in a few hours than the entire solar industry does in a year. And these companies, their workers, and their customers have a natural tendency to favor their ongoing activities.[22]

The government must lead, for the only thing that is going to happen "automatically" in the years immediately ahead is an ever greater stream of imported oil. Some of the most efficient large enterprises in history manage that process in a way to make it seem quite easy, transporting oil halfway around the world and then refining it, all for just a few pennies per gallon.* But government leadership does not mean government management. Rather, it means correcting market defects in a way to create more jobs and more business opportunities for both large established companies and small new firms with a stake in conservation and solar energy.

Are we not talking about a great deal of money? How to pay for this kind of program? Many oil and gas executives realize that a windfall tax on part of the profits that result from deregulation of old oil is likely. We would propose that the windfall tax be specifically assigned to financing—primarily by tax credits, but also by grants and loans—conservation and solar, which are the two most promising alternatives to oil. And as the windfall tax might be self-extinguishing as we move to a free market for domestic oil and gas, so these credits might be self-extinguishing over a period of, say, ten years—an important feature, because it has often proved difficult in the past to stop programs once they are launched. Our proposal would thus respond to two of the most urgent problems in U.S. energy policy—the stimulation of conservation and solar, and the need to gradually free all oil prices, not just those of new oil.[23]

As it is today, the system of price regulation is highly irrational. And if our analysis about the impact of larger, rather than smaller, oil imports is correct, then an irrational American pricing system could be one of the main causes of much higher oil prices in the years ahead, with all that will do to the Western economy. It makes no sense for the United States to be as integrated as it is into the world oil market—by far the largest consumer of OPEC oil—and yet have a pricing system that is partly insulated from the market.

* Most of the price paid at the pump goes in the form of payments to the governments of the producing and consuming countries.

Our proposal would respond to another need, a political need, if the nation is ever to resolve its energy difficulties. It would help to bridge the gap between contending parties in the bitter American energy debate. The various interest groups—oil and gas producers, advocates of conservation and solar, even environmentalists and consumerists—are secret allies, though, to understate the matter, not all by any means would recognize this truth. Oil and gas producers are convinced that they need higher prices in order to maintain production. Advocates of conservation and solar should recognize that they need higher prices to make their programs more attractive. Consumerists need somewhat higher prices now to help protect the public against the awesomely high prices that could eventuate if the United States does end up importing 14 million barrels a day. It is not merely rhetoric, but absolute necessity, to find some ways to make this alliance clear to the various participants. Our proposal reconciles their interests.

Moreover, we wish to stress the need for greater understanding among normally warring parties. For instance, public interest groups must understand the substantial and complex difficulties faced by utilities as they try to adapt to the new energy era. Utilities should be partners in the promotion of conservation and solar. The exclusion of utilities from the conservation business in the 1978 National Energy Act was, in this connection, not some minor mistake but a major blunder, creating a very significant and totally unnecessary barrier to the exploitation of conservation energy.

THE CONSEQUENCE OF A BALANCED PROGRAM

What pattern of energy use will result from the move to a more balanced system? Although the base of experience with incentives for conservation and solar energy is inadequate for detailed predictions, some studies have presented plausible estimates of fairly high energy savings—given propitious conditions—over a period of ten to twenty years. Even after generously discounting such estimates, the potential savings appear attractive. To illustrate, we compare a conventional forecast for the late 1980's with a *possible* pattern that could result from increased usage of conservation and solar energy. Even though the estimates for these two sources are far below some of those appearing in Chapters 6 and 7, it is striking the extent to which conservation, with the aid of solar energy, rather than ever-

Table 8–1

U.S. ENERGY SUPPLY, ACTUAL 1977
AND TWO ALTERNATIVE POLICIES FOR LATE 1980's
(millions of barrels daily of oil equivalent)

	1977 Actuals	Late 1980's Conventional Program	Late 1980's Balanced Program
Domestic (excluding U.S. exports)			
Oil	10	10	10
Natural Gas	9	9	9
Coal ᵃ	7	12	11
Nuclear	1	3	2
Subtotal, "Conventional"	27	34	32
Solar, Including Hydro ᵇ	1	2	4
Total Domestic	28	36	36
Imports			
Oil	9	14	9
Gas	0 ᶜ	1	1
Subtotal	9	15	10
TOTAL	37	51	46
Extra Conservation	–	3	8
GRAND TOTAL	37	54	54

ᵃ Estimates for late 1980's exclude estimated exports of one million barrels per day of oil equivalent. DOE estimate for late 1980's includes new technology—liquefaction and gasification.

ᵇ 1977 excludes biomass, principally wood, used by the forest products industry. Late 1980's includes biomass beyond that used in 1977, all active solar hot water and space heating, passive solar heating, and all hydroelectric.

ᶜ Slightly less than 0.5 mbdoe.

Sources: 1977 from Energy Information Administration, *Annual Report to Congress, Volume III, 1977* (Washington D.C.: Government Printing Office, 1978), pp. 5, 23, 51, 145. Conventional program from middle-demand, middle-supply scenario in *Volume II, 1977,* pp. 28, 119, 206, 216, 229, adjusted for the National Energy Act and more recent nuclear estimates—obtained from DOE news release on National Energy Act, October 20, 1978, and recent discussions with DOE officials. Balanced program developed by authors, based on Chapters 2–7.

increasing oil imports and domestic supplies of coal and nuclear, *could* help meet U.S. energy needs. In our illustration, shown as the balanced program in Table 8–1, conservation and solar energy would provide two thirds of the "increased" energy supplies, compared with only a quarter in the conventional program based on Department of Energy forecasts [24] (that is, 11 million barrels daily in the balanced program, compared with only 4 million daily in the conventional program, out of a total new supply of 17 million.) And imported oil in the balanced program would not increase beyond the 1977 level of 9 million barrels daily, whereas the conventional program foresees an increase of 5 million daily—to a new level of 14 million.[25]

The balanced program would still mean considerable use of traditional energy sources. Our illustration for domestic oil and gas, for example, is the same as the conventional program's. But the balanced program would represent the beginning of a transition to alternatives that pose far fewer problems than increased reliance on imported oil.

Conventional domestic production does not offer the same opportunity. The matter can be viewed thusly: Our conventional energy production—oil, gas, coal, and nuclear—may be thought of as well-explored producing regions. We favor continuing and augmenting production in these terrains. But in terms of allocating resources and effort for further major increments of energy, the evidence strongly suggests that the nation would be better served by concentrating its exploratory and development "drilling" in the partially proven acreage of conservation, and the promising but still largely untested acreage of solar.

What is still missing is an energy policy to guide the transition. What we propose would make possible an economically sound and politically workable transition away from ever-growing dependence on imported oil. No other nation has so great an impact on the international energy system. Now is the time for the United States to come to terms with the realities of the energy problem, not with romanticisms, but with pragmatism and reason. And not out of altruism, but for pressing reasons of self-interest.

SERGIO KOREISHA
AND ROBERT STOBAUGH

Appendix:
Limits to Models

The oil crisis of 1973 attracted intense attention from specialists—
econometricians and technologists—who build formal models in
order to make predictions about the future.

Econometricians build and operate models which consist of
mathematical equations based upon relationships derived from
economic theory and estimates based on historical statistical relationships. The use of such models is often characterized as "looking
forward through a rear-view mirror." These models are usually developed with the aid of a computer. They are used primarily to
make forecasts about *existing* energy systems: How much coal will
be produced at such and such a price? How much oil will be consumed if the price goes up two dollars a barrel?

The technologists, meanwhile, base their analyses mostly on engineering cost estimates. Such models are used extensively to forecast energy supplies that might be forthcoming from *new* sources:
What, for example, is the likely cost of gasifying coal or utilizing
shale oil? They are also used to estimate supplies from conventional
energy sources: What, for example, are the chances of finding new

oil reserves? The results of both types of formal models—econometric and technological—are often modified by personal judgment to make the results correspond more closely to the specialist's understanding of the real world.[1]

A systematic survey of all published formal models is not possible here, for they number in the dozens, perhaps even in the hundreds. But a very important point can still be made. The major studies since 1973 have given us predictions about the U.S. energy situation that have consistently been more optimistic than the reality has proved to be, especially in regard to energy supplies. Some of the models were published without receiving much notice and had virtually no lasting impact. Nevertheless, it seems abundantly clear that some of the optimistic forecasts issued did influence—and mislead—both the energy policymakers and the informed public about the causes and possible solutions for the energy problem.

In a world of contradictory assertions, it is not surprising that, as one leading econometric modeler put it, "Public officials increasingly fall back on the computer model as their ultimate authority." *Fortune* magazine, with some hyperbole perhaps, summarized the situation as follows: "When the history of economic policymaking in the turbulent late 1970's is written, an important part of the story will be about the ever-widening impact of econometric models on federal policy decisions. The wondrous computerized models—and the economists who run them—have become new rages on the Washington scene. These days, it seems, every spending and tax bill is played into mathematical simulations inside a computer. The model managers themselves are star witnesses before congressional committees whose members seek to divine the future. And what these machines and their operators have to say has come to have a significant bearing on what Washington decides to do." [2]

Although other factors, such as the judgments of consultants and industry executives, obviously also affect energy estimates, our discussions with government officials, as well as published evidence, provide support for a conclusion that energy policy indeed has been affected to an important extent by formal models. In November 1974, for example, the federal government released a study that showed how the United States could become self-sufficient in energy by 1985. A *New York Times* article, in reporting the study, quoted an anonymous Administration official, who, citing a large computer model, said, "We expect oil prices to level out between $4 and $6 a barrel." [3] At the time, OPEC prices were about $10 a barrel. In January 1975, Secretary of State Henry Kissinger stated that discoveries of oil and new sources of energy would make it "increasingly difficult for the cartel to operate," and that this would begin to "occur within two or three years." And in order to encourage the development of domestic energy sources in the major industrial nations, Kissinger spent much time trying to arrange a floor price on oil of $7 a barrel. [4]

Models have also had similar impact on some decision-makers in the private sector. In early 1979, Henry Ford II expressed a hope that the federal government "would give up" on efforts to get tighter post-1985 fuel economy standards. He was asked, "Aren't you going to have to develop something in the 1985 period that really copes with the likelihood of oil becoming more and more scarce?" Ford replied, "I thought so once, but I'm not sure I do now. In late 1974 we asked the Stanford Research Institute to study this problem. They said we don't see that as a worry. If it doesn't come out of the ground, it will come from conversion of coal or shale or what Occidental is doing, or gasohol or God knows what." [5]

What went wrong with the models is a question that obviously needs to be pursued. In general, the modeling enterprise needs to be demystified so that a better understanding of both the utility and the limitations of models can be obtained. To try to answer the question of what went wrong with some of the specific predictions, we will show the critical importance of the various assumptions made by the modelers, for these assumptions, in effect, determine the results of the models. This is all too often overlooked.

We also hope to encourage a recent tendency wherein model builders state explicitly, even highlight, what the limitations of their studies are. [6]

From the innumerable studies made about energy prospects, we selected three for detailed examination, precisely because of the authors' extraordinarily high degree of competence and sense of responsibility. We know of no one who could have done better using the tools of the craft. The studies also happened to be three of the most prominent published. They are as follows:

- The Kennedy-Houthakker World Oil Model
- The MIT Energy Self-Sufficiency Study
- The Federal Energy Administration's Project Independence Report

The very word *model* can confuse people. Here it means a representation of some phenomenon or observable system in the real world. Thus, the builder must seek to balance his or her model between two extremes. It must be complex enough to be a reasonable representation of the phenomenon, but it must also be simple enough to be constructed and used.

To fit the complexities of the real world into the variables that can be handled in a model, the builders of energy models are forced to make a number of simplifying assumptions that can lead to inaccuracies. For convenience, we refer to these simplifying assumptions as *red flags,* items to which the model builder and the user of the model's results should pay special attention. The first three red flags discussed are relevant for both econometric and technological models, whereas the last two apply mainly to econometric models.

1. *Exclusion.* The influence of any factor not included in the model is assumed to be unimportant in affecting the conclusions.

In an econometric model, the builder seeks to measure the impact of one or more economic variables on another variable in order to predict future events or to explain some economic process. For example, a very simple econometric model could be constructed on the supposition that demand for oil this year is dependent solely on this year's oil price. In such a case, the model builder would be assuming that natural gas prices had no effect on oil consumption. But in fact, in the real world, one expects that higher natural gas prices lead to increased oil consumption by encouraging consumers to use oil instead of natural gas.

In a technological model, the model builder typically studies the current state of technology, estimates the costs of individual elements that make up the total investment and operating costs, and then estimates the cost of these elements for commercially sized plants. For example, a technological model could be constructed in order to forecast the cost of producing oil from shale rock. One element that would be included in the cost of the plant would be the interest costs incurred during the construction period, costs that would be partially dependent on the number of years needed for construction. If a prototype plant had already been built and was in operation, the model builder might estimate the time period needed to build the commercial plant by extrapolating from the experience obtained during the construction of the prototype. If the prototype plant had not experienced delays due to objections raised by environmentalists, the model builder might make the assumption that environmentalists would not delay the commercially sized plant.

2. *Aggregation.* Data on different subprocesses are often combined, or aggregated, as if they were just one process in order to reduce the number of variables to manageable proportions, or because it is not possible or convenient to measure them separately. For example, U.S. onshore and offshore oil production clearly have different cost structures, yet many econometricians lump them together as though they constituted just one large field. Similarly, for technological models, although workers with different skills have different costs and productivities, cost estimators sometimes lump them together as though they were one homogenous group of workers.

3. *Range.* The data put into an equation used to make forecasts are calculated from observations made within a range of prior experience. If there is no change in the underlying economic behavior, then logic dictates that such data should be useful in making forecasts within this range of prior experience. But if the data are used to make projections outside this range, the logical underpinning of the estimates deteriorates. As one moves further outside the range, the potential for error increases dramatically.[7]

For forecasting purposes in an econometric model, some of the input data consist of economic statistics known as elasticities. In the context of supply and demand, the term *elasticity* can be defined as the responsiveness of the demand or supply of a commodity to changes in factors which influence that demand or supply. The price elasticity of supply relates to the extent to which supplies respond to changes in the price of a commodity. Similarly, the price elasticity of demand is a measure of the extent to which the quantity demanded responds to a change in the price of the commodity. To be more explicit, the price elasticity of demand is the percentage change in the demand that would result from a one-percent change in price. For example, a coefficient of −.5 for the price elasticity of demand for oil means that if the *price* of oil were to rise by one percent, the *demand* for oil would fall by half a percent. That is, the elasticity can be viewed as the percentage change in demand (or supply) divided by the percentage change in price.

Percentage Change in Price	Percentage Change in Demand	Elasticity
↑1%	↓.5%	−.5

If a −.5 elasticity was found to exist when the price of oil rose from $2.90 to $3.00 a barrel, one might assume with some confidence that the same elasticity would apply if the price of oil were to rise from $3.00 to, say, $3.15. But there is no real basis for assuming that the same elasticity also would apply if the price were to rise from $3.00 to $10.00. It would be as though one were predicting the outcome of a marathon among runners whose track record was limited to the 100-yard dash.

The range issue also figures in technological models, as, for instance, when the cost of a commercially sized plant is estimated from the experience gained in building and operating a prototype. Although there is both theory and empirical evidence to suggest that a chemical processing plant twice as large as another plant with identical technology would cost only about 1½ times as much to build, the extrapolation of experience upward by a factor of 10 or 100 produces much more uncertainty.[8]

4. *Reversibility.* This red flag applies primarily to econometric models. For example, if the elasticities used to make forecasts are derived during a demand process in which *reduced* prices are accompanied by *increased* consumption, then, under the reversibility assumption, it is assumed that forecasts can also be made using these elasticities for periods when *increased* prices are expected to be followed by *reduced* consumption.

5. *Time lag.* This red flag also applies primarily to econometric models. Data are seldom available to enable the modeler to make an accurate estimate of the time lag required for the change in one variable, such as price, to achieve any given effect. True, econometric models (using "distributed lags") could take into account slow adjustments to price changes, such as retrofitting furnaces, making cars smaller, and changing commuting habits. But in many cases there are no historical data on which to measure how fast and effective the response is likely to be. This problem is sometimes

handled by using two elasticity estimates: a short-term elasticity to indicate those effects that will appear in a short period of time (often less than a year), and a long-term elasticity to indicate the full effects that will occur (often requiring many years). Thus, estimates of lags generally require considerable personal judgment.

In light of the foregoing, it is not surprising that estimates of similar elasticities can differ significantly from model to model. Table A-1, for example, shows the disparity in agreement for the price elasticity of crude oil supply among the three studies analyzed in this Appendix. The crude oil supply that would result with an elasticity of .87 is very different from the supply that would be forthcoming from an elasticity of .15. The high elasticity implies that a 100 percent increase in price would increase supply by 87 percent, compared with an implied increase of only 15 percent with the lower elasticity.

THE KENNEDY-HOUTHAKKER WORLD OIL MODEL

The World Oil Model, formulated by Michael Kennedy and Hendrik Houthakker * in 1973 and 1974, was the first attempt of which we are aware to set up an analytical structure that could systematically predict long-run levels of consumption, production, and price of oil in various regions of the world based on flexible sets of assumptions.[9] The model consists of four sectors: crude oil production, transportation, refining, and consumption of products. It distinguishes six regions: United States, Europe, Japan, Canada, Latin America, and the Middle East. It uses five commodities: crude oil and four refined products—gasoline, kerosene, distillate fuel, and residual fuel.

The model contains the assumption that the Persian Gulf and North African members of OPEC determine the world oil price, which they do by announcing the revenues that they will receive per barrel and then selling whatever oil is demanded at this revenue plus production costs. (Hereinafter, we refer to this revenue as a

* The model was constructed by Kennedy for his doctoral thesis in economics at Harvard University. He is now at the Rand Corporation. Hendrik Houthakker, a member of Kennedy's thesis committee, is a professor of economics at Harvard University and a world-renowned expert on elasticities and their uses in econometric models; he was formerly a member of the President's Council of Economic Advisors.

Table A–1

ESTIMATES FOR THE PRICE ELASTICITY OF CRUDE OIL SUPPLY

Study	Estimate
Kennedy-Houthakker World Oil Model [a]	
"Pessimistic" Case	.15
"Optimistic" Case	.50
MIT Study (Erickson-Span Model) [b]	.87
Project Independence [c]	.78

[a] Hendrik S. Houthakker, *The World Price of Oil: A Medium-Term Analysis* (Washington, D.C.: American Institute for Public Research, 1976), p. 19.

[b] The Policy Study Group of the MIT Energy Laboratory, "Energy Self-Sufficiency: An Economic Evaluation," *Technology Review,* May 1974, p. 34.

[c] The .78 figure represents the average of .74 and .82, which in turn was derived from data in Federal Energy Administration, *Project Independence Report* (Washington, D.C.: Government Printing Office, November 1974), p. 29.

"tax." [10]) The other members of OPEC are assumed to hold their production constant at 1972 levels; the output of the rest of the world is fixed by the user of the model.

The detailed assumptions made by Kennedy and Houthakker are of particular interest, for they can be used to illustrate some of the red flags discussed above.

First, take the *exclusion* assumption. None of the other energy supply sectors, such as coal, natural gas, and nuclear, were incorporated into the model.[11] Consequently, the model cannot explicitly quantify the effects that changes in the other sectors would have on the oil market.

Second, in order to estimate price and income elasticities, the authors *aggregated* available data for industrialized countries and assumed that these applied for all regions of the world.[12] Furthermore, in three of the four products, data for the United States—consumer of one third of the world's oil—were deficient and hence not included.

Third, let us recall that when elasticity figures are used far outside the *range* of values of the observations used to calculate them, the potential for error is large. In the Kennedy-Houthakker model, data covering the pre-embargo period of 1962 to 1972 were used to estimate the post-embargo period, when prices were several times those of the pre-embargo period. In other words, the historical experiences observed when prices changed from $1.75 to $2.00 a barrel were used to estimate the effects of changing price from $2.00 a barrel up to $10.00 a barrel.

Fourth, the modelers assumed that the processes being modeled were *reversible:* that is, the extent to which reduced oil prices were accompanied by increased consumption would be an accurate guide of the extent to which increased oil prices were accompanied by reduced consumption.

Although this reversibility assumption sounded plausible, in fact we believe that it was not justified. The principal reason is that a fall in the "real" price of oil (that is, the price corrected for inflation) helped oil to replace coal in the 1950's and 1960's. But energy users could not be as responsive to the 1973–74 rise in oil prices as they had been to the earlier drop—that is, they could not switch from oil back to coal as readily as they switched from coal to oil, for a variety of reasons. Many consumers who were allowed to burn coal in earlier years were prevented or discouraged from returning to coal in the 1970's because of laws designed to protect the environment. Whereas the replacement of coal-handling equipment by oil was relatively inexpensive, the reinstallation of the dismantled coal-handling equipment was costly. A boiler designed to burn coal requires a larger firebox per unit of output than one designed to burn oil, so a boiler conversion from coal to oil could be done without any loss of capacity. But if a boiler designed to burn oil was converted to coal, the capacity loss would be substantial. In other words, many factors not included in Kennedy's and Houthakker's equations affected the outcome and undercut the potential reversibility of the economic processes.

Fifth, consider the *time-lag* red flag. The model builders assumed that the full effects predicted by the model would occur by 1980—that is, that six years would be sufficient time for the long-run elasticities to apply. A lag structure had to be *assumed* rather than estimated from the model, because the model was not constructed to trace the pattern of events leading to the conclusions—it was a static model.

Given this list of red flags, it is not surprising that many of the elasticities calculated by the authors differed significantly from what such elasticities were generally believed to be. Consequently, following a practice sometimes used by econometricians, the authors made "some substantial modifications to the . . . [elasticities] for projection purposes." [13]

The projections made by Kennedy and Houthakker for 1980 were very optimistic. For example, Kennedy stated (in 1974) that "price hikes of crude which occurred in late 1973 are not likely to persist in the future." The results of simulations from the model suggested that a host government tax of about $3.50 a barrel, which represented less than one half the then-current tax of about $9.00, was most likely to occur in the long run. But Kennedy used his judgment to select a different scenario as the most likely one: "Changing the assumptions behind the simulations to make them reflect conditions more favorable to a high price *implied* that a value closer to $5.00 was easier to maintain, and thus more likely to occur." (All monetary figures in the Kennedy-Houthakker study were in constant 1972 dollars. Five dollars in 1972 was equivalent to $7.95 in early 1979, when the OPEC price was $13.34.) [14]

Table A-2 (page 244) shows some of the results on which Kennedy based his conclusions. Note that in all cases revenues from oil exports by the Persian Gulf and North African governments are maximized when the tax lies between $3.50 and $5.25 per barrel, and that the *United States becomes an oil exporter* in the "high-supply elasticity case" when the host-government tax reaches $8.75 (or $13.90 in early 1979 dollars).

Noticeably lacking in the model's projections are the likely boundaries, or ranges, associated with each forecast. The absence is actually not too surprising, for it is extremely difficult to make a calculation of likely boundaries for the type of model used by Kennedy and Houthakker. [15] To get around the problem, modelers often construct scenarios typifying the worst and best situations that can occur within reason. These scenarios can be viewed as boundary limits for the forecasts. The wider the range is between the limiting cases, the higher the chances are for the actual figure to fall between the boundary values. Ideally, one would like to specify narrow ranges and still have the actual figures fall within those boundaries.

With these thoughts in mind, we proceed to select two different scenarios as the upper and lower boundaries for the forecasts. We assume that the high-supply elasticity case in Table A-2 represents the *lower limit* for the forecasts of oil exports and oil revenues of

Table A-2

KENNEDY-HOUTHAKKER WORLD OIL MODEL
PROJECTIONS FOR 1980 FOR THREE CASES
(ALL MONETARY FIGURES ARE IN 1972 U.S. DOLLARS)

Host Government Tax at Persian Gulf (dollars per barrel)	High-Supply Elasticity Case			Base Case			High-Income Elasticity Case		
	Persian Gulf & N. Africa		U.S.	Persian Gulf & N. Africa		U.S.	Persian Gulf & N. Africa		U.S.
	Exports (millions of barrels daily)	Revenues (billions of dollars yearly)	Imports (millions of barrels daily)	Exports (millions of barrels daily)	Revenues (billions of dollars yearly)	Imports (millions of barrels daily)	Exports (millions of barrels daily)	Revenues (billions of dollars yearly)	Imports (millions of barrels daily)
$1.75	31	$22	13	31	$22	13	40	$29	n.a.
3.50	19	26	9	21	28	9	27	36ᵃ	n.a.
5.25	13	26ᵃ	4	15	29ᵃ	6	19	36	n.a.
7.00	8	20	1	11	29	4	13	34	n.a.
8.75	3	10	-2	7	24	2	8	26	n.a.

ᵃ Maximum revenue for respective case

n.a.: not available

Source: Constructed from information in Michael Kennedy, "An Economic Model of the World Oil Market," *The Bell Journal of Economics and Management Science*, Autumn 1974, pp. 540–577.

Notes:

1. One dollar in 1972 is equal to $1.59 in early 1979 dollars.
2. The above table represents only part of the information presented by Kennedy. The host-government tax presented by Kennedy ranged from a low of $.80 to a high of $8.75 per barrel.
3. Kennedy refers to the host-government tax as an "export duty"; see Reference 6, this Appendix.
4. Numbers are rounded from those in original source.

the Persian Gulf and North African OPEC members. The reason for our assumption is that a high-supply elasticity implies a large U.S. oil supply and hence relatively low OPEC exports. Similarly, we assume that the high-income elasticity case represents the *upper limit* of that range. The reason for this assumption is that a high-income elasticity implies a strong demand for oil products, resulting in relatively high U.S. oil imports and OPEC oil exports. We then note that in these limiting cases, as values for the host-government tax increase, the forecasts for Persian and North African exports and revenues become quite low. At a time when the host-government tax approximated $8.75 (1972 dollars), the boundary limits of forecasts at a tax of this level indicated that the oil exports of the Persian Gulf and North African countries would be in the range of 3 to 8 million barrels daily; at the time of the forecasts, such exports were approximately 23 million barrels daily.[16]

One can question whether it was reasonable to believe even back in 1974 that forecasts of Persian Gulf and North African exports so much lower than then-existing conditions were likely to prevail in 1980. An observer might conclude either that the assumptions used for these cases were so unreasonable as to preclude the cases from being reported as evidence for what could occur if the supply or income elasticities were at the boundary limits, or that the model was not sufficiently robust so as to be able to handle situations that even then would seem reasonable to consider.

Indeed, in a talk before the Conference Board in April 1974, Houthakker recognized that the results of the model "are subject to a wide margin of error because there is much uncertainty about the exact response of production and consumption to the unprecedently high price levels of the moment." Although Houthakker still made lower estimates of OPEC prices than eventually proved to be the case, he did begin to raise his estimates. The *Wall Street Journal* reported that an "eminent economist" (Houthakker) was relatively optimistic about the world oil situation: "Running his data through a computer, he [Houthakker] came up with a tentative future price of something around $6 a barrel based on 1972 dollars" ($9.50 in early 1979 dollars), a figure lower than in fact proved to be the case, but higher than Kennedy's earlier estimate of $5 a barrel.[17]

Two years later Houthakker published the results of model simulations that contained much greater variations in elasticities for the different regions of the world. He presented two different cases:

1. An "optimistic" case, in which there was considerable response to higher prices in terms of both increasing crude oil production outside of OPEC and decreasing product consumption in the industrial world.

2. A "pessimistic" case, in which the opposite was assumed—that is, neither crude oil production outside of OPEC nor product consumption was very responsive to higher prices.

The results of these simulations differed from the results of the simulations of the earlier model in two important ways. First, the optimum host-government tax, and resulting host-government revenues, for the Middle East and African exporters were substantially higher than in the earlier results.* In the optimistic case, the optimum tax was $7.50 per barrel, while the pessimistic case shows the optimum tax as $20.00 (with Middle East and African annual revenues of $35 billion and $75 billion, respectively, for the two cases). Second, the resulting ranges were much wider than those presented earlier. If we assume, as in the earlier simulations, that the two extreme cases represented the upper and lower bounds of a "reasonable" range, then the range of optimum host-government taxes was from $7.50 to $20.00 per barrel, as shown in Table A-3, instead of the $3.50 to $5.25 per barrel in the two earlier cases that represented the upper and lower bounds (or, in early 1979 dollars in round numbers, $12 to $32 instead of $6 to $8).

Although it is desirable to recognize the width of the range for the estimates of the later simulations, the question arises as to the usefulness of such a model to policymakers. Take the case of U.S. imports. With a host-government tax of $7.50 per barrel, imports are shown as zero for the optimistic case and 10 million barrels daily for the pessimistic case.

In any event, Houthakker ultimately used his judgment instead of the results of the model in order to make a prediction. After explaining the problem of determining the production level of each OPEC country if OPEC's output were to drop substantially, he stated that the optimum export levels shown in Table A-3 "are probably academic. Perhaps an uneasy equilibrium at a total output level not much below that of 1975 [22 million barrels daily

---

* The earlier model simulation included "Persian Gulf and North African OPEC," whereas the later simulation included "Middle East and African OPEC"; the difference is insignificant for our purposes.

Table A–3

KENNEDY-HOUTHAKKER WORLD OIL MODEL RESULTS FOR 1980 FOR "OPTIMISTIC" AND "PESSIMISTIC" ELASTICITY CASES
(ALL MONETARY FIGURES ARE IN 1972 DOLLARS)

Host-Government Tax at Persian Gulf (dollars per barrel)	Optimistic Case			Pessimistic Case		
	Middle East & Africa		U.S.	Middle East & Africa		U.S.
	Exports (millions of barrels daily)	Revenues (billions of dollars)	Imports (millions of barrels daily)	Exports (millions of barrels daily)	Revenues (billions of dollars)	Imports (millions of barrels daily)
$ 5.00	22	$34	3	29	$46	12
7.50	15	35 [a]	0	24	57	10
10.00	10	29	–1	20	65	9
12.50	n.a.	n.a.	n.a.	18	70	9
15.00	n.a.	n.a.	n.a.	15	73	8
17.50	n.a.	n.a.	n.a.	14	75	7
20.00	n.a.	n.a.	n.a.	12	75 [a]	7
22.50	n.a.	n.a.	n.a.	11	74	7

[a] Maximum revenue for respective case

n.a.: not available

Source: Constructed from information in Hendrik S. Houthakker, *The World Price of Oil*, pp. 21, 26.

Notes:

1. One dollar in 1972 is equal to $1.59 in early 1979.
2. The above table presents only part of the results presented by Houthakker. For example, the host-government tax presented by Houthakker ranged from a low of $1.25 to a high of $22.50 per barrel.
3. Houthakker refers to the host-government tax as a "royalty"; see Reference 6, this Appendix.
4. Numbers are rounded from those in the original source.

of exports from the Middle East and Africa] is the most likely outcome for the next several years." Houthakker used his judgment for this volume estimate because the model does not contain a theory of cartels, and he thought that the Middle East and African OPEC members would need at least 20 million barrels daily of output in order for the cartel to hold together. His volume estimates based on this judgment proved to be close—only about 10 percent low—to the actual figures for the next two years.[18]

THE MIT ENERGY SELF-SUFFICIENCY STUDY

In May of 1974, The Policy Study Group of the MIT Energy Laboratory * released a major report, the MIT Energy Self-Sufficiency Study. The study focused on "the question of whether the goal of energy independence [for the United States by 1980] could be achieved and, if so, the price its achievement might entail for the nation." The focus was on price rather than imports, because the study was based on the key assumption that the United States could meet *all* its energy demands from internal sources by 1980.[19]

The group conducted the study in a remarkably short time—in the first few months of 1974—so it was forced to rely mainly on already existing information, much of which was work done earlier at the MIT Energy Laboratory. The group used both econometric models and technological models, but to arrive at a final judgment they checked the results of the models with "judgmental forecasts" that were made separately from the formal models. Such forecasts were made by an expert (or a group of experts) after taking into account past trends and relationships and expected future events that might affect the variables being predicted. The experts included a number of distinguished academicians, consultants, and industry specialists.

The group made estimates of supply and demand at three levels

* Some of the principal members of the Policy Study Group included Morris A. Adelman, professor of economics and author of *The World Petroleum Market*; Paul W. MacAvoy, then professor of management and later a member of the President's Council of Economic Advisors; Henry D. Jacoby, professor of management; David C. White, professor of engineering and director of the laboratory. Much of the group's work was based on studies made by Edward Erickson, Dale Jorgenson, Robert Spann, National Petroleum Council experts, and the staff of National Economic Research Associates.

of energy prices—the equivalent of oil prices of $7, $9, and $11 per barrel in 1973 dollars (or, $10.50, $13.50, and $16.50 in early 1979 dollars).

They estimated the supply of energy for each of five major sources —oil (including natural gas liquids), natural gas, coal, hydroelectric and uranium, and new technology (such as synthetic crude oil, synthetic gas, and shale oil). They then totaled the figures for each of the five energy sources to obtain estimates of total U.S. energy supply at each price level.

They derived demand figures by using a large econometric model of the U.S. economy in conjunction with another econometric model of production and consumer behavior, a model that in turn predicted demand for the output of four sectors—private consumption, private investment, government expenditures, and net exports. The energy input required per unit of output was determined as a function of price by the use of an input-output model. These estimates of energy requirements per unit of output were then multiplied by estimates of the output of each of the four sectors to obtain an estimate of total energy demand.

Using these results to define points on their approximate supply-and-demand curves for total energy, the modelers looked for the point of intersection of the curves—that is, the point where domestic supply and demand were in balance. The price associated with this point was then used as the econometric forecast for what the price of oil would have to be in 1980 for the United States to become energy self-sufficient.

Although the MIT group did not publish as much detail about their models as did Kennedy and Houthakker, sufficient information was available to indicate some of the problems likely to be caused by some of the five red flags.

To illustrate the possible problem encountered with the second red flag, the *aggregation* assumption, we discuss only one issue—the utilization of an econometric model of oil supply that does not make an explicit distinction between onshore and offshore production of oil (that is, onshore and offshore fields are treated as one source).

Within physical limitations, the supply of oil obtained from both offshore and onshore fields should increase as the price for that commodity increases. But start-up costs are generally higher for offshore production. Hence, the threshold price required for production to begin is higher for offshore oil. Further, there is no reason why the rates at which production increases with price would be the same offshore as onshore; in other words, the slopes of the supply curves for the two areas likely would differ. Thus, up to the threshold level of price required for offshore production to begin, the total supply curve would reflect onshore conditions. Beyond that level, the total supply curve would reflect a mixture of onshore and offshore supply curves and hence would differ from that of the onshore curves—that is, it has a kink at the threshold price required for offshore production to begin. Estimating the price elasticity of oil supply using such an aggregated curve is like fitting a straight line through a kinked curve.

We have no way of knowing how much of an error this one aggregation assumption might have caused in the projections. It is possible that the aggregation of onshore and offshore areas into one area may have been an important cause for the grossly overestimated price elasticity value of .87 for the supply of oil. This high elasticity figure, of course, is crucial to the group's conclusion that "crude oil sources are forecast to produce more from onshore and offshore domestic wells in 1980 than at the present [1974], because the incentives of increased price more than compensate for depletion of inground reserves." [20]

Of course, the MIT group faced the same fundamental problem as did Kennedy and Houthakker. They were forced to rely on data obtained under one set of conditions and assumed that it could be used to predict the outcome of processes operating under radically different conditions—the assumptions regarding the applicability of the *range* of the data and the *reversibility* of the supply-and-demand processes. [21]

Although the MIT group indicated, correctly, that energy from new technology would make no contribution to U.S. energy supplies by 1980, their estimates of the prices required for the commercialization of new technologies to produce energy turned out to be much too low. These estimates, which were based primarily on published information, indicated, for example, that synthetic crude oil from shale would become feasible when the price of oil reached $6.80 per barrel (or $10.20 in early 1979 dollars). In contrast, recent reports indicate that even when oil prices reach $16 (in early 1979 dollars), production of oil from shale may still be uneconomical, [22] and the $16 figure is for the in-situ process, which is believed to be more economical than the surface processes assumed in the MIT estimates. The MIT group projected $7.70 per barrel for synthetic crude from coal ($11.60 in early 1979 dollars). In late 1978, the U.S. Department of Energy was projecting about 2½ times that amount—the equivalent of about $30 in 1979 dollars.[23] Indeed, discussions with industry experts suggest that probably all of the estimates in Table A-4 will eventually prove to be too low by a factor of at least two and sometimes three.

Table A–4

ESTIMATES BY MIT ENERGY POLICY STUDY GROUP OF WHAT
THE PRICE OF OIL WOULD HAVE TO BE FOR SOME
"NEW TECHNOLOGY" FUELS TO BECOME ECONOMICAL

Fuel	Cost per Barrel (oil equivalent)	
	(1973 dollars)	(early 1979 dollars)[a]
Synthetic Crude from Oil Shale	$6.80	$10
Synthetic Crude from Coal	$7.70	$12
Synthetic Natural Gas from Coal		
Old Technology	$9.05	$14
New Technology	$7.30	$11
Methanol from Coal	$10.00	$15

[a] Rounded to nearest dollar. One 1973 dollar equals $1.50 in early 1979 dollars.

Source: The MIT Policy Study Group, "Energy Self-Sufficiency."

That the engineering cost estimates used for new technologies by the MIT group were much too low is not surprising. Indeed, for years a narrow focus on technology has resulted in too optimistic a picture of energy supplies that might be forthcoming from new technologies.[24] Some of the lessons of the nuclear story in Chapter 5 of this book are relevant—especially the importance of conflicting political interests, the level and distribution of external costs, and the problems encountered in massive scale-ups of complex systems when working outside of a prior range of experience.[25]

Although the MIT group correctly estimated no energy in 1980 from the new technologies, their estimates of U.S. domestic supplies for 1980 seem high, primarily because of overestimation of new supplies of crude oil, natural gas, and nuclear. As shown in

Table A-5, the econometric models (supplemented by judgmental forecasts when econometric models were not available) predicted 38 million barrels daily of oil equivalent at $7.00 per barrel of oil

Table A–5

ESTIMATES BY MIT ENERGY POLICY STUDY GROUP
OF ENERGY EQUILIBRIUM IN 1980 USING SUPPLY FORECASTS
BASED ON ECONOMETRIC MODELS SUPPLEMENTED BY
JUDGMENTAL FORECASTS
(monetary figures in 1973 dollars)

Fuel	Source of Estimate	Millions of Barrels per Day of Oil equivalent, at prices per barrel of oil		
		$7.00	$9.00	$11.00
Crude oil and natural gas liquids [a]	Econometric models of Erickson-Spann (crude oil) and MIT (natural gas liquids)	10[b]	13	15
Natural gas	MIT econometric model	15	16	17
Coal	Judgmental from MIT analysis	7	8	8
Uranium and hydroelectric	Judgmental from equipment survey	6	6	6
New technology	Judgmental from MIT analysis	0	0	neg.
Total supply		38	43	46
Forecasts of total demand	Econometric model of Hudson-Jorgenson	44	42	41

[a] Estimates for natural gas liquids are 2.1, 2.2, and 2.4 for the three respective prices.

[b] Rounded from 10.5 in the original source, in order to present it closer to, rather than farther from, actual.

Numbers rounded from those in original source.

neg.: negligible.

Source: Constructed from information in The MIT Policy Study Group, "Energy Self-Sufficiency."

Note: A 1973 dollar equals $1.50 in early 1979 dollars.

(or $10.50 in early 1979 dollars), and 43 million barrels at $9.00 per barrel (or $13.50 in early 1979 dollars). In fact, the United States will be fortunate if its domestic production is as high as 30 million barrels daily of oil equivalent in 1980.

True, most of the econometric models used by MIT—with some natural gas projections the principal exception—contain an implicit or explicit assumption that prices would be based on the interplay of market forces rather than controlled by the government; in fact, neither U.S. oil nor natural gas prices have been substantially free of price controls since the time of the MIT study. And even though it is true that by 1978 the price of newly found oil had been allowed to rise to about $9.00 a barrel and newly found gas to a little over $7.00 per barrel of oil equivalent (both values in 1973 dollars), it is possible that the mere existence of price controls might have affected supply and demand independently of the effect on the price level.

Price controls add uncertainty, for example, which might have reduced investment for new supplies below the level that would otherwise have resulted at the uncontrolled price identical to the controlled price. On the other hand, producers might have expected price controls to be eased or removed, thereby allowing prices to rise toward or to the world level. Such an expectation, of course, might have increased investment for new supplies above the level otherwise expected at the controlled price. Thus, it is impossible to say with a high degree of certainty whether the MIT domestic supply estimates were optimistic, given the assumptions of the MIT group.

On the other hand, their rough estimate that the world's productive capacity for oil in 1980 would be 87.4 million barrels daily was clearly too optimistic. This estimate was based on a host-government tax of about $7.00 a barrel in 1973 prices (or $10.50 in early 1979 dollars) and on the group's judgments of the price elasticities of supply. And though recognizing "that the range of conceivable values for the world oil price over the next decade is very large," the group concluded that "it is unlikely—though certainly not impossible —that the price of world oil to the U.S. will rise above this level in the next few years." [26]

The MIT group, of course, recognized shortcomings in their studies. Among other things, they declared that "until more experience is gained at these high price levels, the uncertainty over this range of prices will remain," and "there may be some doubt about the completeness of the reversibility" assumed to exist.[27] And they explicitly stated that the econometric estimates of supply "are very probably optimistic." Indeed, they concluded that the market-clearing price required for U.S. energy independence of $9.00 ($13.80 in late 1979 dollars), as indicated by the econometric models, was "extremely optimistic." In contrast, the judgmental forecasts, based principally on estimates from industry, showed a lower supply estimate and a higher demand estimate than the econometric models. Thus, the judgmental forecasts indicated a far higher market-clearing price than did the econometric models. Therefore, the MIT group backed off from the results of the econometric analyses and stated that, on balance, in order for the United States to achieve self-sufficiency in energy in 1980, "the results point to the conclusion that the price of energy would be from $10.00 to $12.00 a barrel" (or $15.00 to $18.00 in early 1979 dollars).[28]

Fortunately for U.S. imports, U.S. demand in 1980 almost surely will be lower than forecast either by the econometric models or by the judgmental approach—but not so low as to make actual imports as low as the zero to 7 million barrels daily indicated for the $7.00 to $9.00 price range ($10.50 and $13.50 in early 1979 dollars).[29]

PROJECT INDEPENDENCE EVALUATION SYSTEM

The 1974 Project Independence Evaluation System (PIES–74) constitutes a set of models that culminated a colossal effort directed by the Federal Energy Administration to model the U.S. energy system. In fact, it is the most complex, sophisticated set of models yet developed for this purpose. PIES–74 represents the national energy system as a multiregional network that links the producing, refining, processing, converting, distributing, transporting, and consuming sectors of the system. Myriads of models and submodels are interlinked to generate forecasts for the years 1977, 1980, and 1985 under different sets of supply-and-demand scenarios. For explanatory purposes these various submodels have been grouped into four categories: supply, demand, integration, and assessment.

Price-sensitive *supply* curves have been developed for each geographical region by various intergovernmental task groups for eight different energy sources: oil, natural gas, coal, nuclear, synthetic fuels, shale oil, geothermal, and solar. These curves have been developed in order to determine how much production could be achieved for each source of energy at different prices under two alternative policy strategies: Business-as-Usual (BAU) and Accelerated Development (AD). Table A-6 contrasts the assumptions made for each energy source under the two strategies. The supply data provided by the government task groups also contain the requirements for materials, equipment, investment, and manpower associated with the production levels of the various energy sources.

Table A–6

COMPARISON OF BUSINESS-AS-USUAL AND ACCELERATED DEVELOPMENT
ASSUMPTIONS OF PROJECT INDEPENDENCE ENERGY MODELS

Energy	Business-as-Usual (BAU)	Accelerated Development (AD)
Oil	Moderate outer continental shelf leasing program (1–3 million acres per year); Prudhoe Bay developed with one pipeline	Accelerated OCS leasing program, including Atlantic and Gulf of Alaska; expanded Alaskan program, assuming additional pipeline and authority to develop Naval Petroleum Reserve No. 4
Natural Gas	Phased deregulation of new natural gas; liquefied natural gas facilities in Alaska	Deregulation of new natural gas; additional gas pipelines in Alaska; gas produced in tight formations
Coal	Some federal coal land leasing; phased implementation of Clean Air Act with installation of effective stack gas control equipment; moderate strip mining legislation	Same as BAU, with additional leasing and larger new mines

Table A–6 (*continued*)

Energy	Business-as-Usual (BAU)	Accelerated Development (AD)
Nuclear	No change in licensing or regulations; added enrichment and reprocessing capability	Streamlined siting and licensing to reduce lead times; increased reliability; additional uranium availability; material allocation
Synthetic Fuels	No change from current policies	Streamlined licensing and siting; financial incentives; increased water availability
Shale Oil	No change from existing policies	Additional leasing of federal lands; modification of Colorado air-quality standards; financial incentives; increased water availability
Geothermal	Continued R&D and federal leasing programs	Leasing of federal lands; streamlined licensing and regulatory procedures; financial incentives
Solar	Continued R&D program	Additional R&D expenditures and financial incentives

Source: Federal Energy Administration, *Project Independence Report,* pp. 64–65.

Price-sensitive *demand* forecasts for the entire nation have been generated by an econometric simulation model developed by the FEA but based on work done by Data Resources Incorporated (DRI). This model predicts quantities demanded by the major consuming sectors of the economy and by fuel type, conditioned upon expected energy prices and key economic and demographic variables, such as population, employment, GNP, and growth in housing stocks.[30]

The *integration* of the supply-and-demand components of the

U.S. energy system has been accomplished by determining the prices and quantities at which supply and demand are in equilibrium for each energy source for each region. To accomplish this integration, the modelers have derived production levels and prices that meet at minimum cost the demand indicated by the econometric simulation model. The production levels and prices are obtained from a linear programming model that contains resource and economic constraints.[31]

Now, for the *assessment* category: Several models were built to assess the impact that the equilibrium solutions would have on the environment, economy, and international scene. These evaluations were incorporated into the final PIES–74 report in order to help policymakers judge the energy strategies in terms of their influence on other national priorities. Results of the various assessment models, however, were not used as feedback for the other sectors of PIES–74 to refine estimates and reassess forecasts.

More effort has already been made in critiquing this model than perhaps any other in history,[32] and the complexity of PIES–74 precludes a detailed critique here. We do, however, think it worthwhile to comment on PIES–74 in relation to the five red flags mentioned earlier.

PIES–74 handles three of the five issues raised by the red flags more effectively than either the Kennedy-Houthakker model or the MIT study. It includes many more factors—hence the assumption that the effects of factors *excluded* from the model are less likely to cause error. However, the exclusion of factors, such as sulfur levels in coal, and different types of crude oil and certain other fuels, may also, the report states, "make conclusions which appear correct in the aggregate impractical in reality. In particular they may overstate the ability of the supply system to shift to certain domestic fuel sources. For example, anti-pollution standards may preclude the use of coal with high sulfur level." [33] Further, PIES–74 uses data that are not as *aggregated* as those used in the other models; also, PIES–74 estimates *time lags* more formally than do the others. Unfortunately, the handling of these three issues raises another problem—the problem of scale and complexity. As one model analyst says about complicated, large-scale models in general, they "may contain much that is not fully understood and substantiated. The very magnitude of these models may diminish their helpfulness to policymakers." [34] We do not know the extent to which this warning

Figure A-1

ORGANIZATIONAL STRUCTURE OF THE TASK FORCE THAT FORMULATED PIES—74

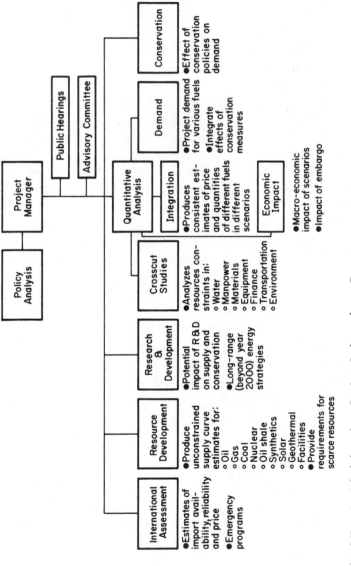

Source: Federal Energy Administration, *Project Independence Report*, p. 4IO.

applies to PIES–74, but it is a complicated, large-scale model, as shown in Figure A-1.

PIES–74, as did all other energy models formulated at that time, contained the problems associated with the other two red flags—the assumptions that *range* of prior experience is relevant and that the processes are *reversible*. The authors of PIES–74 readily state their concerns about these problems. In fact, William Hogan, who directed the PIES–74 models, has been quite articulate in pointing out the limitations of econometric models.[35]

Problems associated with all the red flags led to the determination of several elasticities that were contrary to what theory would presume. For example, the equations, which were derived by fitting curves to historical data, showed that in the household-commercial sector the demand for natural gas would fall as the price of oil increased. According to theory, however, the opposite should be true—that is, the demand for natural gas should rise because higher oil prices should encourage consumers to switch to natural gas. Many other such unexpected elasticities were also obtained from the model.[36] Also, some elasticities declined with time, whereas one would expect responsiveness (and hence elasticity) to increase over time. Furthermore, some elasticities were considered too high. Consequently many judgmental modifications and adjustments were made. Such adjustments, observed the General Accounting Office, "raise questions regarding the accuracy of all elasticities developed by the system." [37]

The PIES authors raised another list of uncertainties. They pointed out that sensitivity analyses on potential rates of domestic oil production "show that, within a range of reasonable assumptions, different values regarding discount rates, financial costs, and finding rates could affect the quantities produced at $4, $7, and $11 per barrel in 1985 by 10 to 40 percent. Other assumptions about drilling costs, effective depletion rates, and co-product prices would affect production levels at these prices by as much as 15 percent . . . The uncertainties inherent in estimating future petroleum production (especially uncertainties having to do with the magnitude of undiscovered resources in as yet totally unexplored provinces and the finding rate per foot of exploratory drilling) are so great that numerical estimates of this type are highly speculative." [38] Similar observations were made for natural gas.

The published projections of U.S. energy supplies made by the

Project Independence Group were quite optimistic. Take 1980, for example. With a $7.00 oil price in 1973 dollars (or $10.50 in early 1979 dollars), domestic supply was projected to be the equivalent of 35 million to 36 million barrels of oil daily (Table A-7). Because their consumption estimates also were higher than is likely to be the case, their projections of 1980 oil imports are much closer to current expectations for 1980 than are their supply figures.

Table A–7

PROJECTIONS BY PROJECT INDEPENDENCE STUDY GROUP
OF U.S. ENERGY SUPPLIES IN 1980
(millions of barrels of oil daily, with oil prices
in 1973 dollars per barrel)

	$7 Oil		$11 Oil	
	Business as Usual	Accelerated Development	Business as Usual	Accelerated Development
Domestic Supply	35	36	37	39
Imports	9	8	4	3
Total	44	44	41	42

Source: Federal Energy Administration, *Project Independence Report,* p. 61 and AI–37–48.
Notes:
1. Quads in original table were converted to millions of barrels daily of oil equivalent by dividing by 2.1, and then rounded.
2. A 1973 dollar equals $1.50 in early 1979 dollars.

Turning to 1985, one finds that the PIES–74 estimates of U.S. supply also seem quite optimistic. As shown in Table A-8, total domestic supplies were estimated as 39 million to 45 million barrels daily at a $7.00 oil price and 46 million to 49 million barrels daily at an $11.00 oil price (or $10.50 and $16.50 in early 1979 dollars). For a reference, our estimate in Chapter 8 is about 32 million barrels daily. Not only do the PIES oil and gas estimates seem especially high, but so do the nuclear projections—presumably because of the *exclusion* of the political forces affecting this issue.

In addition, their projections were too optimistic for the quantity of synthetic fuel production and for the prices at which these fuels could be economically produced. One case, for example, suggested that it would be possible in 1985 to produce 1 million barrels daily

Table A–8

PROJECTIONS BY PROJECT INDEPENDENCE STUDY GROUP
OF DOMESTIC FUEL PRODUCTION BY SOURCE AND IMPORTS, 1985

(millions of barrels of oil daily equivalent, with oil prices
in 1973 dollars per barrel)

	$7 Oil		$11 Oil	
Fuel Source	Business as Usual	Accelerated Development	Business as Usual	Accelerated Development
Domestic				
Oil	11	15	15	18
Gas	11	12	12	12
Coal	9	9	11	10
Nuclear	6	7	6	7
Hydro & Geoth	2	2	2	2
Synthetics	–	–	–	neg.
Subtotal	39	45	46	49
Imports	12	8	3	0
Total	52	53	49	50

neg.: negligible

Source: Federal Energy Administration, *Project Independence Report,*
p. 23, and Appendix A–1, pp. 37, 39, 45, 47.
Notes:
1. A 1973 dollar equals $1.50 in early 1979 dollars.
2. Some columns do not total, due to rounding.
3. Quads in the original table were converted to millions of barrels daily
of oil equivalent by dividing by 2.1, and then rounded.
4. "Oil" apparently includes shale, tar sands, and heavy oil; see *Project
Independence Report,* pp. 46–47.

of oil from shale alone and 1.5 million daily in total from shale, tar
sands and heavy oils, and coal liquefaction [39]—projections that now
seem virtually impossible.

THE DILEMMA OF MODELS

Everyone uses models to predict the future, even if the models
are only implicit mental ones—such as judgmental forecasts.
There is no doubt that in many instances formal models, such as

econometric models, have advantages over implicit mental models, for the formal models provide an explicit, organized framework for the forecaster and thus help to clarify assumptions. They also improve communication between model builders and users, and aid in the accumulation of knowledge and the making of forecasts under different assumptions. Furthermore, formal models can help uncover counter-intuitive results and thus promote deeper understanding.[40] In fact, some model builders state explicitly that the production of individual answers is only a minor element of the contribution of models, and that the purpose of energy modeling should be "insight, not numbers." [41]

The use of formal models to provide insight can be beneficial. The dilemma of formal models, however, is that the scientific aura surrounding them encourages those who use the results of models to expect much more than this.[42] And such expectations at times are encouraged by model builders who get so enthralled with their own models that they promise much more. As we shall show, a model incorrectly used can do more harm than good.

By now it should be clear that the predictions derived from energy models are subject to a great deal of imprecision. The derivation of coefficients that are necessary to generate forecasts often requires the modeler to make assumptions that cannot be completely substantiated, or to make simplifications that may obscure the representation of the system being analyzed. Furthermore, dramatic changes and trend reversals in certain economic processes in the course of the last few years may well mean that assumptions used today, though valid prior to these changes, may no longer apply. The lack of experience with these new economic processes contributes to the general imprecision of results generated by models.

Yet precise—and highly misleading—estimates concerning energy keep appearing. For example, Robert Pindyck, a professor of economics at MIT, writing in the *Wall Street Journal* in late 1977, stated that "OPEC's best price today is between $12.50 and $13.00 —slightly below the actual posted price. . . . In real (constant dollar) terms, this price should grow by no more than 2% per year over the next ten years." [43] How did Pindyck obtain such precise estimates? He used, as he explained, "a small computer model that quantitatively describes the characteristics of total world oil demand (and its response to income growth and price changes), non-OPEC oil supply (and its response to price changes as well as resource depletion), resource depletion within OPEC and the different levels of reserves and different rates of time discounting among OPEC members." [44]

The margin of error for each of those variables is so large that one must question how anyone could suggest, let alone self-confidently state, that OPEC's best price is between $12.50 and $13.00 or that the price should grow by no more than 2 percent per year over the next ten years. In fact, the self-confidence of the assertion seems quite surprising given the uncertainty that exists. Another professor of economics at MIT, M. A. Adelman, states that the optimum price for OPEC is very much above 1977 prices—perhaps more than double—and that if the OPEC "monopoly holds together, then we [can]...predict that the real price of oil (i.e., adjusted for general price-level changes) must rise substantially over the next ten years." [45]

But even though a model's predictions may be quite imprecise and therefore not particularly useful for policymakers, they still can be useful in "organizing the information base and guiding the decisions," as Hogan stated. [46] But even for this purpose, models must be used with great care. When present conditions are far different from the historical experience, it is possible that the models can do more harm than good by directing attention to inappropriate concerns. For example, most models of energy demand until very recently showed, incorrectly, an "iron link" between GNP and energy consumption, thereby focusing undue attention on the trade-off between energy production and loss of GNP. This focus was accentuated by models that, incorrectly, showed high responses of energy supplies to prices. These were no mere intellectual debates. They had serious effects on U.S. energy policy, for these models helped shore up the potential of traditional energy sources at the expense of conservation and solar energy. [47]

There is also a tendency for those working with technological models to concentrate too heavily on technology. An almost comical example of *exclusion* was provided by a recent headline in *Chemical Week*: "Oil from Coal: It could be a gusher by the 1990s." [48] But the opening sentence of the article read, "The technology is now available. The only barriers are political, social and economic."

When facing a future that is likely to be quite a bit different from the past, detailed studies of politics, institutions, and markets are needed along with economics and technology in order to provide the adequate base of data required for plausible answers and realistic frameworks for discussion. In energy, politics is a crucial factor, yet an explicit consideration of it is typically omitted in most formal models. Although the MIT group's report did include a number of suggestions about changing governmental policies, it contained little or no political analysis indicating the extent to which these suggestions were feasible, stating, "Behind every energy bottleneck and in every future decision stand serious societal issues—nuclear power plant safety, environmental protection, and many others. Such issues, though both appropriate and important to the debate which is now in progress throughout the nation, are beyond the scope of this report." [49]

True, one could argue that an exclusion of national politics might be justified if the goal is to help national policymakers comprehend the outcomes of different national policies. But even so, the exclusion of issues such as nuclear safety and environmental protection almost guarantees unrealistic projections and offers solutions that are much further from a theoretical optimum than model builders often acknowledge. In addition, the decentralized nature of political decisions makes it even more imperative to consider the likely actions of regulatory agencies, courts, states, and other local bodies if realistic estimates, especially of energy supplies, are to be obtained. Even if the United States had had no price controls on oil and gas, the actions of different political bodies would still have been especially important in affecting energy supplies from offshore oil, coal, nuclear, and new technologies.*

* We have been discussing why the predictions of energy models often give unrealistic results. Still unanswered is an important question: Why does there seem to be this upward bias in predictions of future energy supplies that is not borne out by subsequent experience? One reason, which we have highlighted throughout the book, is the effect of the political barriers, especially those raised by concerns about environmental issues and income distribution. Another reason, no more than a tentative hypothesis at this point, is consistent with the disappointing results in finding new oil and gas in the United States in spite of the substantially increased exploratory and development activities that resulted from the higher financial incentives. The hypothesis relates to natural resources as a special case. It is this: The relatively slow rate of technological progress normally encountered in older industries does not offset the increasing difficulty of exploitation that results from the

The actual extent to which the results generated by mathematical models influenced decisions is, of course, impossible to tell. But since it is becoming apparent that policymakers are using models to an increasing extent to help them make decisions, it is imperative that the officials find out what lies behind the assumptions made by the modelers before they start using the results. They should also test the validity of the predictions by asking what would happen when certain plausible changes are made to key parameters. Do the predictions remain plausible or do they change so much as to become totally ludicrous? A substantial change may indicate possible faults in the structure of the model. At the same time, of course, the modeler has the burden of thinking of ways to make the model both useful and understandable to the decision-maker.

Once it has been ascertained that the model adequately depicts the situation being analyzed, the results derived from the model should be used in conjunction with the decision-makers' own knowledge of the problem, which often contains factors not explicitly accounted for by the model.

As a final note, we would like to reiterate what we said at the outset of this Appendix: namely, that models can be extremely useful in the formulation of energy policy. They provide a framework for decision-makers to make intelligent choices. They facilitate the evaluations of the influences of the various factors that affect the decision. Furthermore, they allow decision-makers to test their ideas "on paper" without manipulating the actual system. *But a model is not reality.*

---

fact that the reserves easier to develop are exploited first. This slowdown in technological progress, combined with the increasing difficulty of exploitation, could cause a downward-sloping cost curve to change direction and thereby move upward. Any econometric model fitted as a straight line from historical data, as many are, would not pick up the change of direction. A model fitted with a curvilinear equation would show a change in direction. But it would probably not show the change until quite some time after it had occurred.

References

Chapter 1
The End of Easy Oil

1. The price of oil in the Arab/Persian Gulf at the end of the 1960's was $1.00 to $1.20 per barrel. See M. A. Adelman, *The World Petroleum Market* (Baltimore: Johns Hopkins University Press, 1972), pp. 183, 191. The U.S. message was first presented in a meeting of the Oil Committee of the Organization for Economic Cooperation and Development in Paris at the end of 1968 by James Akins, then director of the Office of Fuels and Energy in the State Department.

2. Paul McCracken and Alan Greenspan, quoted in *Wall Street Journal*, November 20, 1978, p. 24, November 21, 1978, p. 22. Of course, there is no reason why over the long run an increase in oil prices should necessarily affect the dollar's equilibrium point in the foreign exchange markets. But with over $300 billion held in the Eurodollar market, the expectations of foreign traders and psychological elements can have an important effect on exchange rates. And our discussions with European government officials, bankers, and business executives lead us to conclude that the dollar is weakened by rising U.S. oil imports. For an example, "the U.S. dollar registered steep declines against most major currencies" because of the possibility of another increase in oil prices. *Wall Street Journal*, December 28, 1978, p. 8. A prominent British banker told us: "As we see it, American oil imports are the main reason for the dollar's weakness. It is not only the oil deficit itself, but also what it symbolizes. Here is a very large and obvious problem that everyone can see—an oil import bill of forty to fifty billion dollars a year—and yet the United States seems completely incapable of doing anything about it. This raises a basic question about the ability of the United States to act effectively in any political or economic area."

3. Department of Energy, Energy Information Administration, *Annual Report to Congress, Volume III, 1977* (Washington, D.C.: Government Printing Office, May 1978), especially p. 61. This represents an increase, in 1979 dollars, of $10 per barrel of the 84 billion barrels of crude oil equivalent that were in U.S. proved

reserves of crude oil, natural gas, and natural gas liquids at the end of 1973. Proved reserves are the estimated quantities that geological and engineering data demonstrate with reasonable certainty to be recoverable in future years from known reservoirs under existing economic and operating conditions. This estimate overstates the value because future income was not discounted to obtain a present value; but it understates the eventual value of all oil and gas in the ground because additions will be made to proven reserves.

4. *New York Times,* June 6, 1978, sec. D, p. 15, and magazine supplement, p. 20.

5. The $120 billion estimate is from Battelle Memorial Institute, *An Analysis of Federal Incentives Used to Stimulate Energy Production* (Springfield, Va.: National Technical Information Service, March 1978). Although we have not yet analyzed the Battelle report in detail, two things seem clear: (1) The estimates will be controversial, and (2) federal subsidies to existing energy sources have been very substantial. The consumer subsidies result mostly at the expense of the owners of "old" oil and "old" gas, and from average pricing for electricity, which masks the incremental costs of electricity from new coal, nuclear, and hydroelectric power plants.

6. For an example of studies that assume away many of the problems that we discuss and regard as most important, see Energy Modeling Forum, *Coal in Transition: 1980–2000, EMF Report 2,* vol. 1, Stanford University, Energy Modeling Forum, July 1978. A series of models analyzed by this forum generally supported the conclusion that President Carter's goal of almost doubling coal production between 1977 and 1985 could be met. But the models specifically excluded a number of environmental, political, and productivity constraints that are important barriers.

Chapter 2
After the Peak: The Threat of Imported Oil

1. Not everyone agrees with the sentiment in this paragraph. Interviews with executives of the major international oil companies, in particular, reveal that some see less problems in relying on imported oil than I do.

Of course, if U.S. import controls had not existed, imports would have become significant a decade earlier. See *The Energy Industry: Organization and Public Policy* (Washington, D.C.: Government Printing Office, 1974).

2. For histories of the U.S. oil industry, see Alfred D. Chandler, Jr., *The Visible Hand* (Cambridge: Harvard University Press, 1977); Ralph W. Hidy and Muriel E. Hidy, *Pioneering in Big Business, 1882–1911* (New York: Harper and Row, 1955); Harold F. Williamson and Arnold R. Daum, *The American Petroleum Industry* (Evanston: Northwestern University Press, 1959); M. A. Adelman, *The World Petroleum Market* (Baltimore: Johns Hopkins University Press, 1972); Anthony Sampson, *The Seven Sisters* (New York: Viking, 1975); Benjamin Shwadran, *The Middle East, Oil and the Great Powers* (New York: John Wiley and Sons, 1973); Raymond Vernon, ed., *The Oil Crisis* (New York: W. W. Norton, 1976); William N. Greene, "Strategies of the Major Oil Companies," Unpublished Ph.D. thesis, Harvard University; Lorenzo Meyer, *Mexico and the U.S. in the Oil Controversy, 1917–1942* (Austin: University of Texas Press, 1977); Mira Wilkins, *The Emergence of Multinational Enterprise* (Cambridge: Harvard University Press, 1970) and *The Maturing of Multinational Enterprise* (Cambridge: Harvard University Press, 1974); U.S. Senate, *Hearings before the Subcommittee on Multinational Corporations of the Committee on Foreign Relations, Multinational Petroleum Companies,* 93 Cong., 2 sess., 1974 (Washington, D.C.:

Government Printing Office, 1975), hereafter referred to as *MNC Hearings;* and U.S. Senate, *Hearings before the Subcommittee on Multinational Corporations of the Committee on Foreign Relations, Multinational Petroleum Companies: Multinational Oil Corporations and U.S. Foreign Policy, Report Together with the Individual Views to the Committee on Foreign Relations,* 93 Cong., 2 sess. (Washington, D.C.: Government Printing Office, 1975), hereafter referred to as *MNC Report.*

3. For statistical sources, see various issues of the British Petroleum Company's *BP Statistical Review of the World Oil Industry;* of U.S. Bureau of Mines, *Minerals Yearbook* (Washington, D.C.: Government Printing Office); of Department of Energy, *Monthly Energy Review;* and of *International Petroleum Encyclopedia* (Tulsa: Petroleum Publishing). See also Energy Information Administration, *Annual Report to Congress, Volume II: Projections of Energy Supply and Demand and Their Impacts;* also *Volume II: Executive Summary; Volume III: Statistics and Trends of Energy Supply, Demand and Prices* (Washington, D.C.: Government Printing Office, April 1978); American Petroleum Institute, *Facts About Oil* (New York: API, 1977) and *Petroleum Facts and Figures* (Washington, D.C.: API, 1971); Leonard M. Fanning, *American Oil Operations Abroad* (New York: McGraw-Hill, 1947); and Central Intelligence Agency, *International Energy Statistical Review* (Washington, D.C.: CIA, April 19, 1978).

4. The change of the United States to a net import position in 1948 was, of course, recognized within the oil industry. The United States now is a permanent importer of oil in the sense of this being likely for the foreseeable future. The Cabinet (advisory) Committee on Energy Supplies and Resources Policy concluded in 1955 that if the ratio of oil imports to domestic production rose above 1954 levels (of about 10 percent), national defense would be endangered, although many observers believe that the politically powerful domestic oil producers played an important role in getting imports restricted. Voluntary controls were instituted in 1957, when the plans of sixty importing companies indicated that imports would reach 17 percent of domestic production for the last half of 1957. Mandatory controls were established in 1959. See Douglas R. Bohi and Milton Russell, *Limiting Oil Imports* (Baltimore: Johns Hopkins University Press, 1978), chaps. 2 and 3.

5. *BP Statistical Review of the World Oil Industry.* Consumption does not equal production plus imports, because exports are omitted. The U.S. Bureau of Mines reports consumption figures higher by 1 to 3 percent, presumably because of processing gains. The official statistics for 1979 do not include imports for the U.S. strategic reserve.

6. Adelman, *The World Petroleum Market.*

7. Between 1900 and World War I, U.S. companies purchased oil leases in Japan and Rumania, and explored for and found oil in Mexico; Hidy and Hidy, *Pioneering in Big Business,* pp. 498, 516–520; Wilkins, *The Maturing of Multinational Enterprise,* pp. 14–15. The Lansing quote is from *MNC Report,* p. 49; the quote by the director of the U.S. Geological Service is from *New York Times,* quoted in *MNC Report,* p. 33. The underlying motivations of the government and the companies, of course, were different. The U.S. government wanted oil in American hands because of security reasons; the companies wanted profits. Throughout this chapter, the companies' names are as of 1978. Of course, Exxon was at one time Standard Oil of New Jersey. For an explicit statement about the State Department's beliefs on the control of oil by U.S. companies, see *MNC Report,* p. 41.

The story that follows mirrors an important thread in history: the rise and fall of American oil power. Two other important threads also exist. First, the major oil companies continued to seek stability in the industry; for example, the big oil companies, Exxon and the foreign-based ones, agreed that non-U.S. market shares that each had in 1928 would be maintained in subsequent years more or less "as is."

Second, in keeping with the American sentiment for the division of power, the U.S. government typically adopted policies that would ensure the survival of the so-called "independent" oil companies. Thus, since the independents felt that they gained by the market stabilization agreement of 1928, the U.S. government made no effort to reject the agreement even though it went against its earlier Open Door policy. Furthermore, at times—especially when conditions were tranquil—the U.S. government seemed to ignore the world oil business. These two themes, along with the theme that I emphasize, are discussed in Raymond Vernon, "The Influence of the U.S. Government Upon Multinational Enterprises—The Case of Oil," in *The New Petroleum Order* (Quebec: Les Presses de l'Université Laval, 1976).

8. *MNC Report*, p. 41.

9. France eventually obtained a 6 percent interest in the Iranian consortium, as discussed later in the chapter. True, the Middle East nations did bargain with the oil companies, but which companies were allowed to begin the bargaining was often decided by the Great Powers and not the host governments (Kuwait is one example).

10. *MNC Hearings*, part 7, pp. 497–500; and Shwadran, *The Middle East, Oil and the Great Powers*, p. 409.

11. *MNC Hearings*, part 7, pp. 84–86.

12. The payments actually were based on pounds sterling—£50,000 and £10,000, respectively—and the £50,000 was an advance against future royalties. See Shwadran, *The Middle East, Oil and the Great Powers*, pp. 304–311. Socal originally operated in Saudi Arabia as owner of the California–Arabian Standard Oil Company, and it was this company in which Texaco bought half interest, in December 1936, for $3 million cash and a promise to pay $18 million out of the oil produced. Also in 1936, Socal put up Bahrain and Texaco put up its marketing facilities east of the Suez to form a fifty-fifty joint venture, the California Texas Oil Company (Caltex). *MNC Report*, p. 37, and Shwadran, pp. 303–312.

13. "Desperate" is a quote by the Socal vice-president in *MNC Report*, p. 37. "American hands" is a quote in Daniel Yergin, *Shattered Peace* (Boston: Houghton Mifflin, 1977), p. 446. In fact, the U.S. government attempted to buy 100 percent of Aramco in 1943, but Socal and Texaco refused to sell. See *MNC Report*, pp. 39–41.

14. This was the Red Line Agreement, so named because of a red line drawn on a map to indicate the territories in which the owners of IPC agreed not to explore on their own for oil. Saudi Arabia was within the red line. The owner of the French share of IPC and Calouste Gulbenkian, who owned 5 percent, sued Exxon and Mobil for violating the Red Line Agreement. The case was settled out of court. See *MNC Report*, pp. 50–55.

15. The State Department's suggestion was consistent with the American concern over the concentration of power and the resulting desire to ensure the survival of the independents; see Reference 7. *MNC Hearings*, part 8, pp. 72–76. The companies received a credit on U.S. tax bills for the taxes paid to Saudi Arabia. Thus, for every dollar paid in income taxes to Saudi Arabia, a dollar could be subtracted from income taxes owed the U.S. government.

16. The Cold War information is from Yergin, *Shattered Peace*, p. 179. This account of the Iranian nationalization crisis comes primarily from Robert B. Stobaugh, "The Evolution of Iranian Oil Policy, 1925–1975," in George Lenczowski, ed., *Iran Under the Pahlavis* (Stanford, Ca.: Hoover Institution, 1978). Also, see the histories in Reference 2. See Adelman, *World Petroleum Market*, chap. 3, for production statistics by company.

17. In a related development, on January 12, 1953, President Truman, acting on General Omar Bradley's assurance that national security called for the action, requested that the Justice Department terminate a grand jury investigation of the U.S. majors for possible criminal violation of the U.S. antitrust laws because of their alleged activities as part of a world oil cartel. Truman's instructions were in accordance with the recommendation of the Departments of State, Defense, and Interior, which had concluded in a report that any attack on the oil companies would be viewed in Europe and the Middle East as a fundamental attack on the whole American system and would pose great potential danger in both Venezuela and the Middle East, which were the only sources from which the free world's import requirements of oil could be supplied. Truman emphasized, however, that he wanted the case pursued vigorously in civil courts. This was done, and the case was settled years later by consent degrees. See *MNC Report*, pp. 60–74.

18. *MNC Hearings*, part 7, p. 297.

19. The Shah expressed his resentment of the market power of the consortium to me in an interview in April 1976. The quote is from *Wall Street Journal*, August 10, 1960, p. 2. See also Sampson, *The Seven Sisters*, pp. 156–158.

20. See Paul Swain, "Discounts Break Middle East Prices," *Oil and Gas Journal*, August 15, 1960, p. 84.

21. *New York Times*, August 10, 1960, p. 41.

22. The words are those of Senator J. W. Fullbright, quoted in Zuhayr Mikdashi, "The OPEC Process," in Vernon, ed., *The Oil Crisis*, p. 214. Some oil executives have told me that OPEC would likely have been formed eventually, and without the price cuts, as Europe and Japan became increasingly dependent on imported oil. But it is not surprising that a price cut triggered the move, because a threat to an established position can be an important galvanizer of action. See Raymond Vernon, *Sovereignty at Bay* (New York: Basic Books, 1971), p. 71.

23. From the author's interviews with Faud Rouhani in Tehran in April 1976 and with Perez Alfonzo in Caracas in February 1974. For a history of OPEC and its methods of operations, see Mikdashi, "The OPEC Process," in Vernon, ed., *The Oil Crisis*, and Mikdashi, *The Community of Oil Exporting Countries* (Ithaca: Cornell University Press, 1972). As of 1978, OPEC membership included Algeria, Ecuador, Gabon, Indonesia, Iran, Iraq, Kuwait, Libya, Nigeria, Qatar, Saudi Arabia, United Arab Emirates, and Venezuela. The reader will recall that BP and Royal Dutch/Shell were international majors along with the U.S. majors. The loss of control by the U.S. majors, of course, was accompanied by a loss of control by BP and Royal Dutch/Shell. See also U.S. Senate, Committee on Interior and Insular Affairs, *United States–OPEC Relations, Selected Materials*, 94 Cong., 2 sess. (Washington, D.C.: Government Printing Office, 1976); and Dankwart A. Rustow and John F. Mugno, *OPEC Success and Prospects* (New York: New York University Press, 1976).

24. The information on Libya is from *MNC Report*, pp. 121–140. The estimate of 1969 oil prices came from Adelman, *The World Petroleum Market*, pp. 183 and 191. Because of lower investment per barrel, however, the return on investment of the companies does not seem to have declined. See Citibank's *Energy Newsletter*, various issues. For a discussion of the difficulty in measuring return on investment, see Thomas R. Stauffer, "Measurement of Corporate Rates of Return and Marginal Efficiency of Captial," Unpublished Ph.D. thesis, Harvard University, 1971.

25. Some oil executives speculate that because Europe seemed to need Libya's oil, whereas Libya had several years of savings in banks, Qaddafi might have still been able to obtain higher prices even if there had been only majors in Libya. Of course, there is no way of determining what would have happened. The Occidental

executive is quoted in Daniel Yergin, "The One-Man Flying Multinational: Armand Hammer Wheels and Deals," *Atlantic Monthly*, June 1975, p. 32.

26. Descriptions of the events between 1970 and mid-1973 are given in a number of places. See, for example, chapters by Robert B. Stobaugh in Lenczowski, ed., *Iran Under the Pahlavis*, and in Vernon, ed., *The Oil Crisis*. Also see chapters in *The Oil Crisis* by Edith Penrose and Mira Wilkins; Sampson, *The Seven Sisters*, pp. 182–184; and *MNC Report*, pp. 121–140. U.S. oil imports, after increasing by 800,000 barrels daily between 1967 and 1970, increased by 2,835,000 barrels daily between 1970 and 1973.

27. Information about Aramco's cushion is from interviews with executives of major oil companies. The U.S. import quota system was abandoned during 1973, and U.S. imports rose 1.5 million barrels daily during 1973, compared with a rise of 800,000 barrels during 1972. For a discussion of market conditions in mid-1973, see Penrose, "The Development of Crisis," in Vernon, ed., *The Oil Crisis*.

28. *MNC Hearings*, part 7, p. 509.

29. For information on the embargo period, see Stobaugh, "The Oil Companies in Crisis," in Vernon, ed., *The Oil Crisis; and .MNC Report*, pp. 144–150.

30. Quoted from Stobaugh, "The Oil Companies in Crisis," in Vernon, ed., *The Oil Crisis*, p. 189. An intermediate meeting of OPEC in October agreed to a unilateral raise in prices that would raise the government income to $3.08 a barrel; but this raise was not agreed to by the companies. For a summary, see Lenczowski, "The Oil Producing Countries," *The Oil Crisis;* and *MNC Report*.

31. Some of the companies that bid very high prices eventually refused to take delivery of the oil and pay such prices. The quote by the refiner is in *Petroleum Intelligence Weekly*, December 3, 1973, p. 3. The majors, through their actions in allocating oil among nations so as to "equalize suffering" during the embargo, helped avoid an unmanageable crisis. See Stobaugh, "The Oil Companies in Crisis," in Vernon, ed., *The Oil Crisis*. The OPEC meeting in Tehran was reported in a number of places, including *New York Times*, December 24, 1973, p. 1. For a discussion of the realization that the problem was one of price, not supply, see *The Oil Crisis*.

32. Libya did not lift its 1973–74 embargo until later, nor was the embargo against countries other than the United States lifted until later. See Lenczowski, "The Oil-Producing Countries," and Appendix A in Vernon, ed., *The Oil Crisis*, pp. 67 and 284.

There was considerable confusion in the world oil market in 1974, and, in fact, the OPEC take never did settle at the $7.00-per-barrel figure announced for January 1, 1974. Negotiations during the year resulted in the following estimated prices —obtained from oil-company files—for Arab Light (also see Reference 26, Appendix):

Date	Average Acquisition Cost, Aramco Partners (including buy-back and equity oil)	Prices Paid by non-Aramco Companies to Saudi Arabia	Prices Paid by non-Aramco Companies to Aramco Companies	Spot Prices
1-1-74	$9.35	$10.83	—	$11.05 (1st Q. avg.)
4-1-74	9.35	10.83	$10.20 (2nd Q. avg.)	10.80 (2nd Q.)

Date	Average Acquisition Cost, Aramco Partners (including buy-back and equity oil)	Prices Paid by non-Aramco Companies to Saudi Arabia	Prices Paid by non-Aramco Companies to Aramco Companies	Spot Prices
7–1–74	9.45	10.46	10.40 (3rd Q.)	10.20 (3rd Q.)
11–1–74	10.24	10.46	10.30 (4th Q.)	10.40 (4th Q.)
1–1–75		10.46	—	
10–1–75		11.51	—	

Carter's toast, *Wall Street Journal*, January 4, 1979, p. 1; disappointments of oil producers about industrial development, Walter J. Levy, "The Years that the Locust Hath Eaten: Oil Policy and OPEC Development Prospects," *Foreign Affairs*, Winter 1978, pp. 287–305; Sheik of Kuwait, *New York Times*, March 11, 1979, p. F17.

33. Statistics on Saudi Arabia's shares of OPEC production and reserves are from the Department of Energy, *Monthly Energy Statistics*, various issues. "Ease of exploitation" refers to the low costs and very large output per well, and hence to the relatively few wells needed. In fact, the geological structure is quite complex, according to geologists I have interviewed. Population estimates vary widely, depending upon the source. Many experts believe that the official statistics overstate the population. The oil-reserves estimate is from Harold J. Haynes, chairman of Socal, who stated that "in each year of its operations, Aramco had added more liquid hydrocarbons to Saudi reserves than it has produced," reaching 151 billion barrels in 1976, in "The Changing Role of Multinational Oil Firms," *Oil and Gas Journal, Petroleum 2000: 75th Anniversary Issue*, August 1977, p. 500. For an explanation of some earlier confusion about a possible decline in Saudi reserves, see *Oil and Gas Journal*, February 14, 1977, p. 62.

Financial reserves figures are from Morgan Guaranty Trust, *World Financial Markets*, various issues. Interviews with U.S. officials and banks indicate that there are also substantial private holdings, perhaps $30 billion to $40 billion. The Department of Energy's *Monthly Energy Review* of August 1978 (p. 88) shows excess capacity held by OPEC nations, some of which has been shut in for internal political reasons and not for lack of a market. Kuwait is an example of a nation that shut in some of its capacity prior to 1973 when it could have sold more oil. The oil executive quote is from an interview with the author. From 1974 to 1975 production was reduced from 8.35 to 6.97 million barrels, or 17 percent, per day in Saudi Arabia, and from 6.06 to 5.39 million barrels, or 11 percent, per day in Iran; see *BP Statistical Review of the World Oil Industry*, 1976. On a monthly basis, the Saudi cut was as deep as 35 percent, falling from a peak of 9.05 million barrels daily in October 1974 to 5.92 million daily in April 1975; see *Petroleum Economist*, May 1976, p. 200.

The Yamani quote was in *New York Times*, December 22, 1977, p. 1. In fact, Saudi Arabia is the only nation that has the freedom to vary production over a sufficiently large range and long time to control prices, essentially. Inside OPEC, Iraq has the ability to expand production, but not enough to keep Saudi Arabia from dominance. Some analysts point out that although Saudi Arabia is clearly the leader within OPEC, it must be joined by others in adjusting output over the long run; see M. A. Adelman, "Constraints on the World Oil Monopoly Price," *Resources and Energy*, 1 (1978), pp. 1–39. Although Saudi Arabia (through OPEC) determines crude oil prices, the companies set product prices.

34. I am aware of eighty-one studies of the world oil outlook. Unless otherwise noted, my conclusion about the world oil outlook came from a review of the better-known ones listed below and from studies by some of the leading consulting firms. For a good summary of the best-known studies, see *Energy: An Uncertain Future*, prepared by Herman Franssen at the request of the Senate Committee on Energy and Natural Resources, publication no. 95–157, December 1978 (Washington: Governmnet Printing Office, 1978). Perhaps the most extensive study was the one by Workshop on Alternative Energy Strategies, *Energy: Global Prospects 1985–2000*, Sponsored by the Massachusetts Institute of Technology (New York: McGraw-Hill, 1977); other well-known studies include Organization for Economic Cooperation and Development, *World Energy Outlook* (Paris: OECD, 1977); Central Intelligence Agency, *The International Energy Situation: Outlook to 1985*, April 1977; U.S. International Trade Commission, *Factors Affecting World Petroleum Prices to 1985*, September 1977; and Petroleum Industry Research Foundation, *The Outlook for Oil to 1990 and After: Overview and Findings* (New York: PIRF, 1978). Other studies include Marcello Colitti, "The Energy Supply Problem Up to and After the Year 2000," Paper presented at the Second International Conference on Environmental Policy and the Fuel Crisis, Turin, April 26–29, 1977; A. Benard, "Prospects for Oil and Gas to the End of the Century," Address to the Commission of the European Communities, Brussels, November 29, 1977; and W. B. Davis, "The World Oil Outlook in 1978," Paper presented at the Young Presidents' Organization Energy Conference, Princeton University, May 11, 1978; Richard Nehring, "Giant Oil Fields and World Oil Resources," Prepared for the Central Intelligence Agency (Santa Monica, Calif.: The Rand Corporation, June 1978). Estimates for the Soviet Union and its allies (Eastern Europe, Mongolia, Cuba, and Vietnam) vary widely; for examples, see Central Intelligence Agency, *Prospects for Soviet Oil Production: A Supplemental Analysis*, ER77–10425, Washington, July 1977; and *International Herald Tribune*, September 14, 1978, p. 4. The North Sea estimate is from D.D.F. Laidlaw, a BP executive, in his "A Global View of the Prospects of the European Oil and Gas Industry in All Its Phases," Paper delivered at the European Petroleum and Gas Conference, Amsterdam, May 1978. Note that these studies consider the subject of "oil proliferation" —discoveries of new fields in non-OPEC areas.

35. See George W. Grayson, "Mexico's Opportunity: The Oil Boom," *Foreign Policy*, Winter 1977/1978, pp. 65–89; "Mexico: A Survey," *The Economist*, April 22, 1978; "Mexico's Reluctant Oil Boom," *Business Week*, January 15, 1979, pp. 64–74; National Economic Research Associates, *Mexico—Potential Petroleum Giant* (Washington, D.C., September 15, 1978); "Mexico: The Premier Oil Discovery in the Western Hemisphere," *Science*, December 22, 1978, pp. 1261–1265; Congressional Research Service, *Mexico's Oil and Gas Policy: An Analysis* (Washington: Government Printing Office, December 1978). Secretary of Energy Schlesinger predicted that Mexico could be producing 4 to 5 million barrels per day by 1985, and the CIA suggested that Mexico could produce 10 million barrels daily by 1990, but the only firm plans are to increase production from 1.5 million barrels daily (including one-half million barrels of exports) at the end of 1978 to 2.25 million at the end of 1989, of which 1.1 million would be exported; *Newsweek*, January 15, 1979, pp. 64, 65. Saudi Arabia's proven reserves of gas are about 16 billion barrels of oil equivalent; *Oil and Gas Journal*, December 25, 1978, p. 102. But any comparison of the ultimate size of Mexico's hydrocarbon reserves with those of Saudi Arabia is highly speculative, since estimates of potential reserves of Saudi Arabia have not been released.

36. See Reference 35. The estimates of Mexico's production and exports were obtained in a confidential interview from a knowledgeable source, but they should be regarded merely as educated speculation. Portillo's quote is from *New York Times*, January 6, 1978, p. 25. The Venezuelan tar sands also represent a potential

new source, but it apparently will be at least the end of the century before major exploitation is being done.

37. Central Intelligence Agency, *China: Oil Production Prospects* (Washington, D.C.: Library of Congress, 1977), p. 11. For a discussion of the theory supporting a gradual decline in the oligopolistic control of multinational corporations in raw-material industries and of the rise in the power of host governments, see Raymond Vernon, *Sovereignty at Bay,* chap. 2; Fariborz Ghadar, *The Evolution of OPEC Strategy* (Lexington, Mass.: Lexington Books, 1977); Daniel Fine, "Multinational Oil, The Energy Crisis, and Government Control in Kenya," Unpublished mimeograph paper, 1978; Irving Kuczynski, "British Off-Shore Oil and Gas Policy," Unpublished D.B.A. thesis, Harvard University, June 1978; Stobaugh, "The Evolution of Iranian Oil Policy," in Lenczowski, ed., *Iran Under the Pahlavis;* and Louis Turner, *Oil Companies in the International System* (London: George Allen and Unwin, 1978). For more optimistic reports, see Selig S. Harrison, "China: The Next Oil Giant, Time Bomb in East Asia," *Foreign Policy,* Fall 1975, p. 25; Choon-ho Park and Jerome Alan Cohen, "The Politics of the Oil Weapon," *Foreign Policy,* Fall 1975, pp. 33, 40; Wang Kung Ping, *The People's Republic of China—A New Industrial Power with a Strong Material Base* (Washington, D.C.: U.S. Bureau of Mines, 1975, p. 38); and *Review of Sino-Soviet Oil,* March 1978, p. 70.

38. For example, a princess was executed on orders of her grandfather for attempting to elope with a man not approved of by her family. The official charge against her was adultery, and her "husband" was stoned to death. See *New Orleans Times Picayune,* January 22, 1978, sec. 1, p. 17. The late King Abdul Aziz Ibn Saud is known in Saudi Arabia as King Abdul Aziz rather than King Ibn Saud, as he was called in the West. For a history of Saudi Arabia, see Charles M. Doughty, *Travels in Arabia Deserta* (New York: Random House, 1946); H. St. John B. Philby, *Saudi Arabia* (London: Been, 1955); Benoist-Mechin, *Ibn-Seoud: Le loup and le léopard* (Paris: Albin Michel, 1955); and for more leisurely reading, parts 4 and 5 of the *Aramco Handbook* (Dhahran: Arabian American Oil Company, 1968).

39. For an alternative view, which assumes that all OPEC nations are profit maximizers, see Basil Kalymon, "Economic Incentives in OPEC Oil Pricing Policy," *Journal of Development Economics* (December 1975), pp. 357–362; Nabil Chartouni, "Optimal Pricing/Investment Decisions for Natural Resources Production," Unpublished D.B.A. thesis, Harvard Business School, June 1978; and Robert A. Marshalla, "An Analysis of Cartelized Market Structures for Non-Renewable Resources," Unpublished doctoral thesis, Stanford University, August 1978. M. A. Adelman states, "Saudi Arabia will produce as much or as little oil as to maximize its revenues," but also recognizes that nations have political objectives. He maintains, however, that "political objectives are served perfectly by economic gain. There is no sacrifice or tradeoff of one for the other." Both of these quotes are in his "Need for Caution Over Prices," *Petroleum Economist,* September 1977, pp. 359–360. He also recognizes the need for current income; see his "Constraints on the World Oil Monopoly Price," *Resources and Energy.* For a discussion of how to make decisions to meet several goals simultaneously, see D. Bell, R. Kenney, and H. Raiffa, eds., *Conflicting Objectives in Decisions* (New York: John Wiley and Sons, 1977). For an explicit discussion of an example of France's not maximizing its income in order to meet international political goals, see *International Herald Tribune,* July 28, 1978, p. 4.

My view is consistent with Yamani's, who said, "The political aspect of the decision [to limit the price increase to 5 percent] was obvious to everybody concerned." Interview in *Al-Medina,* July 10, 1977. My view also is consistent with the fact that the Saudi government has been earning a negative profit (in real terms) on its investments outside the nation, and has a goal of minimizing what would probably be a steady real loss, whereas its leaders state that they believe that the price of oil will rise more rapidly than inflation. For the first point, see the statement of James Akins, former U.S. ambassador to Saudi Arabia, in a footnote in Theodore H. Moran, *Oil Prices and the Future of OPEC* (Washington: Resources for the Future, 1978), p. 19. For second point, see Yamani's statement in *Oil and Gas Journal,* July 3, 1978, p. 28.

40. Some experts on Saudi Arabia have said that King Khalid is enjoying his job more now than he initially did. The press carried a report that Prince Abdullah has asked for a meeting of the full family of perhaps 5,000 to decide the issue, but my interviews with persons who, I believe, have more reliable information indicate that this is not correct. See *The Economist,* July 16, 1977, p. 76, and September 17, 1977, p. 79.

41. For a report on the 1969 attempted coup, see *New York Times,* September 9, 1969, p. 1. The *Petroeconomic File* has contained reports of as many as six attempted coups in the last decade (see, for example, their October 1977 issue), but U.S. company and government officials with substantial experience in Saudi Arabia believe these reports to be highly exaggerated. See also Peter Mansfield, *The Arab World* (New York: Thomas Y. Crowell, 1976). Ian Smart believes that a replacement government might well have the same economic and political goals as the present government; see his "Patterns of Middle East Politics in the Coming Decade," in J. C. Hurewitz, ed., *Oil, the Arab-Israel Dispute, and the Industrial World* (Boulder, Colo.: Westview, 1976).

42. Given to me in interviews.

43. A Libyan cabinet officer told me the story about Qaddafi in Tripoli in 1974.

44. For an analysis of the U.S.S.R.'s aspirations, see John C. Campbell, "The Soviet Union in the Middle East," *The Middle East Journal,* 32 (Winter 1978). The remainder of the information in this paragraph came from interviews with U.S. and European government officials, and Arab individuals and government officials. For an account of a recent assassination of a president of North Yemen, see *New York Times,* June 25, 1978, p. 1.

45. From interviews, as in Reference 44. But recently, Iraq has seemed less pleased with its Soviet connection. For a broad discussion of Middle East politics, see Paul Y. Hammond and Sidney S. Alexander, eds., *Political Dynamics in the Middle East* (New York: American Elsevier, 1972); and Dale R. Tahtinen, *National Security Challenges to Saudi Arabia* (Washington: American Enterprise Institute, 1978).

46. The Shah had expected, for example, that by 1986 Iran would have a real per capita income equal to Canada's in 1976. (The Shah told me this in my interview with him in Tehran, April 1976.) Other information on Iran came from interviews with Iranian individuals and Iranian government officials. Some Saudi Arabian leaders told oil company executives that they were greatly concerned about the military build-up in Iran and found it inconceivable that such a build-up had anything to do with defending Iran against the Russians. Adham's interview is quoted in *MNC Hearings,* part 3, p. 507.

47. Richard C. Steadman, *Report to the Secretary of Defense on the National Military Command Structure,* Department of Defense, July 1978, p. 11.

48. The world-oil-outlook studies are listed in Reference 34.

49. The 60 percent and 230 percent increases in energy demand for the period 1977–2000 is based on the ratio between the annual rate of growth of energy usage to annual rate of growth of the world's economy, estimated to range between .82 and 1.12, and the annual rate of the world's economic growth, estimated to range between 2.5 and 4.5 percent; see studies in Reference 34.

50. *Wall Street Journal*, January 6, 1978, p. 22, and November 28, 1978, p. 2. As of January 1978, Aramco—Saudi Arabia's only producer other than in the Neutral Zone, which is owned jointly with Kuwait—was producing a little over 8 million barrels daily from only fifteen of Aramco's thirty-seven fields. An expansion to 14 million barrels daily would require very major capital expenditures. The estimate of 16 million to 20 million barrels daily is disputed by some in the oil industry, but it comes from discussions with petroleum engineers familiar with the Saudi Arabia oil fields.

51. For a description of Saudi relations with the U.S. government, see the discussion of the issues of selling jet fighters to Saudi Arabia in *Wall Street Journal*, March 13, 1978, p. 1. For a description of U.S. aid to Saudi Arabia, see press reports, such as *New Orleans Times Picayune*, February 9, 1978, sec. 2, p. 12.

52. Some observers have suggested that the U.S. government is encouraging Saudi Arabia to raise prices gradually so as to avoid a sharp upward break later; for speculation in the press on this point, see *Forbes*, March 20, 1978, pp. 31–32, and *Washington Post*, July 10, 1977, p. 1. But U.S. government officials have denied this in interviews with me. In fact, in 1978 the Saudis reduced their projected development plans; see *Wall Street Journal*, January 6, 1978, p. 22, and November 28, 1978, p. 2. Some authors have an opposite view and assume that Saudi Arabia will increase expenditures in geometric progression each year. For an example of an assumption of 50 percent annual increases at current prices in imports for the rest of the 1970's, see James Bedore and Louis Turner, "The Industrialization of the Middle Eastern Oil Producers," *The World Today*, September 1977, pp. 326–334. Also see Theodore H. Moran, "Why Oil Prices Go Up? The Future: OPEC Wants Them," *Foreign Policy*, Winter 1976–77, pp. 58–77, and his *Oil Prices and the Future of OPEC*.

53. From interviews with U.S. individuals and government officials knowledgeable about Saudi Arabia, and with Saudi government officials.

54. In economic terms, this represents a problem of income distribution across generations. For a discussion of the difficulties in handling the problem, see Robert M. Solow, "The Economics of Resources or the Resources of Economics," *American Economic Review*, 64 (May 1974), pp. 1–14.

55. The quote on profit-maximizing is from Adelman, "Need for Caution Over Prices," *Petroleum Economist*. Also see his "Constraints on the World Oil Monopoly Price," *Resources and Energy;* page 16 of this latter reference contains a scenario with a Saudi daily production of 20 million barrels. Both these articles state that current OPEC prices seem to be far below the level that would maximize OPEC revenues. The statement of the Department of Energy official is in *Forbes*, March 20, 1978, p. 32, in response to a suggestion that Saudi Arabia might produce 19 million barrels daily. A recent book disputes the widely held belief that firms maximize profits, stating instead that "firms set a 'comfortable' price in terms of their capacity to survive." The author believes that the fundamental difference between his theory and conventional theory is the fact that decisions are made by agents whose motivations differ from those of an ideal profit-maximizing firm. See Harvey Leibenstein, *Beyond Economic Man* (Cambridge: Harvard University Press, 1977), pp. 216, 271.

56. From confidential reports of a well-known oil consultant, a research institute, and two high-level U.S. government officials. This view also has been reflected in a number of studies in Reference 34.

57. Those believing in a gradual price increase generally place heavy emphasis on the effects of the economic growth of industrial nations on the demand for oil and hence on oil prices; see Adelman in References 2 and 39. Some observers say a fourth "accident" had occurred—the replacement of Libya's King Idris by Colonel

el-Qaddafi. They speculated that if Idris had remained in power, then U.S. political influence (exemplified by the Wheelus Air Force Base) would have deterred him from raising prices; see, for example, Dankwart A. Rustow, "Political Factors Affecting the Price and Availability of Oil in the 1980's," in Petroleum Industry Research Foundation, *Outlook for World Oil into the 21st Century* (Palo Alto, Ca.: Electric Power Research Institute, June 1978), app. A. Others say nonsense, because all OPEC nations always maximize their profits; thus Idris would have taken the same action as Qaddafi.

58. Daniel Yergin, "Killjoy of the Western World," *New Republic,* February 25, 1978, pp. 18–21; Yergin, "West Ready for Another Oil Embargo," *The Boston Globe,* March 21, 1978. There is no legislative authority, however, requiring the United States to live up to its commitments, and some government officials feel that the progress thus far has been "inexplicably slow" (private correspondence). A program to provide 500 million barrels of strategic storage (or two months' supply of 8 million barrels daily of imports) has been underway for some time. In mid-1978, the Department of Energy decided to double the capacity to a billion barrels by 1985; see its *Energy Insider,* July 10, 1978, p. 1.

59. William D. Nordhaus, "Energy and Economic Growth," in Hurewtiz, ed., *Oil, the Arab-Israel Dispute and the Industrial World,* p. 280. Also see estimates in Edward R. Fried and Charles L. Schultze, eds., *Higher Oil Prices and the World Economy* (Washington, D.C.: Brookings Institution, 1975), chaps. 1 and 2; and J.R.B. Associates, Inc., "Estimation of the Short-term Macroeconomic Impacts of Energy Price Changes on the U.S. Economy," Mimeograph report prepared as an account of work sponsored by the U.S. government and Brookhaven National Laboratory, McLean, Va., June 1978. The U.S. share of the economic loss almost surely exceeds $100 billion, with some estimates running into the hundreds of billions of dollars. One study reported that "the large increase in the price of energy in 1974 permanently reduced economic capacity, or the potential output of the U.S. economy, by four to five percent"; see Robert H. Rasche and John A. Tatom, "The Effects of the New Energy Regime on Economic Capacity, Production, and Prices," *Review of Federal Reserve Bank of St. Louis* (May 1977), p. 2. Analyses by the Wharton Econometric Forecasting Associates, Inc., show that the post-1972 increases in world oil price reduced real GNP (and real consumption) by 2 percent in 1978, with the reduction forecast to reach 5 percent by the mid-1980's; private correspondence December 11, 1978. Some economists have told me that economic losses could have been far fewer with proper management by the U.S. government, and that next time the United States might manage the problem much better. For a discussion of why a large change in the price of oil tends to retard economic growth and create inflation at the same time, and of the difficulty of avoiding these conditions, see Nicholas Kaldor, "Inflation and Recession in the World Economy," *The Economic Journal,* 86 (December 1976), pp. 703–714. For a lucid description of the difficulties in managing the U.S. economy, see Lester C. Thurow, "Economics 1977," *Daedalus,* Fall 1977, p. 80. He points out that "the economy is widely perceived as out of control."

60. See Charles Bartlett, "Dollar Decline Now is Nose of the Wolf," *New Orleans Times Picayune,* February 10, 1978, p. 12. Of course, factors in addition to oil imports are causing the U.S. balance-of-payments deficit, and the extent to which a balance-of-payments deficit represents a fundamental weakness in the U.S. dollar is questionable because of the build-up of U.S. assets abroad and other factors. See Raymond Vernon, "A Skeptic Looks at the Balance of Payments," *Foreign Policy,* Winter 1971–72. But it is clear that many people, including many current traders, *believe* that the deficits indicate a weak dollar, and this belief in turn helps *cause* a weak dollar. It is also clear that a number of U.S. industries are seeking greater import protection; see John Nevin, president of Zenith Corporation, "Can U.S. Business Survive our Japanese Trade Policy?" *Harvard Business Review,*

September/October 1978, pp. 165–177; and Bethlehem Steel's advertisement in *Newsweek*, October 2, 1978, following p. 90. Protectionist sentiments also are strong in other nations; see the Japanese example in Nevin, and the French, asking for "organized free trade," in *International Herald Tribune*, June 6, 1978, p. 6.

61. A discussion with almost any European political leader reveals the fears listed here, and by 1978 some Europeans were beginning to issue public warnings to the United States. See the statement by British Prime Minister Callaghan that reduction in U.S. oil imports was an "essential ingredient" for success at the forthcoming summit at Bonn in *International Herald Tribune*, June 7, 1978, p. 3. For a discussion of the European view, see Guy de Carmoy, *Energy for Europe: Economic and Political Implications* (Washington, D.C.: American Enterprise Institute, 1977); Horst Mendershausen, *Europe's Changing Energy Relations* (Santa Monica, Ca.: Rand Corporation, 1976); *International Energy Supply: A Perspective from the Industrial World*, Rockefeller Foundation, May 1978; Romono Prodi and Alberto Clo, "Europe," and Stobaugh's article in Vernon, ed., *The Oil Crisis;* and Louis Turner, "The European Community: Factors of Disintegration—Politics of the Energy Crisis," *International Affairs*, February 1978, pp. 39–49.

62. For example, see Petroleum Industry Research Foundation, *The Outlook for Oil to 1990 and After* (New York: PIRC, 1978).

63. The first two solutions are shown in numerous places. For examples, see statements by Senator Russell Long in the *National Journal*, November 5, 1977, p. 1717; Texaco senior vice-president Alfred C. DeCrane, Jr., "United States Energy Policy, Programs, and Prospects: An Appraisal," Speech, American Bar Association, August 8, 1978; and investment banker John A. Hill, "In Fact, We Do Have an Energy Policy," *New York Times*, May 28, 1978, sec. 3, p. 12. The last-mentioned solution—divestiture—is in John M. Blair, *The Control of Oil* (New York: Pantheon, 1976). A central theme of Blair's book is the alleged withholding of oil production, both domestically and abroad, in order to keep prices high. The estimate of an OPEC-induced increase of $400 billion is in 1978 dollars, and is based on 38.6 billion barrels of proved reserves of crude oil and natural gas liquids existing at year-end 1974 and an assumed increase of $10 a barrel; for estimates of reserves, see Energy Information Administration, *Annual Report to Congress, Volume III, 1977* (Washington, D.C.: Government Printing Office, May 1978), p. 61. As discussed in Chapter 3, the value of U.S. natural gas reserves also increased in potential value by approximately $400 billion.

64. In fact, this is an indeterminate issue. But what does appear to be true is that the supply elasticity in some major oil areas of the world can be quite low with respect to price. A recent estimate indicates that a 33-percent higher price for North Sea oil would result in increased output of 3.2 and 2.3 percent in 1980 and 1985, respectively. See Paul L. Eckbo, Henry D. Jacoby, and James L. Smith, "Oil Supply Forecasting: A Disaggregated Process Approach," *Bell Journal of Economics* (Spring 1978), p. 233. Given that it was neither politically nor economically desirable to allow an immediate jump in domestic oil prices to world levels in 1973–74, a relatively complicated system of controls was necessary to prevent widespread economic dislocations within the industry. Not only were regulations apparently confusing, but some companies stated that the Department of Energy was basing claims on a retroactive interpretation of government rules that differed from the interpretation of rules in effect at the time of the transactions in question. The matter is being litigated in the courts. See Columbia Broadcasting System, News Release, June 17, 1978; *Wall Street Journal*, June 13, 1978, p. 1; and *New York Times*, January 6, 1978, p. 1. The reader should not confuse the cases of retroactive interpretation of rules with the alleged cases of oil-price manipulation by middleman firms; see *Wall Street Journal*, September 22, 1978, p. 1. According to oil industry executives I have interviewed, even if controls were abandoned, the mere fact that controls have been used in the past creates a fear

that they will be used in the future if profits appear too large to the public and politicians; this fact lessens the amount of political security perceived to exist in the United States. This, then, reduces exploration below the level where it would otherwise be at the level of the uncontrolled prices. Of course, if oil companies believed that controls would be eased or removed, thereby allowing prices to move toward or to the world level, then exploration would tend to be higher than it would otherwise be at the level of the controlled price.

65. This conclusion about U.S. production levels does not imply that oil supply is insensitive to price, for without substantially higher prices, oil output would decline because of normal annual depletion. Earlier projections indicated much higher domestic production rates, but more recent estimates approximate 10 mbd. See 1978 Shell and 1978 Exxon estimates in *Energy: An Uncertain Future,* pp. 229, 239. For years the oil companies have been anxious to explore these virgin territories, the property of the federal government. Until a Supreme Court decision in early 1978, lawsuits had prevented drilling on the only East Coast outer continental shelf territory for which leases had been granted (off South Carolina, in August 1976); see *Wall Street Journal,* January 27, 1978, p. 17. And in early 1978 lawsuits brought by the state of Massachusetts and groups of environmentalists and fishermen asking for additional safeguards for the environment and fishing industry prevented the Department of the Interior from granting leases to oil companies to explore on the outer continental shelf off New England. Congress might well pass legislation providing for additional environmental safeguards. The result would be extensive exploration of offshore territories, especially off the East Coast, although the new environmental safeguards might slow the ongoing exploration in the Gulf of Mexico.

66. The U.S. production estimates are based on interviews with company officials and on analysis of literature in Reference 34. The number of wells drilled is from John P. Henry, Jr., "Energy: A World Perspective," Paper presented to the Stanford Research Institute Council, March 3 and 4, 1977, fig. 8. The quote of the oil executive is from an interview with the author.

67. Conventional, or secondary, recovery consists of allowing the natural pressure in the oil field to force the oil out, then installing pumps at the bottom of the well to pump the oil out, and finally, injecting high-pressure natural gas or water into the field to force the oil out.
Even an increase in production from enhanced recovery would require higher oil prices; see U.S. Congress, Office of Technology Assessment, *Enhanced Oil Recovery Potential in the United States* (Washington, D.C.: Government Printing Office, 1978), pp. 7, 19. Also see National Petroleum Council, *Enhanced Oil Recovery* (Washington, D.C.: NPC, 1976).

68. Guy Elliott Mitchell, "Billions of Barrels of Oil Locked Up in Rocks," *The National Geographic Magazine,* February 1918, p. 205.

69. Some observers believe, however, that it is economic at present prices. Occidental Petroleum Chairman Dr. Armand Hammer has stated that his company would earn a 15-percent return on Occidental's investment in a shale-oil project in Colorado, based on the U.S. price paid for OPEC oil in early 1978. Some other companies—Cities Service and Continental Oil, for example—think that Dr. Hammer's estimates of required price are too low. The latter re-estimates Occidental's costs as $16 to $26 a barrel. A shale-oil expert, John J. McKetta, professor of chemical engineering at the University of Texas, estimates costs of about $15 a barrel. See *Business Week,* January 30, 1978, p. 55. A recent Rand study indicates required prices of $23 to $29 per barrel in early 1979 prices for surface retorting. Edward W. Merrow, *Constraints on the Commercialization of Shale Oil* (Santa Monica, Calif.: Rand, September 1978), p. vii. For earlier estimates of the production costs of shale oil, see Federal Energy Administration, *Project Independ-*

ence, Potential Future Role of Oil Shale: Prospects and Constraints, U.S. Department of Interior, November 1974 (Washington, D.C.: Government Printing Office, 1974); Other Energy Resources Subcommittee of the National Petroleum Council's Committee on U.S. Energy Outlook, *An Initial Appraisal by the Oil Shale Task Group: 1971–1985* (Washington, D.C.: NPC, 1972); and various issues of *Oil and Gas Journal, Chemical Engineering,* and *Chemcial and Engineering News.*

70. Although I judge that shale oil is unlikely to make a major contribution to the U.S. supply of oil by the turn of the century, individual companies might earn a good profit from it.

71. For information on the finding of reserves in the last thirty years, see *Energy: An Uncertain Future,* p. 58. These rankings are from *Fortune,* May 8, 1978, p. 240. The rankings of the oil companies would be a little lower if ranked on profits rather than sales, but the point still remains: They are big companies.

72. This is my conclusion after studying the available literature. For those sources tending to support divestiture, see Fred Allvine and James M. Patterson, *Competition, Ltd.: The Marketing of Gasoline* (Bloomington: Indiana University Press, 1972); Blair, *The Control of Oil;* Paul Davidson's statement in *Chemical Engineering Progress,* 73 (April 1977), pp. 14–32; Paul Davidson, "Divestiture and the Economics of Energy Supplies," and Walter S. Measday, "Feasibility of Petroleum Industry Divestiture," in David J. Teece, ed., *R&D in Energy: Implications of Petroleum Industry Reorganization* (Stanford: Institute for Energy Studies, Stanford University, 1977); and Federal Trade Commission proposal to break up oil companies. For those tending not to support divestiture, see Treasury Department analysis of break-up of large oil companies, *Implications of Divestiture* (Washington, D.C.: Superintendent of Documents, 1976); Donald C. Baeder, "Divestiture and the Impact on Pioneering Research in Energy," Thomas Baron, "Consequences of Divestiture for R&D and the Development of Alternative Energy Sources," F. A. L. Holloway, "A View of the Effect of Oil Industry Divestiture on Science and Technology," and David J. Teece and Henry O. Armour, "Innovation and Divestiture in the U.S. Oil Industry," all in Teece, ed., *R&D in Energy;* Edward J. Mitchell, ed., *Vertical Integration in the Oil Industry* (Washington, D.C.: American Enterprise Institute, 1976); David J. Teece, *Vertical Integration and Vertical Divestiture in the U.S. Oil Industry* (Stanford: Institute for Energy Studies, Stanford University, 1976); Frank N. Trager, ed., *Oil, Divestiture and National Security* (New York: Crane, Russak, 1977); W. A. Johnson, R. E. Messick, S. Van Vactor, and F. R. Wyant, *Competition in the Oil Industry,* Energy Policy Research Project, George Washington University, 1976; Edward W. Erickson, "The Energy Crisis and the Oil Industry," Statement before the U.S. Senate, Committee on the Judiciary, Subcommittee on Antitrust and Monopoly, January 27, 1976; Barbara Hobbie and Richard Mancke, "Oil Monopoly Divestiture: A Clash of Media Versus Expert Perceptions," *Energy Policy,* September 1977; Exxon, "Competition in the Petroleum Industry," Submission before the Senate Judiciary Subcommittee on Antitrust and Monopoly, January 21, 1975. The estimate of the tank of gasoline per capita is about one-half the extreme estimate appearing in print; an executive of one major estimated a loss (with no strong evidence for the statement) in GNP of $6 billion yearly (or about 2½ cents per gallon of oil used). On the other hand, independent refiners have estimated that, based on their experience, they can compete with the majors when their crude oil costs do not exceed the major's costs by 50 cents to $1 a barrel (or about 1¼ to 2½ cents a gallon), thereby implying a greater efficiency than that of the majors; see Robert Yancey, president of Ashland Oil, in U.S. Congress, Senate, *Hearings Before the Subcommittee on Antitrust and Monopoly of the Committee on the Judiciary of United States Senate, S.1167, Part 8,* August 6–9, 1974, p. 5918. For other testimony on this subject, see *S.1167, Part 9,* January 21, 22 and 30, 1975; *S.2387* and related bills *S.739, S.745, S.756, S.1137,* and *S.1138* in

Part 1, September 23, 26, October 29, 31, November 12, 19, 1975, and *Part 3*, January 21, 22, 27, 28, 30, and February 3, 18, 1976. The text discusses divestiture from the standpoint of domestic supply. There are also arguments in the above sources about the extent to which the majors help OPEC maintain price; but as indicated in the text, I believe Saudi Arabia's position is by far the dominant influence.

73. Quote from Trager, ed., *Oil, Divestiture, and National Security*, back cover. Some members of the financial community suggest that the shareholders of at least some of the majors would be financially better off if management dismantled the company by selling the individual operations. As one analyst wrote, "As preposterous as it sounds, the idea of voluntary self-liquidation by Exxon Corporation is not without economic and strategic logic insofar as the shareholders' interest is concerned." (William L. Randall, "A Case Study in the Limits to Corporation Growth," Pamphlet published by Blyth Eastman Dillon and Company, June 1978, p. 6.) John S. Herold, Incorporated, in *Petroleum Outlook*, June 1978, reports that Exxon shares were selling at a 46-percent discount from the value as appraised by Herold. Many other oil companies, including a number of majors, were selling at comparable discounts. But I doubt whether purchasers could absorb a number of liquidations simultaneously without a substantial fall in value, although the shares of the individual operations could be spun off to existing shareholders.

74. A number of polls have shown that Americans place heavy blame for energy problems on the oil companies. For evidence that a positive noneconomic effect could result just from the mere fact of breaking up a concentration of power, see Richard Hofstadter, *The Paranoid Style in American Politics and Other Essays* (New York: Alfred A. Knopf, 1966), p. 205; R. A. Dahl, *Pluralist Democracy in the United States: Conflict and Consent* (Chicago: Rand McNally, 1967), p. 24; S. P. Huntington, "Political Modernization: America vs. Europe," *World Politics* (April 1966), pp. 378–414; Bernard Bailyn, *Ideological Origins of the American Revolution* (Cambridge: Harvard University Press, 1967), pp. 53–93, 273–301; and Raymond Vernon, "The Influence of the U.S. Government Upon Multinational Enterprises: The Case of Oil," in *The New Petroleum Order* (Quebec: Les Presses de l'Université Laval, 1976). For literature stressing the benefits of more equal income distribution, see Tibor Scitovsky, *The Joyless Economy* (New York: Oxford University Press, 1976); and Fred Hirsch, *Social Limits to Growth* (Cambridge: Harvard University Press, 1978). Those readers doubting that this is a serious issue should read the description by Daniel Yergin of a meeting in the United Kingdom in "The Great Fritter Debate," *New Republic*, August 5–12, 1978, pp. 16–17.

75. The $15 billion estimate is based on the difference between the price of imported crude oil acquired by refiners ($14.53 in 1977) and the price for domestic oil ($9.55) times the annual production of domestic oil (8.2 million barrels daily of crude oil, ignoring 2.0 million barrels of natural gas liquids), or $4.98 × 8.2 × 365 = $14.9 billion. Consumption data is from the Department of Energy, *Monthly Energy Review*, various issues.

76. U.S. imports of crude oil, products, and natural gas liquids were 9 million barrels daily in 1977. As Alaskan production began in late 1977, imports declined by about a million barrels daily but began to climb again in mid-1978; see Department of Energy, *Monthly Energy Review*. There are likely to be imports in the region of 9 million barrels daily in 1979. An Administration official was quoted as saying, "We expect oil prices to level out between $4 and $6 a barrel"; see *New York Times*, November 13, 1974, p. 1. Also see Federal Energy Administration, *Project Independence*. The price of newly found oil, adjusted by the GNP deflator to correct to 1973 dollars, was $9.94 in 1975, $9.16 in 1976, $8.22 in 1977, and about $8.26 in 1978; see Federal Energy Administration, *Monthly Energy Review*,

July 1978, p. 61; and *International Financial Statistics,* July 1978, pp. 380–381. These prices were far above the $3.89 that existed in 1973, or the $3.60 in 1972 (in 1973 dollars); see Federal Energy Administration, *Annual Report to Congress,* p. 39.

77. See Reference 58. Anthony Copp, vice-president, Energy and Resource Development Group, Salomon Brothers, estimates that the investment cost of the oil storage for the United States for 500 million barrels will be some $10.7 billion, or almost 35 percent greater than anticipated in 1977 by the Federal Energy Administration; see his "Strategic Oil Storage Requirements and Costs for the United States: Some Economic Observations," Unpublished mimeograph paper, April 14, 1977. Dr. Copp says that is "peanuts" compared with the possible damage to the U.S. economy of an Arab oil boycott. To obtain my estimate of $2 billion annually, I assumed U.S. imports of 9 million barrels daily, assumed the cost of capital to the U.S. government of about 10 percent, and ignored annual operating costs, which are very minor. M. A. Adelman had recommended strategic storage for the Eastern Hemisphere noncommunist countries in 1967; see his "Security of Eastern Hemisphere Fuel Supply," Unpublished mimeograph paper M.I.T. Economics Department, December 1967.

78. Refer to Stobaugh, "Comments on Recovery and Beyond," Interview on energy outlook in *Saturday Review,* July 12, 1975, pp. 28–30; "Gas Rationing is No Panacea," *Wall Street Journal,* January 4, 1974, p. 4; "The Hard Choice on Energy," *Wall Street Journal,* December 9, 1974, p. 22; and "For Oil and Gas Compromises," *New York Times,* December 2, 1977, p. A27.

79. For example, Richard N. Cooper, under-secretary of state, stated that "reducing energy demand . . . should be judged by a test of cost-effectiveness" without giving any indication that the world price of oil does not reflect the risks associated with its use; see *Department of State Bulletin,* February 1978, p. 29. Many opponents of the Carter Administration's energy plan hold, of course, that the world price of oil is too high for the United States.

80. For a discussion of externalities and social costs, see Reference 6, Chapter 8.

81. Although 1978 imports might be closer to 8 million barrels daily than to 9 million, in 1979 U.S. imports are likely to be in the region of 9 million. Other than our efforts at Harvard Business School, the only attempt that I have seen to quantify the incremental costs of imported oil are in a series of unpublished memoranda by Harvard economist Thomas Schelling as part of his activities as a member of the project entitled "Energy: The Next Twenty Years," of the Resources for the Future. (I also am a member of this project.) I am grateful for discussions with Schelling and other members of the RFF project, although I am solely responsible for my estimates.

82. A recent OECD study assumes a positive and direct relationship between U.S. oil imports and other OECD oil imports, with a variation in U.S. imports of 8 million barrels daily being accompanied by a variation in the remainder of OECD of 7 million daily. See OECD, *World Energy Outlook.*

83. If it is assumed that the 5 mbd of imports are before the effects of the price, rather than after, as I assume here, the cost of marginal imports still exceeds $30 a barrel. Similarly, other reasonable assumptions, such as a price of $20 a barrel moving to, say, $26, also would result in costs of marginal oil of more than $30. Some sets of plausible assumptions show costs as high as $80 a barrel. The assumptions in the text are consistent with estimates by consultants and by officials of the U.S. government, primarily in Departments of Energy and State. For example, one official estimated a 50-percent probability of a doubling of price in real terms between 1978 and 1985. The printed record of the expected increase is not quite so clear as to whether the expected doubling is in real terms or in

current dollars (that is, not corrected for inflation). For examples, see *Capital Energy Letter, Special Supplement,* March 6, 1978; *Forbes,* March 20, 1978, pp. 31–32; *Washington Post,* July 10, 1977, p. 1; and Richard N. Cooper's statement in *Department of State Bulletin,* February 1978, pp. 26–29. Congress was being told by its Office of Technology that by 1985 imported oil may cost about thirty dollars a barrel; see *Chemical Week,* November 2, 1977, p. 17. For ease in exposition, I ignored the effect of the higher prices on U.S. demand, the small element of price (perhaps 2 percent) representing profits earned by U.S. companies that might be returned to the United States, and the fact that transportation costs, which are about 10 percent of total costs, might not increase proportionately, because only a proportion of transport costs are fuel-related. Any error caused by these assumptions would not affect the basic conclusion that incremental oil imports might be very costly to the United States. Readers interested in the price elasticity of demand inherent in these calculations can write me for a working paper.

84. See References 59 and 60. The exact amount of lost economic activity would depend on (1) whether oil prices rose gradually throughout the period, rose in several larger steps, or rose in one giant step reminiscent of 1973–74; (2) the fiscal and monetary responses of other governments to higher oil prices; and (3) the fiscal and monetary response of the U.S. government to (1) and (2). I am grateful for discussions with Professor Francis Bator on this subject, although I am solely responsible for my estimates.

85. The Japanese are especially dependent on imported oil, having less domestic coal than the European countries. The Japanese (uncharacteristically) leveled their sharpest criticism at the United States since World War II, expressing "severe disappointment" with U.S. leadership; see *International Herald Tribune,* July 11, 1978, p. 7; *The Economist,* July 8, 1978, p. 67.

86. Although these costs can be thought of as "external costs" in the sense of the phrase as used here, economists sometimes refer to direct payments for the oil as pecuniary costs.

87. Estimating the likelihood of different events occurring during different time periods, the costs associated with the events and the risk-averseness of U.S. society as a whole are not easy tasks; neither is estimating an appropriate discount rate for consumption for society as a whole. See Solow, "The Economics of Resources or the Resources of Economics," *The American Economic Review.*

88. This assumes, of course, that the consumer is basing investment decisions on current costs—an assumption in line with analyses that I have seen on conservation and solar energy. Companies, on the other hand, are likely to be assuming some rise in costs, but nothing that can approach the estimate of the incremental costs to the nation of imported oil. For discussions of why innumerable small decisions might result in a different outcome for society than the decision-makers would want, see Thomas Schelling, "On the Ecology of Micromotives," in Robin Marris, ed., *The Corporate Society* (London: Macmillan, 1974), pp. 19–64; and Alfred E. Kahn, "The Tyranny of Small Decisions: Market Failures, Imperfections, and the Limits of Economics," *Kyklos* (1966), pp. 23–46.

89. This seems obvious, since oil prices in the domestic market are being held below world price, by approximately $15 million yearly.

90. Many observers, including some environmental experts with whom I have spoken at universities, believe that current environmental rules are satisfactory. Also see Reference 66.

91. In theory, new oil should be priced substantially above the deregulated price (that is, substantially above the imported price) in order to give domestic oil an equal chance with imported oil. I judge that this would not be acceptable politically.

(See Reference 89.) It must be recognized that many observers question whether even world prices, much less a financial incentive in addition to world prices, would be justified by the incremental domestic energy forthcoming. This is especially so given the lack of success in keeping U.S. oil and gas reserves from declining even after the dramatic price jumps that have occurred since 1973.

92. Payments from one region to another are also involved, thereby presenting another political problem to be solved in any elimination of price controls. For testimony suggesting an investigation of the antitrust implications of the majors' controlling vast oil supplies abroad in joint ventures, see Robert B. Stobaugh, *MNC Hearings*, part 9, July 25, 1974, p. 191. The bankruptcy of Commonwealth Oil Refining Company in Puerto Rico, involving hundreds of millions of dollars, is an example of what can occur when changes in competitive positions are caused by changes in the U.S. government's oil import program. (An earlier advantage granted by the U.S. government disappeared, only to be replaced by a disadvantage.) Other causes were also at work. See *Wall Street Journal*, various issues in 1977 and early 1978, for further details. For an indication of the concern of some of the oil companies that are smaller and less vertically integrated than the majors, see the letter of March 22, 1978, from Clyde A. Wheeler, Jr., Sun Company, to the Honorable John F. O'Leary, deputy secretary, Department of Energy, endorsed by twenty companies, two industry associations, and two labor unions.

Chapter 3
Natural Gas: How to Slice a Shrinking Pie

1. U.S. Department of Energy, *Monthly Energy Review*, November 1978, pp. 4 and 32. Congressional staff member interview by authors. For background on regulation, see Stephen G. Breyer and Paul W. MacAvoy, "Regulating Natural Gas Producers," in Robert J. Kalter and William A. Vogely, eds., *Energy Supply and Government Policy* (Ithaca: Cornell University Press, 1976), pp. 161–192. See also Breyer and MacAvoy, *Energy Regulation by the Federal Power Commission* (Washington, D.C.: The Brookings Institution, 1974); and U.S. Senate, Committee on Interior and Insular Affairs, *Natural Gas Policy Issues*, 92 Cong., 2 sess. (Washington, D.C.: Government Printing Office, March 1972).

2. Newspapers carried reports of $100 billion at stake in the debate over decontrol of natural gas prices in 1978. Our estimate of $400 billion is based on a notion of economic values of natural gas at world energy prices rather than value with prices controlled by law or long-term contracts. For example, U.S. proved reserves of natural gas were 250 trillion cubic feet (tcf), or 44 billion barrels of crude oil equivalent, at the end of 1973. The OPEC price rises of 1973–74 increased the world oil prices, in round numbers, by about $10 a barrel (in 1978 dollars). Thus, the value of U.S. proved reserves of natural gas increased, based on a world oil price, by some $400 billion. This has not been corrected by discounting future cash flows to obtain present value, nor does it consider enlargement of reserves. See Department of Energy, Energy Information Administration, *Annual Report to Congress, Volume III, 1977* (Washington, D.C.: Government Printing Office, May 1978), p. 61. A second way to view the potential magnitude of the transfer of wealth to producers from total deregulation of all natural gas is to compare the revenues received by producers in 1977 for interstate gas with what they would have received if interstate gas had been priced the same as intrastate gas—apparently 27 cents per mcf higher (Reference 10, this chapter). So the apparent price subsidy was about $3.5 billion (or about 13 tcf × $.27 per mcf × billion mcf/tcf). Some industry observers say, however, that the mere existence of the controlled inter-

state market caused intrastate prices to be lower than they otherwise would have been. Also, average prices in the intrastate market are held down by long-term contracts that will gradually expire. Thus, another way to view the issue is to compare the actual prices received in 1977 for all gas—interstate and intrastate—with the revenues if natural gas prices had been equal to world oil prices for comparable heat contents. (Implicit in this comparison is an assumption of deregulation of domestic oil prices as well as the renegotiation of existing gas contracts upward to the equivalent value of world oil.) Since approximately 5,800 cubic feet of gas equals the heat content of a barrel of oil, if oil sells for $14.50 per barrel delivered to the United States, natural gas would sell for $2.50 per thousand cubic feet (mcf). (For oil prices, see *Monthly Energy Review*, November 1978, p. 58.) In 1977, the average wellhead price of all U.S. gas was $.78 per mcf (*Annual Report to Congress*, p. 59). Total volume of gas consumed in the United States was 19.5 tcf (Reference 1). Therefore, the difference between U.S. price and a price equivalent to world oil was 19.5 × ($2.50 − .78), or $34 billion each year. We recognize that the economic value of gas, when compared with the price of world oil, would be more than this in some cases and less in others, depending upon a number of factors, including the comparative costs of delivering gas versus oil products, the particular oil product in question, and the premium value of gas because of its cleanliness. Regardless of assumptions, the main conclusion is clear: U.S. consumers are receiving a large subsidy, at the expense of producers, in using natural gas compared with the world price of oil.

3. Proved reserves were 293 tcf at the end of 1967 and 216 tcf at the end of 1976; see *Annual Report to Congress*, pp. 53, 61. For information about possible U.S. production rates, see Paul W. MacAvoy and Robert S. Pindyck, *Price Controls and the Natural Gas Shortage*, National Energy Study 7 (Washington, D.C.: American Enterprise Institute for Public Policy Research, 1975), pp. 4 and 5; and Federal Power Commission, Bureau of Natural Gas, *Natural Gas Supply and Demand, 1971–1900* (Washington, D.C.: FPC, February 1972), p. 3; and U.S. Senate, Committee on Interior and Insular Affairs, *Natural Gas Policy Issues and Options, A Staff Analysis*, 93 Cong., 1 sess. (Washington, D.C.: Government Printing Office, 1973). But even on the decline of reserves, some critics dispute published estimates. See Bethany Weidner, "What Natural Gas Shortage," *The Progressive*, April 1977, pp. 19–23.

4. This account of the history of the American manufactured gas business draws heavily on Paul J. Garfield and Wallace F. Lovejoy, *Public Utility Economics* (Englewood Cliffs, N.J.: Prentice Hall, 1964), chap. 4.

5. National Economic Research Associates, "The Natural Gas Industry: An Overview," Aide memoir submitted to the U.S. Senate, Committee on Energy and Natural Resources, June 16, 1977; "The Natural Gas Pipeline Industry," in Salomon Brothers' *A Quarterly Review*, March 29, 1977; U.S. Congress, Committee on Commerce Hearings, U.S. Senate, *Hearings on Interstate Pipelines and Transmission Companies*, 94 Cong. (Washington, D.C.: Government Printing Office, 1975); James T. Jensen, *Energy Company Strategies for the 1970s* (Cambridge, Mass.: Arthur D. Little, 1969). Useful articles on various aspects of the natural gas industry appear regularly in the *Bell Journal of Economics*, published in New York. Additional industry data appear regularly in the *Oil and Gas Journal*, published in Tulsa, Oklahoma.

6. Garfield and Lovejoy, *Public Utility Economics*, chap. 4. While this chapter primarily focuses on price control, regulations also include certification and dedication of reserves.

7. *Phillips Petroleum Co.* v. *Wisconsin*, 347 U.S. 622, 1954. The 1954 decision traced back to the 1877 decision in *Munn* v. *Illinois*, which allowed state legislatures to fix the maximum price for grain storage, and to *Nebbia* v. *New York* in

1934, which allowed legislatures to impose price controls on any business within their jurisdiction, "where in their judgment it would serve the public interest, provided only that they did not do so in an utterly capricious or discriminatory manner." Quoted in Garfield and Lovejoy, *Public Utilities Economics*, p. 113.

8. *New York Times*, February 7, 1956, pp. 1, 22, and February 18, 1956, p. 1.

9. Breyer and MacAvoy, in *Energy Supply and Government Policy*, pp. 161–192; and MacAvoy and Pindyck, *Price Controls and the Natural Gas Shortage*, p. 12. Most observers believe that this decision helped speed the development of the transcontinental pipelines because it facilitated financing of the lines by ensuring them of low-priced, long-term supplies (dedicated reserves were an item regulated), just as the 1938 Act had helped their development by giving them public-utility status. In doing so, the Federal Power Commission used various formulas. During the first few years after the 1954 Supreme Court decision, the FPC tried to control wellhead prices for individual producing companies in much the same way that individual electric utility companies have been controlled—with a price intended to cover the costs of production, including an adequate return on investment. The large numbers of producers and gas fields led to an administrative nightmare. By the late 1950's, the commission had become inundated with some 2,900 applications for price review—a logjam that would have taken decades to free. In 1961 the agency switched to area rate regulation. On the basis of historical average production costs in five gas-producing regions, the FPC set a ceiling price for gas within each one. Although the switch to area rates hardly settled the question of what a "just and reasonable" price for gas should be, the technique did prove administratively workable. See Breyer and MacAvoy, *Energy Regulation by the Federal Power Commission*, p. 174.

10. It is possible to calculate that the difference was about 27 cents per mcf in 1977. The 27 cents represents the difference between the average interstate price of 69 cents per mcf in Table 3–1 and an estimated average price of 96 cents for intrastate gas, which is the price intrastate gas had to sell at in order for the average price of all gas to be 78 cents. Interstate is approximately two thirds and intrastate one third of the total $[(\frac{1}{3} \times 96) + (\frac{2}{3} \times 69) = 78]$, which is reported in *Annual Report to Congress, vol. III*, p. 59.

11. Subcommittee on Energy and Power, Committee on Interstate and Foreign Commerce, House of Representatives, *Long-Term Natural Gas Legislation*, 94 Cong., 2 sess. (Washington, D.C.: Government Printing Office, 1976), p. 469.

12. Federal Energy Administration, *Annual Report to Congress, Volume II, 1977: Projections of Energy Supply and Demand and Their Impacts* (Washington, D.C.: Government Printing Office, 1978), p. 166.

13. Texas governor and New York commissioner, *Natural Gas Pricing Proposals of President Carter's Energy Program*, 95 Cong., 1st sess. (Washington, D.C.: Government Printing Office, June 1977), pp. 113 and 150.

14. Many studies have been done about the cost of gas production. For a particularly lucid discussion on the various viewpoints, see Clark A. Hawkins, *The Field Price Regulation of Natural Gas* (Florida: Florida State University Press, 1969), chap. 3; for Joskow and Steele, see Hawkins, pp. 88–90.

15. In the 1950's and early 1960's a much greater percentage of gas was associated with oil than in the 1970's. For gas production associated with oil, see *National Energy Outlook, 1976*, pp. 142–143, 145; for an excellent discussion of the complicated problem of "joint costs," see Alfred E. Kahn, *The Economics of Regulation: Principles and Institutions*, vol. 1 (New York: John Wiley and Sons, 1970), pp. 77–83. For alternative studies placed in evidence before FPC, see National Academy of Sciences, National Resource Council, Panel on Gas Reserve Estimation, Washington, D.C., 1975.

16. James W. McKie, "Market Structure and Uncertainty in Oil and Gas Exploration," *Quarterly Journal of Economics,* 16 (September 1972), pp. 543–571.

17. Jackson quote is in *Natural Gas Pricing Proposals of President Carter's Energy Program,* p. 506.

18. Quote from O'Leary in *Natural Gas Pricing Proposals of President Carter's Energy Program,* p. 160.

19. In 1974, two MIT economists, Paul W. MacAvoy and Robert S. Pindyck, constructed an econometric model that predicted that if the government stopped controlling the wellhead price of interstate gas, domestic discoveries would climb to 33 tcf per year by 1980 at a price of $1.00/mcf. (At the time of their estimate, the national average of controlled prices for new discoveries was $.54/mcf.) See MacAvoy and Pindyck, *Price Controls and the Natural Gas Shortage,* p. 57. For ERDA study, see Energy Research and Development Administration, *Market Oriented Program Planning Study—I* (Washington, D.C.: Government Printing Office, 1976). Also, see U.S. Senate, *Hearings before the Committee on Energy and Natural Resources: Market Oriented Program Planning Study,* 95 Cong., 1 sess. (Washington, D.C.: Government Printing Office, June 23 and 27, 1977), p. 95; and Comptroller General of the United States, Report to the Committee on Government Operations, House of Representatives, *Implications of Deregulating the Price of Natural Gas* (Washington, D.C.: General Accounting Office, January 14, 1976).

20. The U.S. Geological Survey conducts periodic studies to update natural gas supplies. A summary of 1975 estimates appears in the *U.S. Geological Survey Circular,* 725, 1975; also see U.S. Central Intelligence Agency, *The International Energy Situation: Outlook to 1985* (Washington, D.C.: Library of Congress, April 1977), p. 6. A recent USGS estimate indicates recoverable natural gas reserves of some 330 to 660 tcf, or the equivalent of 16 to 33 years at a consumption rate of 20 tcf. These estimates include the 205 tcf of proven reserves existing at the end of 1978. The USGS estimate is in *Energy: An Uncertain Future,* prepared by Herman Franssen at the request of the Senate Committee on Energy and Natural Resources, Publication 95-157 (Washington, D.C.: Government Printing Office, December 1978), p. 59.

21. Since the 1973 oil embargo, many of the major oil companies have published reserve estimates of oil and gas in the United States. Most of the estimates are revised periodically and appear in the firms' annual reports. One of the more comprehensive sets of estimates is in *Energy Outlook 1976–1999,* published by the Exxon Corporation. Other similar estimates are published periodically by the American Gas Association in Arlington, Va., and the Independent Petroleum Association of America in Washington, D.C.

22. For Canada, see *Annual Report to Congress, Volume III,* p. 51; *Annual Report to Congress, Volume II: Projections of Energy Supply and Demand and Their Impacts,* pp. 161–171. An unpublished report from Stanford Research Institute, December 1978, contains estimates for Canada and Mexico. Also, for Mexico see "Supplemental Natural Gas Sources: Factors and Policy Issues," Report prepared by the Congressional Research Service, Library of Congress, for the use of the Subcommittee on Energy and Power, Committee on Interstate and Foreign Commerce, U.S. House of Representatives, June 1978, p. 11.

23. U.S. Department of Commerce, *Natural Gas from Unconventional Geologic Sources* (Washington, D.C.: National Academy of Sciences, 1976), sec. IV. True, geopressured brine can be found at depths less than 15,000 feet, but the geologists that we interviewed believe that most of it is likely to be found at depths greater than 15,000 feet.

24. See Chapter 4 of this book, and National Petroleum Council, *U.S. Energy Outlook—Coal Availability* (Washington, D.C.: NPC, 1973).

25. Speech appearing in Sonya Marchand, ed., "Energy—The Global Challenge," Proceedings of a conference sponsored by California State University at Northridge, May 6–7, 1977.

26. The most authoritative study that we have seen is U.S. Congress, Office of Technology Assessment, *Transportation of Liquefied Natural Gas* (Washington, D.C.: Government Printing Office, 1977); our account of the LNG situation draws heavily on this source. Also see Edward K. Fariday, *LNG Review: 1977* (Energy Economics Research Limited, 1978). Ironically, the oldest operating marine LNG project in the United States is one from which Phillips Petroleum and Marathon Oil export Alaskan LNG to Japan. This project started operation in 1969. Predictions of OPEC prices are from interviews by the authors with executives of interstate pipeline companies. Information on the Energy Regulatory Agency's El Paso refusal from news release, Office of Public Affairs, Department of Energy, Washington, D.C., December 21, 1978, and interviews with officials of the Department of Energy.

27. "Independents Vital to U.S. Petroleum Supply," *Oil & Gas Journal*, October 24, 1977, reprint; "Natural Gas Pipeline Industry," *A Quarterly Review;* and *Annual Report to Congress, Volume III*, p. 51.

28. Interview with Ben Cubbage, June 1978.

29. Interview with Jim Daugherty, president, Weal Drilling Company, June 1978. Other interviews with producer companies indicate that besides objections to the regulated price, companies cite instances of long delays in FERC's licensing processes.

30. This section is based heavily on interviews by the authors with officials of several interstate pipeline companies, particularly Mr. Hayward Coleman and his associates at the Southern Natural Gas Company, and Mr. Ed Najaiko and his associates of El Paso Natural Gas. Also see *Natural Gas Pricing Proposals of President Carter's Energy Program*, p. 62.

31. Transportation cost estimate is from Independent Petroleum Association of America, "This is a Bargain?" 1978; also see *Wall Street Journal*, June 21, 1977, p. 22.

32. Interview with Howard Boyd, June 1978.

33. From authors' interview with Robert Herring, chairman, Houston Natural Gas Company, July 1978.

34. Interview by authors with Ed Najaiko, vice-president, El Paso Company, July 1978.

35. Testimony of John F. O'Leary, administrator, Federal Energy Administration, *Natural Gas Pricing Proposals of President Carter's Energy Program*, pp. 45ff.

36. Information on transfer of wealth is given in Department of Energy, Energy Information Administration, "An Evaluation of Natural Gas Pricing Proposals," Analysis memo AM/IA, 7802, June 14, 1978. Lee White quote from *Natural Gas Pricing Proposals of President Carter's Energy Program*, pp. 517–518.

37. Interview by authors with the Gallup Poll firm, Princeton, N.J., July 1978.

38. U.S. Congress, Senate Conference Report 95–1126 (to accompany House Report 5289), 95 Cong., 2 sess., 1978. See also U.S. Congress, U.S. House of Representatives, Committee on Interstate and Foreign Commerce, and U.S. Senate, Committee on Energy and Natural Resources, *Natural Gas Pricing Agreement Adopted by the Conferees on H.R. 5289*, 95 Cong., 2 sess. (Washington, D.C.: Government Printing Office, June 1978). In addition to provisions discussed

in the text, the Act provides that FERC must issue regulations so that certain incremental gas costs for interstate pipelines will be passed through to industrial consumers. But even without considering any possible controversy over these forthcoming regulations, many companies that we interviewed estimate litigation of up to five years because of the Natural Gas Policy Act. In addition, many companies and congressional staff members question whether the bill does in fact give Washington more flexibility, as we state.

Chapter 4
Coal: Constrained Abundance

1. Gordon Young, "Will Coal be Tomorrow's 'Black Gold,'" *National Geographic*, August 1975, pp. 234–259. "The Great Black Hope: Coal," *Colorado Business*, September/October 1976, pp. 33–34. "Old girl" is from Edmund Faltermayer, "Clearing the Way for the New Age of Coal," *Fortune*, May 1974, pp. 215–338. The idea of coal as a "transition" or "swing" fuel is widespread: see, for instance, R. L. Gordon, "Coal—Swing Fuel," in R. J. Kalter and W. A. Vogely, eds., *Energy Supply and Government Policy* (Ithaca: Cornell University Press, 1976); D. Meadows and J. Stanley-Miller, "The Transition to Coal," *Technology Review*, 75, No. 1 (October/November 1975), pp. 19–29; A. Ford, "Environmental Policies for Electricity Generation: A Study of the Long-Term Dynamics of the SO_2 Problem," *Energy Systems and Policy*, 1 (1975), pp. 287–304.

2. The President's Address, "National Energy Program," *Presidential Documents: Jimmy Carter*, 13, No. 17, Delivered before a joint session of Congress, April 20, 1977; Executive Office of the President, *The National Energy Plan*, April 29, 1977. The 1976 coal production figure comes from Energy Information Administration, *Annual Report to Congress, Volume III–77* (Washington, D.C.: Government Printing Office, May 1978), p. 77.

3. David J. Goerz, "Coal: Presentation to Harvard Energy Seminar," Speech, May 1, 1978; Federal Energy Administration, *Final Task Force Report on Coal, Project Independence* (Washington, D.C.: Government Printing Office, November 1974); *1977 Keystone Coal Industry Manual* (New York: McGraw-Hill, 1977), pp. 777–778.

4. For various estimates for 1985 coal production, see Earl T. Hayes, "Energy Resources Available to the United States, 1985 to 2000," *Science*, vol. 203, January 19, 1979, p. 236; for 800 million tons 1985 estimate and for coal executive quote, see *Wall Street Journal*, February 21, 1979, pp. 1, 35. For Carter Administration policy change favoring switch from oil to natural gas, see U.S. Department of Energy news release, N–79–003, January 25, 1979; "ERA Seeks State Backing for Switch to Gas Plan," *Oil and Gas Journal*, January 22, 1979, p. 40; *Oil and Gas Journal*, February 5, 1979, newsletter. For "new fears," see Tom Alexander, "New Fears Surround the Shift to Coal," *Fortune*, November 20, 1978, pp. 50–60. For "there is a lot of it" quote, see General Accounting Office, *U.S. Coal Development —Promises, Uncertainties*, EM–77–43, September 22, 1977, p. 1.1. For factors working to stimulate the use of coal in the short term, see "New Developments Revealed in Moves to Hike U.S. Coal Use," *Oil and Gas Journal*, January 22, 1979, p. 40; U.S. Department of Energy news release, "National Energy Act to Conserve Energy, Accelerate Shift to Coal and Reduce U.S. Oil Import Needs," R–78–413, October 20, 1978.

5. For coal-consumption patterns for the 1950–73 period, see Energy Information Administration, *Annual Report to Congress, Volume III–77*, pp. 78–79; Na-

tional Electric Reliability Council, *7th Annual Review,* July 1977, p. 8; General Accounting Office, *U.S. Coal Development,* chap. 2; Gordon, "Coal—Swing Fuel," *Energy Supply and Government Policy;* and *1977 Keystone Coal Industry Manual.* For ESECA, see Stuart M. Rosenblum, "The Future of the Coal Substitution Option," *Duquesne Law Review,* 3 (1975), pp. 581–622; Ronald F. Ayres, *Coal: New Markets/New Prices—Ramifications of the Federal Coal Conversion Program* (New York: McGraw-Hill, 1977).

6. General Accounting Office, *U.S. Coal Development,* chap. 2; and ICF, Inc., *Final Report: Production and Consumption of Coal, 1976–1980,* May 1976. The Federal Energy Regulatory Commission found for the four-state region of Texas, Arkansas, Louisiana, and Oklahoma that the total coal demand for new units which were scheduled to start operating between 1977 and 1986 would be about ten times greater than total coal shipments to utilities in those states in 1976. See Department of Energy, Federal Energy Regulatory Commission, *Status of Coal Supply Contracts for New Electric Operating Units, 1977–1986, First Annual Supplement,* May 1978, p. 6. For 1978 National Energy Act, see U.S. Department of Energy news release, "National Energy Act to Conserve Energy, Accelerate Shift to Coal and Reduce U.S. Oil Import Needs," R–78–413, October 20, 1978. For Texas Railroad Commission, see *Wall Street Journal,* February 21, 1979, p. 35.

7. National Electric Reliability Council, *8th Annual Review,* August, 1978, p. 9. For statistics and reasons for decline in growth rates of electricity, see Federal Energy Regulatory Commission, *Status of Coal Supply Contracts,* p. 1; *Electrical World,* September 15, 1976, p. 54, and September 15, 1977, pp. 45, 54; Energy Information Administration, *Annual Report to Congress, Volume II–77,* p. 215.

8. Fred J. Abbate, "Kilowatts and Morality: The New Criticism of Utility Decision-Making," *Electric Perspectives,* 78/1, pp. 2–7.

9. National Electric Reliability Council, *7th Annual Review,* July 1977, pp. 1, 6–7; *Electrical World,* September 15, 1976, p. 54. Planner in "Planning for Uncertainty", *EPRI Journal,* May, 1978, pp. 6–11.

10. National Electric Reliability Council, "The Coal Strike of 1977–78: Its Impact on the Electric Bulk Power Supply in North America," 1978. Analysts' comments in A. D. Rossin and T. A. Rieck, "Economics of Nuclear Power," *Science,* August 18, 1978, p. 583; National Coal Association, *Study of New Mine Additions and Major Expansion Plans of the Coal Industry and the Potential for Future Coal Production,* November 1977; and "Awash in Coal" in *Wall Street Journal,* May 8, 1978, p. 28. For continued market softness, see *Wall Street Journal,* February 21, 1979 pp. 1, 35.

11. For information on the rise of and causes for long-term coal delivery contracts, see ICF, Inc., *Final Report: Coal Mine Expansion Study,* May 1976; General Accounting Office, *U.S. Coal Development,* chap. 4, pp. 1.41–1.42; Federal Power Commission, *Status of Coal Supply Contracts for New Electric Generating Units, 1976–1985,* Staff Report by the Bureau of Power, January 1977; Marvin H. Kahn and Robert Hand, *Implications of Ownership Patterns of Western Coal Reserves and their Impact on Coal Development* (McLean, Va.: The Mitre Corp., 1976); Richard L. Gordon, *U.S. Coal and the Electric Power Industry* (Washington: Resources for the Future, 1975). For 1978 estimates, see Federal Energy Regulatory Commission, *Status of Coal Supply Contracts.*

12. For the geographic distribution of coal and for statistical trends according to coal production methods, see Energy Information Administration, *Annual Report to Congress, Volume III–77,* pp. 76–92. But uncertainty exists over coal reserve and "heat value" estimates; see Francis X. Murray, ed., *Where We Agree: [Final] Report of the National Coal Policy Project* (Boulder: Westview Press, 1978),

pp. 76–99; General Accounting Office, *Inaccurate Estimates of Western Coal Reserves Should Be Corrected,* EM–78–32, July 11, 1978.

13. Federal Power Commission, *Status of Coal Supply Contracts for New Electric Generating Units, 1976–1985;* General Accounting Office, *U.S. Coal Development,* chap. 5.

14. Federal Energy Regulatory Commission, *Status of Coal Supply Contracts,* pp. 31–37. For an estimate of capital requirements, see *U.S. Coal Development,* chap. 5.

15. Congressional Research Service, *National Energy Transportation: Issues and Problems,* vol. 3, March 1978, pp. 473–476.

16. National Coal Association, *Coal News,* No. 4403, January 20, 1978, p. 2; U.S. Congress, Office of Technology Assessment, *A Technology Assessment of Coal Slurry Pipelines* (Washington, D.C.: OTA, March 1978); "Why Coal Gasification is Leaving the West," *Business Week,* September 13, 1976, pp. 76J–76K; "Coal Transporters Face Challenge," *Coal Age,* January 1977, pp. 13–15. One coal slurry pipeline clearly was a deterrent to high rail rates. In 1957 Consolidation Coal Company built a 108-mile coal slurry pipeline in Ohio. But in 1963 Consolidation stopped operation of the slurry pipeline after the railroads lowered their freight rates. Today in the West the railroads are attempting to capitalize on their seeming monopoly position in coal transportation. But again, the threat of coal slurry pipelines, along with other factors, such as the inertia of the regulatory process, tends to keep rail rates lower than they otherwise would be. See Martin B. Zimmerman, *Rent and Regulation in Unit-Train Rate Determination: Regional Discrimination and Inter-Fuel Competition,* MIT Energy Laboratory Working Paper, MIT–EL 78–010WP, revised June 1978.

17. *Coal Age,* August 1978, p. 45; *Oil and Gas Journal,* July 21, 1978, newsletter; "Slurry Line Clears Kansas Right-of-Way Hurdle," *Oil and Gas Journal,* March 5, 1979, p. 71.

18. Private communication, Harry Perry to author, July 1978; "Revised Air Standards Will Influence Coal Mining and Burning," *Coal Age,* September 1978, p. 9; "EPA Proposes Strict New Sulfur Curbs." *Oil and Gas Journal,* September 18, 1978, p. 69. For leasing, see "Major Decisions Near on Federal Leasing," *Coal Age,* November 1978, pp. 11–13; U.S. Department of Energy, Leasing Policy Development Office, *Federal Coal Leasing and 1985 and 1990 Regional Coal Production Forecasts,* June 1978; Congressional Research Service, *The Coal Industry: Problems and Prospects—A Background Study* (Washington, D.C.: Government Printing Office, 1978), pp. 67–81.

19. For the environmental effects of coal, see General Accounting Office, *U.S. Coal Development;* Richard L. Gordon, "The Hobbling of Coal: Policy and Regulatory Uncertainties," *Science,* April 14, 1978, pp. 153–158; *Report of the Committee on Health and Environmental Effects of Increased Coal Utilization,* December 27, 1977 (henceforth called *The Rall Committee Report,* after the committee's chairman, David P. Rall, director of the National Institute of Environmental Health Sciences); C. E. Chrisp, G. L. Fisher, and J. E. Lammert, "Mutagenicity of Filtrates from Respirable Coal Fly Ash," *Science,* 199 (January 6, 1978), pp. 73–75.

20. For the two suits, see *The Boston Sunday Globe,* February 26, 1978, p. 14; *New York Times,* April 30, 1978, p. 26.

21. For information on carbon dioxide emissions from coal (and other fossil fuels) see National Academy of Sciences, *Energy and Climate: Outer Limits to Growth?* (Washington, D.C.: NAS Geophysics Board, 1977); *New York Times,* June 10, 1977, p. A28; *New York Times,* July 25, 1977, pp. 1, 37; George M. Woodwell, "The Carbon Dioxide Question," *Scientific American,* January 1978,

pp. 34–43; Richard A. Kerr, "Carbon Dioxide and Climate: Carbon Budget Still Unbalanced," *Science,* September 30, 1977, pp. 1352–1353; U. Sigenthaler and H. Oeschger, "Predicting Future Atmospheric Carbon Dioxide Levels," *Science,* January 27, 1978, pp. 388–395; "Energy Agency to Study Carbon Dioxide," *Chemical and Engineering News,* December 5, 1977, p. 4; William W. Kellogg, "Is Mankind Warming the Earth," *The Bulletin of the Atomic Scientists,* February 1975, pp. 10–19.

22. *The Rall Committee Report.* For the Massachusetts utility, *The Boston Globe,* May 21, 1978, p. 21. However, the Federal Energy Regulatory Commission estimated in 1978 that 43 percent of new coal-fired generating capacity between 1977 and 1981 will use scrubbers to meet air-quality standards. See Federal Energy Regulatory Commission, *Status of Coal Supply Contracts,* pp. 38–42.

23. For the early twentieth century history of labor relations in the U.S. coal industry, see United Mine Workers of America, *It's Your Union, Pass It On,* September 1976; Hoyt N. Wheeler, "Mountaineer Mine Wars: An Analysis of the West Virginia Mine Wars of 1912–1913 and 1920–1921," *Business History Review,* 50 (Spring 1976), pp. 69–91; Alicia Tyler, "Dust to Dust," *Washington Monthly,* January 1975, pp. 49–58; Melvyn Dubofsky and Warren Van Tine, *John L. Lewis: A Biography* (New York: Quadrangle, 1977).

24. Figures on the U.S. coal industry employment and union membership come from the following sources: *1977 Keystone Coal Industry Manual,* p. 550; United Mine Workers of America, *It's Your Union, Pass It On; Wall Street Journal,* October 5, 1977, p. 44; and *New York Times,* February 26, 1978, sec. III, pp. 2, 16. The "new breed" in *New York Times,* September 1, 1977, pp. 33, 39; for attraction of skilled labor, see *Wall Street Journal,* November 10, 1978, p. 1; "wildcat strike" in *New York Times,* February 3, 1978, p. 43.

25. General Accounting Office, *U.S. Coal Development,* chap. 4.

26. For an in-depth discussion on the results and implications of the diverse attempts by certain representative coal companies to innovate in the area of management organizational structure, personnel, and labor relations, see Balaji Chakravarthy, *Adapting to Changes in the Coal Industry: A Managerial Perspective,* D.B.A. dissertation, Graduate School of Business Administration, Harvard University, 1978. For the overall need and opportunities for improving the management of the coal work force, see Congressional Research Service, *The Coal Industry,* chap. 7.

27. *Kentucky Coal Mining Corporation (A-C),* ICH 9–678–101–103 (Boston: Intercollegiate Case Clearinghouse, 1977).

28. For observers stressing coal's negative social effects, see Harry M. Caudill, *Night Comes to the Cumberlands* (Boston: Atlantic, Little, Brown, 1962); Dick Kirschten, "Troubles of the 'Eastern Tilt,'" *National Journal,* 10, No. 12 (March 25, 1978), pp. 461–464; Kai T. Erikson, *Everything in Its Path: Destruction of a Community in the Buffalo Creek Flood* (New York: Simon and Schuster, 1976). For the positive and adaptive aspects of the communities in the Eastern coal-field regions, see the review of *Everything in Its Path* by Dwight Billings and Sally Maggard in *Social Forces,* vol. 52:2, December 1978, pp. 722–723; and Harry K. Schwarzweller, James S. Brown, and J. J. Mangalam, *Mountain Families in Transition* (University Park: The Pennsylvania State University Press, 1971). For the miners' new prosperity, see *Wall Street Journal,* November 10, 1978, p. 1.

29. Information on petroleum-firm activity in the U.S. coal industry comes from the following sources: U.S. Bureau of Mines, *The State of the U.S. Coal Industry,* Information Circular IC 8707, 1976; *Note on the U.S. Coal Industry,* ICH 9–676–133 (Boston: Intercollegiate Case Clearinghouse, 1976); U.S. Congress, Senate, *Interfuel Competition: Hearings Before the Subcommittee on*

294 — REFERENCES

Antitrust and Monopoly of the Committee on the Judiciary (Washington, D.C.: Government Printing Office, June 17, 18, 19, July 14, October 21 and 22, 1975); U.S. Congress, Senate, *Petroleum Industry Involvement in Alternative Sources of Energy,* Prepared at the request of Frank Church for the Subcommittee on Energy Research and Development of the Committee on Energy and Natural Resources, Publication No. 95-54, September 1977; *Submission of Exxon Company, U.S.A., before the House Judiciary Committee on Monopolies and Commercial Law,* September 11, 1975.

30. For arguments in favor of divestiture, see Herbert S. Sanger, Jr., and William E. Mason, *The Structure of the Energy Markets: A Report of TVA's Antitrust Investigation of the Coal and Uranuim Industries,* and *Appendices A–H,* June 14, 1977; Walter Adams, "Horizontal Divestiture in the Petroleum Industry: An Affirmative Case," in Edward J. Mitchell, ed., *Horizontal Divestiture in the Oil Industry* (Washington, D.C.: American Enterprise Institute for Public Policy Research, 1978), pp. 7–19; *Coal Patrol,* 33 (August 5, 1977), pp. 8–10.

31. Evidence against divestiture can be found in U.S. Federal Trade Commission, *Concentration Levels and Trends in the Energy Sector of the U.S. Economy,* March 1974; Jesse W. Markham, Anthony P. Hourihan, and Francis L. Sterling, *Horizontal Divestiture and the Petroleum Industry* (Cambridge, Mass.: Ballinger, 1977), p. 105; U.S. Department of Treasury, *Implications of Divestiture,* June 1976; Donald Norman, *The Performance of Oil Firm Affiliates in the Coal Industry,* American Petroleum Institute, Research Study #004, March 1977; National Coal Association, *Implications of Investments in the Coal Industry by Firms from Other Energy Industries,* September 1977; General Accounting Office, *The State of Competition in the Coal Industry,* EMD–78–22, December 30, 1977; *Oil and Gas Journal,* April 10, 1978, newsletter. Actually, economists usually measure market concentration rather than concentration of production. But in the case of coal, production and market concentration are so congruent that production concentration is an accurate indicator of market concentration and can be substituted as the measure used. For data on control of reserves, see Hossein G. Askari, Timothy W. Reufli, and Michael P. Kennedy, *Horizontal Divestiture of Energy Companies and Alternative Policies* (Austin: Graduate School of Business, The University of Texas at Austin, undated), p. 13. For data on control of production, and on coal-company shares, see National Coal Association, *Implications of Investment in the Coal Industry,* pp. 2, 22–24; unlike the National Coal Association, I used 438 billion tons, the total amount of coal in the United States that is suitable for mining using current methods, as the total reserve figure in calculating oil-firm shares of reserve holdings. The National Coal Association used 219 billion tons, a more likely figure for total recoverable reserves using current methods and economics. My reason for selecting the higher figure is that the firms that reported their reserve holdings did not differentiate between recoverable and in-place reserves. For oil-company share of total research and development, see Norman, *The Performance of Oil Firm Affiliates in the Coal Industry.*

32. A recent case suggests that the threat or possibility of legal or legislative considerations like horizontal divestiture may already be influencing the actions of top management at coal companies. In early September 1978 Amax rejected an offer to merge with Standard Oil of California (Socal), the fourth largest U.S. petroleum company. Socal, which already owned 20 percent of Amax's common stock, proposed acquiring an additional 25 to 39 percent of Amax's common stock. But Amax claimed that the offer raised "serious and substantial antitrust questions." Its chairman added that such antitrust considerations "raise grave doubt as to the feasibility of any such proposal and would in any event involve long delays and great uncertainties." On the other hand, Socal's chairman commented that it

was "unfortunate" that the Amax board "had taken refuge in supposed antitrust problems." See *Wall Street Journal*, September 8, 1978, p. 10.

33. For ways to preserve competition short of divestiture, see *New York Times*, April 23, 1977, pp. 25, 27; Askari *et al., Horizontal Divestiture of Energy Companies; Oil and Gas Journal*, April 10, 1977, newsletter. For the estimate of the federal share of Western coal reserves, see "Major Decisions Near on Federal Leasing," *Coal Age*, November 1978, pp. 11–12.

34. For General Electric, utilities, and U.S. Steel, see National Coal Association, *Implications of Investment in the Coal Industry; 1977 Keystone Coal Industry Manual*, pp. 546, 770; Louis Kraar, "General Electric's Very Personal Merger," *Fortune*, August 1977, pp. 187–194.

35. For the establishment of the Peabody Holding Company and on the members of this holding company, see *1977 Keystone Coal Industry Manual*, pp. 546, 770; "Splicing Together the Peabody Deal," *Business Week*, November 1, 1976, p. 24; Carol J. Loomis, "Down the Chute with Peabody Coal," *Fortune*, May 1977, pp. 228–248; *New York Times*, June 8, 1977, pp. 47, 56; "FTC Approves Sale of Peabody," *Coal Age*, July, p. 27; Western Gasification Company, *Coal Gasification: A Technical Description*, revised April 1, 1974. The composition of the shares in the Peabody Holding Company was as follows: Newmont Mining Corporation—27.5 percent, Williams Company—27.5 percent, Bechtel Corporation—15 percent, the Boeing Company—15 percent, Fluor Corporation—10 percent, and Equitable Life Assurance—5 percent.

36. Multiorganizational and multisector enterprises are discussed in Mel Horwitch and C. K. Prahalad, "Managing Technological Innovation: Three Ideal Modes," *Sloan Management Review*, Winter 1976, pp. 77–89; and Mel Horwitch, "Uncontrolled Growth and Unfocused Growth: Unsuccessful Life Cycles of Large-Scale, Public-Private, Technological Enterprises, With Special Reference to the United States SST Program and the United States Attempt to Develop Synthetic Fuels from Coal," Paper presented at the Symposium on the Management of Science and Technology, in Rio de Janeiro, June 22–23, 1978, to be published in *Interdisciplinary Science Reviews*.

37. For large-scale mining/synthetic fuels projects in the West, see Federal Energy Administration, *Western Coal Development Monitoring System*, Summer Quarter, FEA/G-77-306, August 1, 1977; *1977 Keystone Coal Industry Manual; Pacific Lighting: Wesco*, ICH 9–377–889 (Boston: Intercollegiate Case Clearinghouse, 1977); *The ConPaso Coal Reserve Project*, ICH 4–678–152 (Boston: Intercollegiate Case Clearinghouse, 1978); "Two Gas Companies Plan Commercial Coal-Gas Plant," *Coal Age*, May 1977, p. 17.

38. For the traditional technological innovation process in the coal industry, see E. Mansfield, "Firm Size and Technological Change in the Petroleum and Bituminous Coal Industries," in Thomas D. Duchesneau, ed., *Competition in the U.S. Energy Industry* (Cambridge, Mass.: Ballinger, 1975), pp. 317–345; *Coal Mining: Research and Development*, Society of Mining Engineers, February 28, 1978; N. P. Chironis, "Consol's Record Production Runs Give Impetus to Use of Shield Supports for U.S. Longwalls," *Coal Age*, August 1976, pp. 92–97; "Productivity Perspective," "Underground Mining," and "Surface Mining" in *Coal Age*, July 1976, pp. 63–169. For fluidized-bed combustion, see "Energy Department Will Emphasize Direct Coal Combustion," *Coal Age*, January 1979, p. 9; U.S. Department of Interior, Office of Coal Research, *Clean Energy from Coal Technology* (Washington, D.C.: Government Printing Office, 1974), p. 41; U.S. Energy Research and Development Administration, Office of Fossil Energy, *Power and Combustion*, October-December 1975 (Washington: ERDA, 1976), pp. 1–13, 25–33, 51–53; U.S. Federal Power Commission, *The Status of Flue Gas Desulfurization Applications in the United States: A Technological Assessment*, a Staff

report by the Bureau of Power, July 1977, pp. III–56 to III–65; U.S. Department of Energy, *International Coal Technology Summary Document* (Washington, D.C.: Office of Technical Programs Evaluation, December 1978), pp. 19–29.

39. For South Africa, see Roger Vielvoye, "South Africa's Optimistic Outlook," *Oil and Gas Journal,* January 15, 1979, p. 41; "Liquid Coal Looks Good as Hunt Heats Up for Other Oils," *Financial Times World Business Weekly,* February 26, 1979, pp. 9–11; "U.K., South Africa Boosts Oil-from-Coal Work," *Oil and Gas Journal,* March 5, 1979, p. 78; *New York Times,* March 8, 1979, pp. D1, D4. For the history of pre-1960 U.S. synthetic fuels policy, see Richard E. Vietor, "The Synthetic Liquid Fuels Program: Energy Politics in the Truman Era," Harvard Business School Working Paper 78–54, 1978. The brief history of U.S. synthetic fuels development in the sixties comes from the following sources: *Chemical Week,* April 13, 1963, p. 69; August 3, 1963, pp. 63–64; November 18, 1964, p. 23; May 28, 1966, p. 110; December 3, 1966, p. 76; December 30, 1967, p. 40; March 2, 1968, p. 50; March 23, 1968, p. 67; May 11, 1968, p. 59; June 15, 1968, p. 41; June 29, 1968, p. 13. *Coal Age,* April 1965, p. 55; January 1966, pp. 64–71; April 1966, p. 46; January 1967, pp. 90–92; March 1967, p. 48; September 1967, pp. 64–68; March 1968, pp. 40–42; December 1968, pp. 26–27; May 1969, p. 50; July 1969, p. 44.

40. For synthetic gas cost estimates, see Roger Detman, C. F. Braun and Company, *Preliminary Economic Comparison of Six Processes for Pipeline Gas from Coal,* Presentation to the Eighth Synthetic Pipeline Gas Symposium, Chicago, Illinois, October 18–20, 1976; *Chemical Week,* December 23, 1970, p. 23. For synthetic crude estimates, see Earl T. Hayes, "Energy Resources Available to the United States, 1985 to 2000," *Science,* vol. 203, January 19, 1979, p. 238. For Coalcon, see Luther J. Carter, "Synfuels: Data Gap Imperils 'Coalcon' Demonstration," *Science,* vol. 195, January 7, 1977, p. 38; General Accounting Office, *First Federal Attempt to Demonstrate a Synthetic Fossil Energy Technology—A Failure,* EMD–77–59, August 17, 1977. For the need for a liquefaction capability, see L. E. Swabb, Jr., "Liquid Fuels from Coal: From R&D to an Industry," *Science,* vol. 199, February 10, 1978, pp. 619–622. For the shifts in government coal technology strategies, see "Carter Budget Cuts Coal, Oil Outlays," *Oil and Gas Journal,* January 29, 1979, pp. 94–96; "Coming This Spring: Son of National Energy Plan" and "Energy Department Will Emphasize Direct Coal Combustion," *Coal Age,* January 1979, p. 9; U.S. Department of Energy news release, "DOE Authorizes Both Competitors to Proceed with Coal-to-Gas Demonstration Plant Designs," R–79–035, January 23, 1979.

41. Certain recent studies indicate that the cost of high-BTU synthetic gas may be economically competitive with the delivered cost of coal-fired electric residential heating; see Patrick Crow, "Prospects Seem Brighter for U.S. LNG and SNG," *Oil and Gas Journal,* January 16, 1978, pp. 23–28; William F. Hederman, Jr., *Prospects for the Commercialization of High-BTU Coal Gasification,* R–2294, Rand Corp., April 1, 1978, pp. 23–25, append. D.

42. Federal R&D in U.S. Congress, Senate, *Energy Research and Development— Problems and Prospects,* Prepared at the request of Henry M. Jackson, chairman, Committee on Interior and Insular Affairs, Serial No. 93–21 (92–56), (Washington, D.C.: Government Printing Office, 1973); General Accounting Office, *U.S. Coal Development,* app. III; National Coal Association, *Special Analysis: Federal Funding for Activities Concerned with Coal—Fiscal Years 1978 and 1979,* March 27, 1978. Industry R&D in Markham, *et al., Horizontal Divestiture and the Petroleum Industry;* Norman, *The Performance of Oil Firm Affiliates in the Coal Industry;* and U.S. Congress, Senate, *Petroleum Industry Involvement in Alternative Sources of Energy.* Synthetic fuels projects are identified in U.S. Bureau of Mines, *Projects to Expand Fuel Sources in Western States,* IC 8719, 1976.

43. For the 1977–78 coal strike, see *New York Times*, various issues, October 1977–March 1978; *The Boston Evening Globe*, March 9, 1978; Thomas N. Bethell, "UMW in the Pits," *The New Republic*, April 1, 1978, pp. 7–9; *Coal Patrol*, February 15, 1978, and March 1, 1978.

44. *Where We Agree: [Draft] Report of the National Coal Policy Project*, February 9, 1978, p. 2 (the final report of this group is Francis X. Murray, ed., *Where We Agree: Report of the National Coal Policy Project*); *The Boston Globe*, February 10, 1978, p. 23; Tom Alexander, "A Promising Try at Environmental Détente for Coal," *Fortune*, February 13, 1978, pp. 94–102; National Coal Association, *Coal News*, no. 4451, December 22, 1978, p. 2; and personal interviews.

Chapter 5
The Nuclear Stalemate

1. For Project Independence, see U.S. Federal Energy Administration, *Project Independence Report* (Washington, D.C.: Government Printing Office, 1974). Chirac's statement was part of his remarks opening the First Conference of the European Nuclear Society, held in Paris, April 1975. See also The Energy Policy Project of the Ford Foundation, *A Time to Choose: America's Energy Future, Final Report* (Cambridge, Mass.: Ballinger, 1974); and U.S. Atomic Energy Commission, Office of Planning and Analysis, *Nuclear Power Growth, 1974–2000*, WASH-1139(74) (Washington D.C.: Government Printing Office, February 1974).

2. On March 28, 1979, an apparent series of mechanical failures and human errors caused massive damage to the radioactive core of the most recently commissioned nuclear reactor at the Three Mile Island plant near Harrisburg, Pennsylvania.

3. Robert Smock, "Nuclear Executives Question Industry's Viability," *Electric Light and Power*, December 1979, p. 39. For the U.S. picture, see Alvin M. Weinberg, "Energy Policy and Energy Projections: The Case of a Nuclear Moratorium," Address to the Atomic Industrial Forum Conference on United States Energy Policy, Washington, D.C., January 1977; and Citibank, "Nuclear's 1985 Goal— An Impossible Dream," *Energy Newsletter*, vol. XV, no. 2, 1978. For the situation abroad, see Louis Puiseax, *Le Babel Nucléaire* (Paris: Editions Galilée, 1977); and *Proceedings of the Colloquium on the Socio-psychological Implications of the Development of Nuclear Energy*, Published by the French Society for Radioprotection, Paris, January 1977.

4. The "anti-nuclear" literature is enormous. One of the best books is McKinley C. Olsen's *Unacceptable Risk* (New York: Bantam, 1976). See also Joel Primack and Frank von Hipple, "Challenging the Atomic Energy Commission on Reactor Safety," in *Advice and Dissent: Scientists in the Public Arena* (New York: New American Library, 1974). For a more complete set of references to the literature, see Irvin C. Bupp and Jean-Claude Derian, *Light Water: How the Nuclear Dream Dissolved* (New York: Basic Books, 1978); and Amory B. Lovins, *Soft Energy Paths* (San Francisco: Friends of the Earth, 1977).

5. This conclusion is based on the author's interviews and informal discussions with prominent nuclear critics belonging to organizations such as the Union of Concerned Scientists, The National Resources Defense Council, and the Friends of the Earth.

6. Calculated as follows:
   a. Operation at 100 percent of capacity = (624 million barrels per year of oil, most residual fuel oil ÷ 365 days per year) ÷ (358,000 GW hours per year via oil ÷ 8,760 GW hours per year of potential output per each GW of capacity) = 42,000 barrels of oil per day

b. At actual 1977 operating rate = 42,000 barrels of oil per day per each GW operating at 100 percent of capacity × (2,125,000 GW hours per year total actual ÷ 515 GW × 8,760 hours possible output per year) = 42,000 × .47 = 20,000 barrels of oil per day

Source: U.S. Department of Energy, *Monthly Energy Review*, June 1978, pp. 33–34.

7. Statistics on nuclear generating capacity in operation, under construction, or "on order" can be found in numerous trade journals. Especially comprehensive is the compilation published annually in April by *Nuclear Engineering International*.

8. From telephone surveys of major electric utility companies, conducted by author and research assistant Frank Schuller during summer of 1978.

9. For example, by 1980 half or more of the electricity produced in New England and in some southern Atlantic states will be nuclear. Comprehensive statistics are available in *Moody's Public Utilities*, 1977.

10. This section draws heavily on Bupp and Derian, *Light Water*. See also Richard G. Hewlett and Francis Duncan, *Atomic Shield 1947–1952: A History of the USAEC*, vol. 2 (University Park: Pennsylvania State University Press, 1969); and Richard G. Hewlett and Francis Duncan, *Nuclear Navy, 1946–1962* (Chicago: University of Chicago Press, 1974).

11. Bupp and Derian, *Light Water*, chap. 2.

12. "The Jersey Central Report," *Atomic Industrial Forum Memo*, 11, no. 3 (March 1964).

13. Bupp and Derian, *Light Water*, chap. 2.

14. See, for example, *Nuclear News* (formerly *Atomic Industrial Forum Memo*), 11 (January 1968). For Weinberg quote, see U.S. Congress, Joint Committee on Atomic Energy, *Nuclear Power Economics—1962 through 1967*, 90 Cong., 2 sess. (Washington D.C.: Government Printing Office, February 1968), p. 5.

15. Irvin C. Bupp, "Priorities in Nuclear Technology: Program Prosperity and Decay in the United States Atomic Energy Commission, 1956–1971," Ph.D. thesis, Harvard University, 1971.

16. Harold P. Green and Alan Rosenthal, *Government of the Atom: The Integration of Powers* (New York: Atherton Press, 1963).

17. Bupp and Derian, *Light Water*, chap. 1.

18. U.S. Congress, Joint Committee on Atomic Energy, *AEC Authorizing Legislation Fiscal Year 1968*, 90 Cong., 1 sess. (Washington, D.C.: Government Printing Office, March 1967), part 2, pp. 660–662, 667–900, and throughout.

19. Bupp and Derian, *Light Water*, chap. 1.

20. Bupp and Derian, chap. 11.

21. Bupp and Derian, chap. 9. See also Spurgeon Keeny *et al.*, *Nuclear Power: Issues and Choices* (Cambridge, Mass.: Ballinger, 1977); and U.S. Congress, Committee on Government Operations, *Nuclear Power Costs*, 95 Cong., 2 sess. (Washington, D.C.: Government Printing Office, April 1978). For an industry defense of nuclear power's recent economic performance, see A. D. Rossin and T. A. Rieck, "Economics of Nuclear Power," *Science*, vol. 201, no. 18, August 1978, pp. 582–589.

22. David L. Bodde, "Regulation and Technical Innovation: A Study of the Nuclear Steam Supply System and the Commercial Jet Engine," Doctoral thesis, Graduate School of Business Administration, Harvard University, 1975.

23. H. E. Van, M. J. Whitman, and H. I. Bowers, "Factors Affecting the Historical and Projected Capital Costs of Nuclear Plants in the USA," *Proceedings of the Fourth International Conference on the Peaceful Applications of Atomic Energy*, vol. 2 (Geneva: September 1971), pp. 21–43.

24. William E. Mooz, "Cost Analysis of Light Water Reactor Power Plants," Rand Corp., R–2304–DOE, June 1978. See also Bupp and Derian, chap. 9; Duncan Burn, *Nuclear Power and the Energy Crisis* (London: Trade Policy Research Centre, 1978), chap. 4; U.S. Nuclear Regulatory Commission, "Coal and Nuclear: A Comparison of the Cost of Generating Baseload Electricity by Region," NUREG–0480 (Washington, D.C.: U.S. Government Printing Office, December 1978); U.S. Congress, Committee on Government Operations, *Nuclear Power Costs*, 23rd report, April 1978.

25. Bupp and Derian, part II, chaps. 6–8.

26. This was true before the accident at Three Mile Island. It will be some time until the events at Harrisburg in March–April 1979 can be fully assessed. But the preliminary reports and evidence certainly appeared to vindicate many specific technical criticisms by the Union of Concerned Scientists about reactor design, operating, and inspection standards and practices.

27. Personal communication with the author. I am indebted to Professor Brooks for his advice and comments on this chapter. Naturally, this does not imply that he necessarily agrees with all of the analyses or conclusions.

28. This proposition, stated rather baldly here, is the central theme of Bupp and Derian, *Light Water*.

29. *The Risks of Nuclear Power Reactors: A Review of the Nuclear Regulatory Commission's Reactor Safety Study* (Cambridge, Mass.: The Union of Concerned Scientists, August 1977).

30. This argument is developed at length in Bupp and Derian, *Light Water*, chap. 9.

31. State of Wisconsin, Public Service Commission, *Advance Plans for Construction of Facilities—Findings of Fact, Conclusion of Law and Order*, August 17, 1978, p. 15; also, considerable quantitative support for this finding is contained in the appendix to the order. See also State of New York, Public Service Commission, Case 26974, "Proceeding on Motion of the Commission as to the Comparative Economics of Nuclear and Fossil Generating Facilities," Recommended decision by Administrative Law Judge Reed, Albany, December 18, 1978, p. 171; see also pp. 175–181.

32. For striking empirical evidence of these assertions, see Alan S. Manne and Richard G. Richels, "Probability Assessments and Decision Analysis of Alternative Nuclear Fuel Cycles," Unpublished manuscript, Stanford University, January 1979.

33. Alan Jakimo and Irvin C. Bupp, "Nuclear Waste Disposal: Not in My Backyard," *Technology Review*, March/April 1978. See also Charles L. Hebel, *et al.*, *Report to the American Physical Society* by the Study Group on Nuclear Fuel Cycles and Waste Management (Washington, D.C.: American Physical Society, 1977); and Richard G. Hewlett, "Federal Policy for Disposal of Radioactive Wastes from Commercial Nuclear Power Plants," U.S. Department of Energy, March 1979.

34. Theodore B. Taylor and Mason Willrich, *Nuclear Thefts: Risks and Safeguards* (Cambridge, Mass.: Ballinger, 1975); John McPhee, *The Curve of Binding Energy* (New York: Farrar, Straus and Giroux, 1974); and Daniel Yergin, "The Terrifying Prospect: Atomic Bombs Everywhere," *Atlantic Monthly*, April 1977.

35. Michael Brenner, "Carter's Non-Proliferation Strategy: Fuel Assurances and Energy Security," Center for Arms Control and International Affairs, Univer-

sity of Pittsburgh, September, 1977, cited by permission of the author. See also Victor Gilinsky, "Plutonium, Proliferation and The Price of Reprocessing," *Foreign Affairs*, January 1979, pp. 374–386.

36. Data provided to the author by the Office of Nuclear Policy, U.S. Department of Energy. See also statement of Dr. John P. Cagnetta on behalf of the Atomic Industrial Forum before the Subcommittee on Science, Technology, and Space of the Committee on Commerce, Science and Transportation, U.S. House of Representatives, August 10, 1978.

37. Testimony to the U.S. House of Representatives, Committee on Government Operations, Subcommittee on Environment, Natural Resources and Energy, September, 22, 1977.

38. Letter from the Steering Committee of the Keystone Radioactive Waste Management Group to Dr. John Deutsch, assistant secretary of Energy for Energy Technology, U.S. Department of Energy, and Dr. Frank Press, director, Office of Science and Technology Policy, Executive Office of the President, January 26, 1979. See also Jakimo and Bupp, "Nuclear Waste Disposal."

39. Statement of A. Z. Roisman and S. J. Scherr on behalf of the Natural Resources Defense Council before the House Committee on Interior and Insular Affairs, January 25, 1979. This section also draws upon interviews by author with Department of Energy officials.

40. Private communication from Dr. Peter Montague, Southwest Research and Information Center. This is a public interest organization, based in Albuquerque, New Mexico, that has opposed the Department of Energy's plan for WIPP. It has developed an impressive set of arguments against the WIPP facility.

41. Executive Office of the President, Office of Science and Technology Policy, "Isolation of Radioactive Wastes in Geological Repositories: Status of Scientific and Technological Knowledge," Working paper prepared for the Interagency Review Group on Nuclear Waste Management, July 3, 1978. See also U.S. Department of Energy, Directorate of Energy Research, *Report of Task Force for Review of Nuclear Waste Management*, DOE/ER–0004/D, February 1978; and Executive Office of the President, Office of Science and Technology Policy, *Report of the Subgroup on Alternative Technology Strategies*, Interagency Review Group on Nuclear Waste Management, August 7, 1978.

42. Office of Science and Technology Policy, *Alternative Technology Strategies for the Isolation of Nuclear Wastes: Report of Subgroup One*, September 8, 1978; see also U.S. Department of Energy, *Final Report of the Task Force for Review of Nuclear Waste Management*, March 1979.

43. Keeny, *et al., Nuclear Power: Issues and Choices*, chaps. 8, 11, and 12.

44. Letter from the Steering Committee of the Keystone Radioactive Waste Management Discussion Group to Dr. John Deutsch, assistant secretary for Energy Research, U.S. Department of Energy, and Dr. Frank Press, director, Office of Science and Technology Policy, Executive Office of the President, September 19, 1978. As this is written, the Carter Administration appears to be moving toward a multi-site program based on "intermediate scale facilities." Such a program is recommended in the final report of the interagency Waste Management Task Force, March 1979.

45. Some nuclear critics might argue that an alternative way to avoid this kind of regional catastrophe would be to strengthen the national electricity transmission network so that the loss of nuclear capacity could be shared across all regions, and thus would have a minimum impact on any one of them. We have no specific evidence on the cost and scope of such an effort, but it does not seem likely to be practical by the early 1980's.

46. Brian Flowers, "Nuclear Power: A Perspective of the Risks, Benefits, and Options," *Bulletin of the Atomic Scientists,* vol. 34, no. 3, March 1977, pp. 21ff.

47. Letter from the Keystone Group to Drs. Deutsch and Press, January 26, 1979.

Chapter 6
Conservation: The Key Energy Source

1. Planners at British Petroleum call conservation "non-use." See Robert Belgrave, "An Analysis of the Energy Balances for the U.K. and Western Europe to the Year 2000," Paper presented at the Royal Institute for International Affairs, London, March 1977. Roger Sant, as assistant FEA administrator for conservation, made the case for conservation as a legitimate energy source at a Camp David meeting to develop President Ford's energy plan in December 1974. For general surveys, see Robert Socolow, "The Coming Age of Energy Conservation," *Annual Review of Energy: 1977,* vol. 2, pp. 239–289; Lee Schipper and Joel Darmstadter, "The Logic of Energy Conservation," *Technology Review,* January 1978, pp. 41–50; Schipper, "The Best Energy Policy: Waste Not, Want Not," *Washington Post,* Outlook sec., April 10, 1977; Schipper, "Raising the Productivity of Energy Utilization," *Annual Review of Energy: 1976,* vol. 1, pp. 455–517; and Denis Hayes, *Energy: The Case for Conservation,* Worldwatch Paper 4, January 1976. One of the most important of all energy studies conducted so far is that of the MIT Workshop on Alternative Energy Strategies, which concluded, "Energy conservation may well be the very best of the alternative energy choices available." Carroll Wilson, ed., *Energy: Global Prospects 1985–2000* (New York: McGraw-Hill, 1977). The toast example is borrowed from Grant Thompson.

2. The often ignored but indeed substantial job-creating potential of conservation is discussed in U.S. Congress, Joint Economic Committee, *Creating Jobs Through Energy Policy: Hearings,* 95 Cong., 2 sess. One of the most detailed efforts to investigate job-creating potential of various energy sources was carried out for Long Island. Its findings: A six-billion-dollar investment in conservation and solar provided twice as much energy for end use as a seven-billion-dollar investment in nuclear power. Over a thirty-year period, the conservation and solar package created 178,000 jobs, as opposed to 72,000 jobs created by the nuclear investment; *Creating Jobs Through Energy Policy,* pp. 28–39.

3. Schipper and Darmstadter, "The Logic of Energy Conservation," p. 42. The imagery of curtailment and overhaul are powerful. In 1977, 48 percent of those questioned in a national survey agreed that "conservation is not a realistic solution to the energy crisis unless we are all prepared to accept a much lower standard of living." Forty-two percent disagreed. This result, from *Cambridge Report,* is cited by René Zentner, "The Myth of Energy Conservation," in Ragaei and Dorothea el-Mallakh, *Energy Options and Conservation* (Boulder, Colo.: International Research Center for Energy and Economic Development, 1978), pp. 147–48.

4. A Massachusetts poll found that half of the respondents expected a technological production fix to solve the problem. Letter from Henry Lee, director, Massachusetts Energy Office, to author, July 24, 1978.

5. The proceedings of the conference are in Sonja S. Marchand, ed., "Energy— The Global Challenge," Proceedings of a Conference sponsored by California State University of Northbridge, May 6–7, 1977. The senator's aide's comment is from an interview with the author. At a recent national governor's conference, there were six energy panels. Only two governors bothered to show up at the single

panel concerned with conservation. The other forty-eight hurried off to hear about supply.

6. Scientist is Socolow, "The Coming Age of Conservation," in *Annual Review of Energy: 1977*, p. 252. For a critique of the lack of support for applied research on conservation, see Terry Grew, George W. Sutton, and Martin Zlotnick, "Fuel Conservation and Applied Research," in *Science*, April 14, 1978, pp. 135–142.

7. The first prominent statement encouraging energy conservation was the Ford Foundation's Energy Policy Project, *A Time to Choose: America's Energy Future* (Cambridge, Mass.: Ballinger, 1974). The reader is advised to note the searing dissents of those associated with supply on the advisory committee. As visible has been Amory Lovins, "Energy Strategy: The Road Not Taken?" *Foreign Affairs*, October 1976, pp. 65–96. For the Carter Administration, see *National Journal*, August 22, 1978, p. 1286; Daniel Yergin, "U.S. Energy Policy: Transition to What?" *World Today*, March 1979, pp. 81–91 (also in *Europa Archiv*, February 1979). The Carter pollster Pat Caddell reported that in 1976 no more than 6 percent of the public regarded energy as one of the nation's most important problems. That figure rose to 35 percent in the spring of 1977, when Carter introduced his National Energy Plan, but had dropped back to 20 percent by the summer of 1978. For the utility, see W.R.Z. Willey, "Alternative Energy Systems for Pacific Gas & Electric Co: An Economic Analysis," Testimony submitted to the California Public Utilities Commission by the Environmental Defense Fund, 1978, pp. 8–9.

8. A study by the Conference Board found the reductions of energy use after the 1973–74 price hikes were not "as much as would be expected from a review of published econometric price elasticity estimates." One suggested reason "for the apparent weak response to real energy price increases is the leveling off in new capital investment during the 1970s." Bernard A. Gelb, "U.S. Energy Price and Consumption Changes in the Mid-1970s," Conference Board Information Bull. No. 38, March 1978. The inadequacies—and great social strains—of trying to depend exclusively upon price signals to stimulate conservation are stressed in Massachusetts Energy Office, *The New England Policy Alternatives Study: Final Report*, October 1978, pp. 41–44.

9. *Energy Report from Chase*, September 1976. Oil company in an internal memorandum. The Texas Railroad commissioner was quoted in *Newsweek*, April 18, 1977, p. 73. The Federal Energy Agency's 1976 outlook assumed that the connection between energy consumption and economic activity was relatively fixed for industry, which may be one of the reasons conservation was so low a priority in the Nixon and Ford administrations. Federal Energy Agency, *National Energy Outlook: 1976* (Washington, D.C.: Government Printing Office, 1976), p. 27.

10. Energy to GNP data is from John G. Myers, "Energy Conservation and Economic Growth—Are They Incompatible?" *The Conference Board Record*, February 1975, pp. 27–32. Also see Sam H. Schurr and Joel Darmstadter, "Some Observations on Energy and Economic Growth," Resources for the Future, 1977. Some of the factors that have to be considered in evaluating how elastic the energy–GNP link has become since 1973 include economic recession, weather, differential growth in electricity demand among different countries, structural shifts in industry, rate of investment in capital stock that embodies energy savings, and so on and on. No wonder the task is so difficult. Thus, at a recent meeting the chief economist of one major oil company postulated energy savings per unit of GNP in the OECD countries during the period 1973–77 as 4 percent; another oil company economist with no less certainty more than doubled the estimate to 9 percent. No one could deny that energy conservation has occurred since 1973, but the data for the 1970's would seem much too uncertain to make any bold claims about "conservation having really caught on"; the ratio between growth rates in energy consumption and growth in GNP in the United States has been erratic, as the following table indicates:

	Energy Growth	Economic Growth	Ratio Energy Growth/ Economic Growth
1970	+3.3	− .3	—
1971	+1.8	+3.0	.60
1972	+4.8	+5.7	.84
1973	+4.1	+5.4	.76
1974	−2.6	−1.4	—
1975	−2.8	−1.3	—
1976	15.3	+5.7	.93
1977	+2.0	+4.9	.41
1978 (est.)	+2.4	+3.8	.63

11. For 1972 data (though slightly revised since then), see Joel Darmstadter, Joy Dunkerley, and Jack Alterman, *How Industrial Societies Use Energy: A Comparative Analysis* (Baltimore: Johns Hopkins University Press, 1977), p. 5. For an in-depth study of Sweden, see Lee Schipper and A. J. Lichtenberg, "Efficient Energy Use and Well-Being: The Swedish Example," *Science*, December 3, 1976, pp. 1001–1013.

12. The complexities involved in these comparisons have been much illuminated in the important new study by Darmstadter, Dunkerley, and Alterman, *How Industrial Societies Use Energy*. Some have leaped upon this work to argue, as *The Oil and Gas Journal* of March 6, 1978 (pp. 30–31) did in a headline, "Study Explodes U.S. Energy-Waste Myth." On the contrary, it does not. Despite the rather cautious conclusions by the authors of *How Industrial Societies Use Energy*, their evidence and arguments do affirm that the United States is more intensive in its energy use.

13. The similarities are noted in Socolow, "The Coming Age of Conservation," *Annual Review of Energy: 1977*, p. 247. For a particularly sophisticated but not persuasive attempt to deny the value of energy–GNP ratios, see Chauncy Starr, "Energy Use: An Interregional Analysis with Implications for International Comparison," in an important volume, Joy Dunkerley, ed., *International Comparison of Energy Consumption* (Washington D.C.: Resources for the Future, 1977). Starr warns that GNP–energy ratio comparisons "can lead to wrong conclusions"; he argues that such ratios reflect "the most efficient way to do things for that society" (p. 36). That is not at all clear: How does such a measure of efficiency factor in nuclear proliferation or the dangers of imported oil? Moreover, he sets up a straw man. No one has said these are anything other than rough, but suggestive, indications. Nor, as the volume seems to suggest, has anyone suggested that these comparisons in themselves are "a useful guide to policy formulation" (p. xxvii). But if the "more disaggregated studies" called for (p. xxxi) do end up pointing in the same direction as the general comparisons, then the general comparisons are helpful indicators. The disaggregated studies do so point, and thus it is appropriate to find in the general comparisons suggestive evidence for greater flexibility than traditionally assumed in the relation between economic activity and energy consumption. Indeed, one might ask why the correlation between energy and GNP should be any more fixed than that between GNP and the size of the labor force. Increased productivity of labor—not just a growing labor force—has been a major source of economic growth. See Robert Solow, "Technological Change and the Aggregate Production Function," *Review of Economics and Statistics* 39 (April 1957), pp. 312–320. The same might be true, to some degree, for an increasing "productivity" of energy. Although some might reply that energy is a trade-off for human labor, the evidence suggests otherwise: that even in industry, most energy is used for process and space heat, not power. Darmstadter, Dunkerley, and Alterman, *How Industrial Societies Use Energy*, pp. 116–117.

14. S. M. Lambert of Shell U.S.A., quoted in *New York Times,* February 11, 1978. W. W. Rostow has pointed out that a radical change in the price of a commodity such as oil sets off a dynamic process that has far-reaching consequences difficult to capture in conventional models—direct income effects that differ by regions, changing investments and population flows, and alterations in the pace and character of development. "Energy-economy models are substantially misleading," he observes, because "these models mask out by assumption the critical features" in the current U.S. situation. Rostow, "Energy, Full Employment, and Regional Development," Paper delivered at the American Association for the Advancement of Science, February 14, 1978. Suddenly, the economic life of the entire capital stock is foreshortened.

15. The discussion of the Los Angeles Plan draws on three Rand Corporation studies: Jan Paul Acton and Ragnhild Mowill, "Regulatory Rationing of Electricity under a Supply Curtailment," P–5624; Acton and Mowill, "Conserving Electricity by Ordinance: A Statistical Analysis," R–1650–FEA; and Acton, M. H. Graubard, and D. J. Weinschrott, "Electricity Conservation Measures in the Commercial Sector: The Los Angeles Experience," R–1592–FEA. Interview with Harold Williams by author in January 1975; Harold Williams, "Why We Must Break the OPEC Cartel: A Response to President Ford's Energy Program," Statement, October 10, 1974. (Williams, now chairman of the Securities and Exchange Commission, was Los Angeles' energy coordinator during the crisis.) U.S. Congress, Joint Economic Committee, Subcommittee on Energy, *Energy Conservation: Hearings,* 94 Cong., 2 sess., p. 272; and *Los Angeles Times,* various articles, November–December 1973 and January 1974.

16. *Monthly Energy Review,* March 1975, pp. 11–12, and January 1978, p. 47; Motor Vehicles Manufacturers Association, *Motor Vehicle Facts & Figures '77,* p. 32.

17. *Motor Vehicles Facts & Figures '77,* p. 67. Forty percent of all car miles in 1970 were devoted to commuting. Three quarters of all workers drove to work, and four fifths did so alone. See Dorothy Newman and Dawn Day, *The American Energy Consumer* (Cambridge, Mass.: Ballinger, 1975), p. 78.

18. Jeffrey S. Milstein, "Energy Conservation and Travel Behavior," Department of Energy paper, October 1977, pp. 3–4; letter from Henry Lee, director, Massachusetts Energy Office, to author, July 24, 1978.

19. See Jerry Ward and Norman Paulhus, *Suburbanization and Its Implications for Urban Transportation Systems* (Washington, D.C.: Department of Transportation, 1974); and Congressional Budget Office, *Urban Transportation and Energy: The Potential Modes of Different Savings* (Washington, D.C.: Government Printing Office, 1977), pp. 42–51.

20. Average fuel-economy loss has been estimated at between 4 and 6 percent. J. L. Duda *et al., Program Evaluation Support for the Motor Vehicle: Diagnostic Inspections, and Demonstration Program, Volume 2, Costs and Benefits,* Department of Transportation Report HS–802406 (Falls Church, Va.: Computer Science Corporation, 1977); Ted Baylor and Leslie Eden, *Fuel Economy Improvement through Diagnostic Inspection,* DOT–NHSTA Report HS–802284, March 1977. Synthetic motor fuel oils can improve miles per gallon by more than 4 percent. J.A.C. Krulish, H. B. Lowthen, and B. J. Miller, "An Update on Synthesized Engine Oil Technology," Paper prepared for Fuels and Lubricants Meeting, Society of Automotive Engineers, June 1977. Also see Daniel Yergin, "France's Tough Energy Program Puts the Heat on the Admen," *Fortune,* June 17, 1978, pp. 106–112. For speed limits, see General Accounting Office, *Speed Limit 55: Is it Achievable,* CED–77–27, 1977. Low-grade tires may impose a 15 percent penalty on fuel economy. *Ward's Automotive Reports,* December 18, 1978, p. 404.

21. U.S. Federal Task Force, *Motor Vehicle Goals Beyond 1980,* vols. 1 and 2 (Washington, D.C.: Energy Resources Council, 1976), chap. 8

22. Society of Automotive Engineers, *Passenger Car Fuel Economy Trends Through 1976,* Paper 750957, October 1975; Newman and Day, *The American Energy Consumer,* p. 72; Robert Williams, ed., *The Energy Conservation Papers* (Cambridge, Mass.: Ballinger, 1975), pp. 23, 312–314; U.S. Congress, Senate, Finance Committee, *Energy Conservation and Conversion Act of 1975: Hearings,* 94 Cong., 2 sess., p. 465; John Tien, Ray W. Clark, and Mahendra K. Malu, "Reducing the Energy Investment in Automobiles," and Joseph Kummer, "The Automobile as an Energy Converter," both in *Technology Review,* February 1975, pp. 26–43. Between 1968 and 1973, pollution standards may have imposed as much as 12 percent penalty on efficiency on a sales-weighted average. However, improvements in emissions-control devices had reduced the penalty by 1975.

23. A senior executive with one of the top three manufacturers recalled, "The low priority assigned to fuel efficiency before 1973 was associated with ever-rising world oil reserves and a real decline in the price of gasoline when measured in minutes of labor at the average wage rate" (private communication with author). And the minutes of a meeting involving senior officials of another of the top three manufacturers in 1972 point to some suspicion of the oil industry: "We must evaluate realistically the basic energy supply data and energy demand data. Almost all studies are based on [American] Petroleum Institute statistics, an unneutral body. It is likely safe to say that proven reserve statistics are not overstated; the question is to evaluate how much they are understated."

24. Ward and Paulhus, *Suburbanization and Its Implication for Urban Transportation Systems;* James J. Mutch, *Transportation Energy Use in the United States: A Statistical History: 1955–71,* Rand Corporation, R–1391–NSF, December 1973; Eric Hirst, "Transportation Energy Conservation: Opportunities and Policy Issues," in U.S. Department of Transportation, *Energy Primer.*

25. According to the Commission of the European Community, *Rational Use of Energy: Second Periodical Report of Sub-Group C—Road Transport Vehicles* (Draft), "There is a strong evidence that total fuel consumption is rather insensitive to fuel prices; only major increases in fuel taxation are likely to have a significant effect." Of course, gas prices are generally higher in Europe to begin with. For influence of gas prices on auto purchases, see Daniel Yergin, "France's War on Energy Waste," Harvard Business School Energy Research Project paper, June 1978. Renault introduced the R5 GTL after the 1973 crisis, a variant on its 950 ("Le Car" in the United States). It cost 10 percent more. The only advantage was greater fuel economy. Thirty-five percent of the people buying this popular car in France have opted for the more efficient model.

26. Milton Russell, "Energy," Resources for the Future reprint, No. 145, p. 331. *Monthly Energy Review,* October 1978, p. 65.

27. Darmstadter, Dunkerley, and Alterman, *How Industrial Societies Use Energy,* pp. 92–93; General Accounting Office, *U.S. Energy Conservation Could Benefit from Experiences of Other Countries,* ID–78–4, January 10, 1978, p. 10–11; Organization for Economic Cooperation and Development, *Energy Conservation in the International Energy Agency: 1976 Review* (Paris: OECD), pp. 42–43.

28. Woodcock in U.S. Congress, Senate, Finance Committee, *Energy Conservation and Conversion Act of 1975: Hearings,* 94 Cong., 1 sess., p. 473. For down-sizing, see Tien, Clark, and Malu, "Reducing the Energy Investment in Automobiles," *Technology Review,* February 1975, pp. 38–43; U. S. Congress, Senate, Finance Committee, *Energy Conservation and Conversion Act of 1975: Hearings,* 94 Cong., 2 sess., pp. 173–174. Three quarters of GM's weight reduction came from resizing and engineering improvements, between 10 and 15 percent from

smaller engines, drivetrains, and accessories, and the rest from use of premium materials. *Automotive News,* August 28, 1978, p. 43.

29. For possible effects on market share, see William Abernathy, and Balaji S. Chakravarthy, "Technological Change in the U.S. Automobile Industry: Assessing the Federal Initiatives," Paper prepared for the Department of Transportation, December 1977, pp. 1, 45. For the competing demands, see Abernathy and Chakravarthy, p. 59; Raymond E. Good, "The Automobile: Interaction of Energy, Safety, Environment and the Economy," in U.S. Department of Transportation, *Regulation and Transportation: Report of the Third Workshop on National Transportation Problems* (Washington, D.C.: DOT, 1975). "One element of uncertainty is the success of specific technology we are now developing," said Henry Duncombe, chief economist for General Motors in 1977. "However, the range of this uncertainty is relatively small. Technical feasibility is not the key issue here today—cars on the market already exceed 27.5 mpg. The major uncertainty will be the potential losses of auto sales caused by fuel economy standards." Cited in Richard John, Philip Coonley, Robert Ricci, and Bruce Rubinger, "Mandated Fuel Economy Standards as a Strategy for Improving Motor Vehicles' Fuel Economy," Paper presented at Symposium on Technology, Government, and the Future of the Automobile Industry at the Harvard Business School, October 19, 1978, p. 29.

30. General Motors in *Energy Conservation and Conversion Act of 1975: Hearings,* p. 139; Henry Ford II, Speech to the White House Conference on Balanced National Growth and Economic Development, January 30, 1978.

31. *New York Times,* "National Economic Survey," January 8, 1978, p. 41; Milstein, "Energy Conservation and Travel Behavior," p. 3; Richard Strombotne, "Transportation: Energy Outlook," Speech at Conference Board, December 7, 1977; *Ward's Automotive Yearbook: 1976,* p. 20. The cumulative fuel savings in Richard John *et al.,* "Mandated Fuel Economy," p. 4. For rising gasoline sales, see American Petroleum Institute, *Monthly Statistical Report,* November 1978; Department of Energy, "Petroleum Demand Watch," Release, December 12, 1978, and "DOE Plans Conference to Assess Middle Distillate and Gasoline Supply Situation," Release, November 30, 1978. Some possible reasons for the surge in gasoline sales, despite the fuel standards, are discussed in Roger Sant, "Annual Demand Assessment," Paper prepared for Aspen Institute Energy Committee, 1978, p. 4. On light vans, see Robert F. Hemphill, Jr., "Energy Conservation in the Transportation Sector," in John Sawhill, ed., *Energy Conservation and Public Policy* (New York: Prentice Hall, 1979).

32. For importation of technology and $80 billion capital estimate, see Richard John *et al.,* "Mandated Fuel Economy." The $5 billion to $10 billion figure is from U.S. Federal Task Force, *Motor Vehicle Goals Beyond 1980,* vol. 1, p. 29, and vol. 2, pp. 6–6, 6B–1–2. Also see Abernathy and Chakravarthy, "Technological Change in the U.S. Auto Industry," p. 60; "Machine Tools: Uproar over a Bottleneck," *New York Times,* February 26, 1978; Robert Irvin, "Big Four Spending Jacked Up Again," *Automotive News,* June 5, 1978, p. 1.

33. For future mileage, see Charles Cohn, "Improved Fuel Efficiency for Automobiles," *Technology Review,* February 1975, pp. 45–53; *Economist,* January 20–27, 1978, pp. 76–77; Charles Deutsch, *Economies d'Energie par la Conception des Voitures Particulières, Les Dossiers de l'Energie* 10 (Paris: Ministry of Industry and Commerce, 1977); *Motor Vehicle Goals Beyond 1980;* and Joan Claybrook, "The Snail's Pace of Innovation," Speech at *Automotive News* World Conference, June 13, 1977; Marion Meader, *Seminar on Automobile Fuel Efficiency: Proceedings,* vol. 2 (McLean, Va.: Mitre Corp., 1978), p. 64; Gerald Leach *et al., A Low Energy Strategy for the United Kingdom* (London: Science Reviews, 1979), pp. 156–164. The innovation problem, as well as fuel economy potentials, is discussed in Abernathy and Chakravarthy, "Technological Change in the U.S. Auto Industry," pp. 28–29, 60–65, and in Richard John *et al.,* "Mandated

Fuel Economy." On the unexpected export market, see "To a Global Car," *Business Week,* November 20, 1978, pp. 102–113. The 1979 mileage results for the Rabbit are in the Environmental Protection Agency's revised city driving test; see *Automotive News,* September 18, 1978, p. 1.

34. On inattention to materials-saving innovation in the United States in the postwar years, see William H. Davidson, "Patterns of Factor-Saving Innovation in the Industrialized World," *European Economic Review 8* (1976), pp. 207–217.

35. Aluminum, in John Myers and Leonard Nakamura, *Saving Energy in Manufacturing: The Post-Embargo Record* (Cambridge, Mass.: Ballinger, 1978), p. 122. The energy-intensive paper industry gets a growing part of its fuel needs, now 45 percent, from burning "hog fuels"—that is, such wastes as sawdust, chips, barks—and in the process reducing environmental pollution; see *Energy User News,* June 6, 1977, December 29, 1977, and February 13, 1978; see also Roger Slinn, "The Conflicts: Ecology, Jobs, Regional Diversity," Speech at Conference Board, December 7, 1977.

36. From *Energy User News,* April 11, 1977, April 18, 1977, June 13, 1977, October 31, 1977, and February 20, 1978.

37. *Energy User News,* April 10, 1977, June 13, 1977, and February 27, 1978.

38. Energy–GNP ratio (thousand BTUs per 1972 dollar of GNP) from Table 2–2 in John G. Myers, "Industrial Energy Demand 1976–2000," Draft study prepared for the U.S. General Accounting Office, January 31, 1979. The qualifications are stressed in this paper and in an earlier paper by Myers, "Energy Conservation in Manufacturing Since the Embargo," Paper presented at Conference on the Economic Impact of Energy Conservation, July 1978. For criticism of the voluntary industrial energy use reporting scheme, see Myers and Nakamura, pp. 11–13, and *Energy User News,* June 20, 1977, October 17, 1977, and October 30, 1977. Obviously, associations and firms eager to avoid further regulation will tend to put the best interpretation on ambiguous data. Another rather critical energy audit, this with twenty major industrial firms, also found a gap between claims and results. See General Accounting Office, *Federal Agencies Can Do More to Promote Energy Conservation by Government Contractors,* EMD–77–62, September 30, 1977, pp. 5–19. A more recent report points to the confusion between energy saving and production runs. General Accounting Office, *The Federal Government Should Establish and Meet Conservation Goals,* EMD–78–38, June 30, 1978, p. 32.

39. Myers and Nakamura, *Saving Energy in Manufacturing,* pp. 7–9, 24, 121–23. Some possible reasons for the rise in the energy-output ratio in the five industries would include lower use of capacity (generally meaning less efficiency), the effects of safety and pollution regulations, lack of funds available for conservation investments—and lack of sufficient attention.

40. Goodyear in *Energy User News,* November 28, 1977. See also Neil DeKoker, "Energy Conservation in Management," *Industrial Engineering,* December 1977; Allied Chemical, "Energy Conservation Program," Unpublished paper, June 1978. The absolute requirement of senior management commitment is stressed by the chairman of British Petroleum and an executive of Courtaulds. David Steel, Speech, and J.R.S. Morris, "Implementing an Energy Management Policy," Paper, both presented at National Energy Management Conference, Birmingham, England, October 1978. As Morris said, "Everything starts from the top." For a study of the need for commitment of senior management and methods used to induce organization-wide change in activities that often do not impact heavily on profits, see Robert Ackerman and Raymond Bauer, *Corporate Social Responsiveness: The Modern Dilemma* (Reston, Va.: Reston, 1976). For Armco, see Henry Miller, "User Requirements and Conservation," Speech at Conference Board, December, 7, 1977, p. 2, 8.

41. Henry Miller, "User Requirements and Conservation"; *Energy User News,* April 11 and 25, 1977.

42. G. N. Hatsopoulos, E. P. Gyftopoulos, R. W. Sant, and T. F. Widmer, "Capital Investment to Save Energy," *Harvard Business Review,* March–April 1978, pp. 111–122. Also Roger Sant to author, October 26, 1978. Courtaulds, an international chemical company with major operations in the United States, initially established a three-year payback criteria, then shortened it to two. This was partly because there were "ample opportunities" at two years—apparently much in excess of the allocated capital. J.R.S. Morris, "Implementing an Energy Management Policy." Many companies fail to note the "insurance valve" of conservation. See Robert C. Lind, "The Rationale for Federal End Use (Conservation and Solar) Programs: Implications for Policy and Program Evaluation," Draft of paper prepared for Office of Conservation Planning and Policy, Department of Energy, pp. 5–6, 10–11. For a survey of how different firms calculate returns on conservation investments, see Barnaby J. Feder, "Energy Paybacks Draw More Attention," *Energy User News,* November 27, 1978, p. 1.

43. Michael Tenebaum, "Reflections on Steel's Energy Maze," Speech to the American Iron and Steel Institute, May 25, 1977. It is generally not recognized that the prices of some of the most important fuels have actually declined in real terms. Between August 1975 and August 1978, the real price of Number 6 fuel oil, the main industrial fuel oil, declined by 15 percent. During that same period, interstate natural gas used by industrial consumers increased in price by 69 percent. However, in intrastate markets, natural gas prices for industrial users decreased by about 5 percent in constant dollars. *Monthly Energy Review,* November 1978, pp. 75–78.

44. Letter from Henry Lee to author, July 24, 1978; Massachusetts Energy Office, "Survey of New England Business Attitudes Toward Energy and Energy Conservation," Mimeograph memo; National Federation of Independent Business, "Fifth Energy Report for Small Business," June 1977, p. 30; General Accounting Office, *Federal Agencies Can Do More to Promote Energy Conservation by Government Contractors,* pp. 19–22, 40; "Carpet Firms on Fuel Crisis: Is it Real, Can We Cope?" *Energy User News,* December 1977, p. 7; Samuel I. Doctors, Liam Fahey, and G. Richard Patton, "The Response of Small Manufacturers to the Energy Environment," Working Paper 248, Graduate School of Business, University of Pittsburgh. A detailed survey of potential for industrial conservation in Texas pointed out several constraints that "will reduce the impact of economics as a stimulus for conservation." In larger firms, conservation projects must compete with other often higher-priority activities for investment capital. In smaller firms, the requisite knowledge for economic and engineering analysis may be lacking. "Another disincentive for conservation is that in industries where competition is low or demand is insensitive to price increases, the increased cost for energy can be passed on to the consumer through increased product price. In many cases also, the costs of energy are only a small fraction of the total production costs. In this case small savings may be ignored." William Cepeda, *Potential for Energy Conservation in Texas: Report to Governor's Energy Advisory Council,* Number 77–001, April 1977, pp. 67–8. Here we see numerous examples of Alfred Kahn's "tyranny of small decisions"; see "The Tyranny of Small Decisions: Market Failures, Imperfections, and the Limits of Economics," *Kyklos* (1966), pp. 23–46.

45. *Energy User News,* May 14, 1978, p. 9.

46. Thermo-Electron, *A Study of Inplant Electric Power Generation in the Chemical, Petroleum Refining and Paper and Pulp Industries: Final Report,* Prepared for the Federal Energy Administration, Contract CO–04–50224–00.

47. Kjell Larsson, "District Hearing: Swedish Experience of an Energy Efficient Concept"; Commission of the European Community, *Rational Use of Energy,*

Second Periodical Report of Sub-Group G: Conversions in Power Stations, pp. G–33–36; U.K. Department of Energy, *District Heating Combined with Electricity Generation in the United Kingdom,* Energy Paper No. 20; Robert Williams, *The Potential for Electricity Generation as a Byproduct of Industrial Steam Production in New Jersey* (Princeton: Center for Environmental Studies, 1977), p. 24.

48. In the 1920's and early 1930's, several major paper companies went into the electricity business via cogeneration. And a very profitable business they found it. In the 1930's, the Justice Department took an interest in their activities, and in a series of court suits the companies were forced to decide whether they were in the paper business or the electric power business. They chose paper. See Charles Berg, "Conservation in Industry," *Science,* April 19, 1974, p. 268.

49. U.S. Congress, Senate, Energy Committee, Subcommittee on Conservation, *Status of Federal Energy Conservation Programs: Hearings,* 95 Cong., 1 sess., p. 196.

50. Williams, *The Potential for Electricity Generation as a Byproduct of Industrial Steam Production in New Jersey,* pp. 2–4; Thermo-Electron, *A Study of Inplant Electric Power Generation,* p. 2–2; U.K. Department of Energy, *District Heating Combined with Electricity Generation in the United Kingdom,* p. 3; Federal Energy Agency, *Comparison of Energy Consumption Between West Germany and the United States,* Conservation Paper No. 33.

51. Dow Chemical, *Energy Industrial Center Study,* June 1975; Thermo-Electron, *Summary Assessment of Electricity Cogeneration in Industry,* March 15, 1977; Hatsopoulos, Gyftopoulos, Sant, and Widmer, "Capital Investment to Save Energy," *Harvard Business Review,* p. 115; Williams, *The Potential for Electricity Generation as a Byproduct of Industrial Steam Production in New Jersey,* pp. 14–15. For Massachusetts, see Resource Planning Associates, *The Potential for Cogeneration Development in Six Major Industries by 1985: Executive Summary,* December 1977, p. iv; and Robert Elgin, "Some Implications for Cogeneration in New England from the Resource Planning Associates Study," Paper prepared for Governor's Commission on Cogeneration, 1978. These studies discuss return on investment.

52. The author, having sat for a year on the Governor's Commission on Cogeneration in Massachusetts, speaks from experience on the difficulties in analyzing the problem. The Massachusetts study, intended as a guide for the rest of the country, is a thorough investigation of the economic, regulatory, environmental, and market potential for cogeneration in New England. Governor's Commission on Cogeneration, *Cogeneration: Its Benefits to New England* (Boston: Commonwealth of Massachusetts, 1978). The report and its preparations provide the background for the discussion of cogeneration. It is a basic study of the real-world implementation for cogeneration. Also see John Belding, "Alternatives to Oil and Gas Through Energy Management," Paper prepared for New Options in Energy Technology Conference of the American Institute of Aeronautics and Astronautics, 1977.

Increasing attention is now being focused on proposals to free potential cogenerators from federal and state utility regulations and to ensure that utilities take their surplus power. The cogeneration question can be seen as part of what has been called "the great rate debate" (W. Donald Crawford, "An Electric Utility Perspective on Rate Design Revision," *Public Utilities Fortnightly,* September 14, 1978, pp. 15–19). The 1978 National Energy Act, as well as some state regulatory bodies, seeks to change utility-pricing systems to encourage conservation through such measures as peak-load pricing and the phasing out of declining block rates. Very significant energy and capital cost savings could result, for instance, from peak-load pricing. See Jan Paul Acton, Bridger M. Mitchell, and Willard G. Manning, *Projected Nationwide Energy and Capacity Savings from Peak-Load Pricing of Electricity in the Industrial Sector,* Rand Corporation, R–2179–DOE, June 1978. The shift of emphasis represented in such measures involves "a basic change in the electric utility industry from one of supplying energy to one of encouraging conservation and efficiency in its use while, at the same time, maintaining reliable service and meeting the energy needs of its customers. This is a demanding challenge." (Hamilton Treadway, "Energy Conservation Rates for an Effective Conservation Program," *Public Utilities Fortnightly*, August 17, 1978, pp. 16–20.) A case study of utility fears about cogeneration is in Tom Alexander, "The Little Engine That Scares Con Ed," *Fortune*, December 31, 1978, pp. 80–84. A contrary case is that of Southern California Edison, which is working with customers and regulatory agencies to promote cogeneration. "It's a new ball game," observed Arthur Blake, the utility's supervisor of the load management project. Rising marginal cost for new capacity and uncertain and expensive energy supplies encouraged the utility to welcome cogeneration as a "means to defer construction of expensive new generating plants as well as conserve natural resources"; Arthur J. Blake, "Utility Looks at Cogeneration to Manage Peaks," *Energy User News*, October 30, 1978, pp. 22–23.

An example of the kind of cooperative exchange required of normally warring parties if such problems are to be sensibly resolved for a reasonable transition, is provided in an admirable document, New England Energy Congress, *New England Blueprint for Energy Action* (Somerville: 1979). As is evident from the coal and nuclear chapters in this book, utilities now find themselves trying to make long-term decisions in a very uncertain environment. Coping with the energy transition certainly does impose heavy burdens on utilities, forcing them to face several competing demands at once, and could well bring about substantial changes in their role and orientation. The question of their future role is one of the most important in the entire energy field, but is one that has only begun to be addressed by research.

53. Hatsopoulos, Gyftopoulos, Sant, and Widmer, "Capital Investment to Save Energy," *Harvard Business Review*, pp. 111–122; Thermo-Electron, *Summary Assessment of Electricity Cogeneration in Industry*, p. 2–3.

54. Hatsopoulos, Gyftopoulos, Sant, and Widmer, "Capital Investment to Save Energy," *Harvard Business Review;* letter from Roger Sant to author, October 26, 1978.

55. This point was made after a survey of firms by the General Accounting Office in *The Federal Government Should Establish and Meet Energy Conservation Goals*, p. 38.

56. Myers and Nakamura, *Saving Energy in Industry*, p. 45; Myers, "Conservation in Manufacturing Since the Embargo," p. 7.

57. See General Accounting Office, *The Federal Government Should Establish and Meet Energy Conservation Goals*, p. 34.

58. The discussion about Dow is based on interviews with several Dow executives. Also see J. C. Robertson, "Energy Conservation in Existing Plant," *Chemical Engineering*, January 21, 1974, pp. 104–22; W. A. Rollwage, "Energy Conservation in Chemical Plants," *Chemical Engineering Progress*, October 1975, pp. 44–49; Gerald Decker, "Energy Conservation in Industry," Unpublished paper, September 27, 1976, and Decker, "Energy Conservation at the Dow Chemical Company," Unpublished paper, December 1976; *New York Times*, February 5, 1978, sec. 3.

59. The data on residential energy use (in 1975) from Eric Hirst and Jane Carney, *Residential Energy Use to the Year 2000: Conservation and Economics*, Oak Ridge National Laboratory/Con 13, September 1977. Also Fred Dubin, "New Energy Conservation Ideas for Existing and New Buildings," *Specifying Engineer*, January 1976; U.S. Congress, Senate, Interior Committee, *Energy Conservation Act of 1976: Hearings*, 94 Cong., 2 sess., p. 109. New York office building information is cited in Richard G. Stein, *Architecture and Energy* (New York: Doubleday, 1977), pp. 60–61. For the trend in private dwellings, see Eric Hirst and Jerry Jackson, "Historical Patterns of Residential and Commercial Energy Uses," *Energy*

2 (June 1977). Also see Newman and Day, *The American Energy Consumer,* pp. 39–43. Of course, new sealed buildings have one advantage: Dirt does not blow in the window.

60. Stein, *Architecture and Energy,* pp. 215, 292.

61. IBM in Claire Stegmen, "Not Bad, But Still Not Good Enough," *Think,* October–November 1976; U.S. Congress, Joint Economic Committee, Subcommittee on Energy, *Energy Conservation: Hearings,* pp. 67–74; U.S. Congress, Senate, Interior Committee, *Energy Conservation Act of 1976: Hearings,* p. 182; and Letter, John Honeycomb, IBM, to author, August 4, 1978.

62. GSA in Fred Dubin, "Energy Management for Commercial Buildings," Paper presented at Lawrence Berkeley Laboratory, July 1976, p. 25; and Dubin, "Energy Conservation Studies," *Energy and Buildings,* vol. 36, 1977. Ontario Hydro in *Energy User News,* March 20, 1978.

63. H. C. Fischer *et al., The Annual Cycle Energy System: Initial Investigations,* Oak Ridge National Laboratory ORNL/TM-5525, October 1976. For a practical viewpoint from inside the construction industry, see Ray Harrell, "Can We Build a More Efficient House?" in Paul Hendershat, ed., *Energy Conservation and Development* (Murfreesboro, Tenn.: Middle Tennessee State University, 1977).

64. This viewpoint is eloquently expressed by Stein in *Architecture and Energy.*

65. City of Seattle, Energy Office, *First Quarter's Report,* January–March 1978. For a description of the process whereby HUD is tightening standards, see *New York Times,* March 18, 1978, p. 25.

66. Newman and Day, *The American Energy Consumer,* pp. 40–42; Robert Rosenberg, "Energy Usage in the Home—Consumption and Conservation," in George Morganthaler and Aaron N. Silver, eds., *Energy Delta: Supply vs. Demand* (Tarzana, Calif.: American Astronautical Association, 1974), pp. 76–77.

67. U.S. Congress, Senate, Commerce Committee, *Energy Conservation Act of 1976: Hearings,* 94 Cong., 2 sess., p. 126; U.S. Congress, Senate, Interior Committee, *Energy Conservation Act of 1976: Hearings,* pp. 667–668; Harold B. Olin, "Put Some Sunshine into Your Mortgage Portfolio," *Savings and Loan News,* January 1977; "Energy Loan Programs Square Off against the Elements," *Savings and Loan News,* October 1977.

68. U.S. Congress, Senate, Interior Committee, *Energy Conservation Act of 1976: Hearings,* p. 117.

69. U.S. Congress, Joint Economic Committee, Subcommittee on Energy, *Energy Conservation: Hearings,* p. 70; IBM, *Annual Report: 1977,* p. 33. In IBM world-trade countries, the energy reduction over preconservation levels in 1973 to 1977 was 30 percent. Between 1973 and 1977, AT&T reduced its energy use per square foot by 30 percent, also with very little investment. See AT&T, "Energy Conservation in the Bell System," Mimeographed paper, December 1977, p. 7.

70. For 3M, see U.S. Congress, Joint Economic Committee, Subcommittee on Energy, *Energy Conservation: Hearings,* pp. 74–104; U.S. Congress, Senate, Interior Committee, *Energy Conservation Act of 1976: Hearings,* pp. 174–77; 3M, "Plant Energy Optimization Guidelines," January 14, 1977, and "Engineering Energy Conservation Standards," Mimeographed papers, July 14, 1977. Ashland case described in letter from Paul Chellgren to author, April 28, 1978. Bell Telephone, in its effective conservation program, also found outside air to be one of the biggest villains. AT&T, "Energy Conservation in the Bell System," Unpublished paper, December 1977, p. 8. These various cases stand out more as examples of what is possible than of what is being generally done. While the design of new commercial buildings emphasizes energy saving, there is a considerable body of evidence that conservation energy in the existing commercial stock is not being tapped anywhere

near its potential. One major reason is that a two- to three-year payback is often considered too long. Also, expenditures for improvement in management and maintenance of buildings lag when the economy is uncertain or weak. These are the conclusions of a DOE-sponsored study of the commercial building sector. See Joseph H. Newman, "Commercial Buildings: Retrofit and Other Energy Opportunities and Strategies," in Richard F. Hill, ed., *Energy Technology V: Challenges to Technology* (Washington, D.C.: Government Institutes, 1978).

71. W. R. Godwin, "Energy Conservation," *National Journal*, April 3, 1976, p. 456; General Accounting Office, *National Standards Needed for Residential Energy Conservation*, RED–75–377, June 20, 1975; Ralph Johnson, "Retrofit: a New Business Opportunity for Remodelers and Homebuilders," Speech at National Association of Home Builders Convention, January 22, 1975; Federal Energy Agency, "Retrofitting Homes for Energy Conservation," Energy Conservation Paper No. 23, pp. 3–4; Rosenberg, "Energy Usage in the Home—Consumption and Conservation," in Morgenthaler and Silver, eds., *Energy Delta: Supply vs. Demand*, p. 72; U.S. Department of Commerce, *Annual Housing Survey: 1975*, sec. A, p. 1.

72. *Monthly Energy Review*, March 1976, pp. 4–6; George A. Tsongas, *Home Energy Conservation Demonstration Project: Final Report for Chevron USA*, August 1977. Johns-Manville came to similar conclusions in applying a basic retrofit package to a standard ranch-style home in Belleville, Illinois. See Johns-Manville, "Residential Energy Savings: Retrofit," and John C. Moyers, *The Value of Thermal Insulation in Residential Construction: Economics and the Conservation of Energy*, ORNL–NSF–EP–9. So did the TVA: W. C. Whisenant, to author, May 23, 1977.

73. U.S. Congress, Senate, Commerce Committee, *Energy Conservation Act of 1976: Hearings*, 94 Cong., 2 sess. (Washington, D.C.: Government Printing Office, 1974), pp. 69–70; Donald C. Navarre, Washington Natural Gas, "Conservation and Utility Marketing," Mimeographed paper, November 1, 1977. The company estimates that its Super Heat Keeper package reduces by 70 percent the energy that an uninsulated house would consume for space heating.

74. Letter from W. C. Whisenant, TVA, to author, May 23, 1977. This is true for other utilities as well. See U.S. Congress, Senate, Energy Committee, *Energy Conservation Provisions of President Carter's Energy Program*, 95 Cong., 1 sess., p. 216. Conservation as an alternative to new capacity is discussed in Willey, "Alternative Energy Systems for Pacific Gas and Electric Co.: An Economic Analysis." Also see Willey, Prepared testimony submitted to the Arkansas Public Service Commission, on behalf of Attorney General Bill Clinton, May 15, 1978. "It is our hope," Attorney General Clinton notes, that the Arkansas Public Service Commission "will decide to embark on this innovative path to meeting Arkansas' power and energy needs for the 1980's." Letter to author, June 16, 1978.

75. Robert Socolow, ed., *Saving Energy in the Home: Princeton's Experiments at Twin Rivers* (Cambridge, Mass.: Ballinger, 1978), pp. 63–64, 100. But it is still doubtful whether the skills and knowledge represented in the Twin Rivers study would have become part of the repertoire of the typical insulation contractor. The best guide available for the layman is John Rothchild and Frank Tenney, *The Home Energy Guide: How to Cut Your Utilities Bill* (New York: Ballantine, 1977).

76. Moyers, *The Value of Thermal Insulation in Residential Construction: Economics and The Conservation of Energy;* NAHB, "Retrofitting Homes for Energy Conservation," p. 4; Johns-Manville, "Residential Energy Savings: Retrofit."

77. General Accounting Office, *National Standards Needed for Residential Energy Conservation*, p. 17.

78. Moyers, *The Value of Thermal Insulation in Residential Construction*, p. 1. For mobility, see Newman and Day, *The American Energy Consumer*, p. 45. There may, however, be a growing perception that the cost of energy-saving investments can be passed on to the buyer.

79. Florence Leyland discusses her case in U.S. Congress, Joint Economic Committee, Subcommittee on Energy, *Energy Conservation: Hearings,* p. 26.

80. Cepeda, *Potential for Energy Conservation in Texas,* p. 148. The serious effects of lack of knowledge are discussed, on the basis of a series of studies, in Lind, *The Rationale for Federal End-Use (Conservation and Solar) Programs,* p. 9. The institutional and political problems faced by the Energy Extension Service are perceptively analyzed by Robert Reisner, "The Federal Government's Role in Supporting State and Local Conservation Programs," in John Sawhill, ed., *Energy Conservation and Public Policy.*

81. Ralph Johnson, "Retrofit: A New Business Opportunity," Speech to NAHB Convention, January 22, 1975. For the warning signs about consumer dissatisfaction, see Sheilah Kast, "Insulation: What You Need to Know," parts 1 and 4, *Washington Star,* June 5 and 8, 1977.

82. One method to involve the utilities is through the Michigan Plan, where utilities take responsibility for retrofit. See William G. Rosenberg, "Conservation Investments by Gas Utilities as a Gas Supply Option," *Public Utilities Fortnightly,* January 26, 1977; Detroit Edison, "Detroit Edison: Home Insulation Finance Man," brochure; Joel Sharkey, "Report to the Michigan Public Service Commission on the Home Insulation Promotion and Financing Program," Mimeographed paper.

83. U.S. Senate, Committee on Energy, *Energy Conservation Provisions of President Carter's Energy Program,* parts A, B, C, and G of S. 1469, p. 164.

84. Leon R. Glicksman, "Heat Pumps: Off and Running . . . Again," *Technology Review,* June/July 1978, pp. 64–70; General Accounting Office, *The Federal Government Should Establish and Meet Energy Conservation Goals,* pp. 60, 88–89. Industrial applications of the heat pump are discussed in R.L.J. McLaren, "Heat Pumps," Paper presented at National Energy Management Conference, Birmingham, England, October 1978.

85. Eric Hirst and Robert Hoskins, "Residential Water Heaters: Energy and Cost Analysis," *Energy and Buildings I* (1977–78), pp. 393–400; Hoskins, Hirst, and W. S. Johnson, "Residential Refrigerators: Energy Conservation and Economics," *Energy,* 3 (August 1978), pp. 43–49; Socolow, ed., *Saving Energy in the Home,* pp. 3, 13; C. Franklin Montgomery, "Product Technology and the Consumer," *Scientific American,* December 1977, pp. 47–53.

86. Eric Hirst, Janet Carney, Dennis O'Neal, "Feasibility of Zero Residential Energy Growth," *Energy,* 3 (August 1978).

87. For temperature settings, see General Accounting Office, *The Federal Government Should Establish and Meet Conservation Goals,* pp. 60–61; and Robert Socolow, ed., *Saving Energy in the Home,* pp. 207, 232. Some aspects of the behavioral issues are surveyed, including a warning against coercion, in Paul C. Stern and Eileen M. Kirkpatrick, "Energy Behavior," *Environment,* December 1977, pp. 10–15. Psychological issues involved in energy conservation are surveyed in "Motivating the Troops for the Energy War," a special section in *Psychology Today,* April 1979, pp. 14–33.

88. For limitation of economic models in this regard, see Otto Eckstein and Sara Johnson, "Forecast Summary," in Federal Energy Administration, *1976 National Energy Outlook;* also see William Hogan, ed., *Energy and the Economy,* vol. 1 (Stanford, Calif.: Energy Modeling Forum, 1977), for the difficulty of including "non-market behavior. . . . The effects of regulation, industrial organization, or the expectations created by government's future role are not well understood . . . nor the impacts of unexpected embargoes."

89. The results are reported in Demand and Conservation Panel of the Committee on Nuclear and Alternative Energy Systems, "U.S. Energy Demand: Some Low

Energy Futures," *Science*, April 14, 1978, pp. 142–152. The four scenarios projected total primary energy consumption in 2010 of 58, 74, 94, or 136 quads. Consumption in 1975 was 71 quads. Price assumptions were a quadrupling in real terms of 1975 prices in the first two scenarios, a doubling in the third, and unchanged in the last. The panel stressed the important conceptual point—that energy conservation is, in effect, a source of energy: "A major slowdown of demand growth can be achieved simultaneously with significant economic growth by substituting technological sophistication for energy consumption." In other words, after capitalizing on the relatively simple steps that can save considerable energy, conservation technologies can compete with supply technologies for the energy "services" market.

90. This study was undertaken by Marc Ross and Robert Williams, and is reported in "The Potential for Fuel Conservation," *Technology Review*, February 1977, pp. 49–57. See Marc Ross and Robert Socolow, eds., *Efficient Use of Energy, American Institute of Physics Conference Proceedings, No. 25* (New York: AIP, 1975). The date of substitution was very generalized, and they emphasized physical possibilities, without attempting to calculate a cost. The purpose was to stake out a potential.

91. *Monthly Energy Review*, May 1977, p. 44.

92. Indeed, one of the most compelling pieces of evidence challenging the need for reprocessing of nuclear waste during Britain's Windscale Inquiry was a physical modeling effort similar to the American Physical Society study. Its conclusions for Britain paralleled the American study. See Gerald Leach, "Written Evidence to the Windscale Inquiry on Behalf of Friends of the Earth Limited," Mimeographed paper, 1978. Also see his "Energy for the Built Environment," Paper to the Social Science Research Council Seminar on Energy Utilisation, April 26, 1978.

93. General Accounting Office, *U.S. Energy Conservation Could Benefit from Experience of Other Countries;* International Energy Agency, *Energy Policies and Programs of IEA Countries: 1977 Review* (Paris: OECD, 1978). Daniel Yergin, "France's War on Energy Waste," Energy Research Project paper, June 1978; German Information Center, *Economic Reports,* June 23, 1978. For Canada, see Ministry of Energy, Mines, and Resources, *Energy Conservation in Canada: Programs and Perspectives,* Report EP–77–7, 1977.

94. Yergin, "France's War on Energy Waste."

95. Roger Sant, formerly FEA assistant administrator for conservation and now directing Carnegie Mellon Institute of Research's Energy Productivity Center, has estimated on the basis of preliminary data that over $220 billion of capital investment in conservation would be cost effective at *current* marginal costs of energy, in "A Preliminary Assessment of the Potential to Adjust Capital Stock to Higher End Use Efficiencies," Paper at American Association for the Advancement of Science, February 14, 1978. Applied between 1978 and 1985, this investment, Sant estimates, would save about 10 million barrels a day in 1986 energy consumption. The paper draws on the work of Eric Hirst for the residential sector; J. R. Jackson, for the commercial sector; Sant's own work, for the industrial sector; and, for transportation, on the U.S. Federal Task Force, *Motor Vehicle Goals Beyond 1980.* The caveat against conservation is in *Citibank Energy Newsletter* 4 (1978), p. 4. The value of conservation as a way to buy time and promote flexibility is stressed in a paper by a very shrewd energy analyst: Mons Lönnroth, "Nuclear Power in the Swedish Energy Futures—Commitments and Alternatives," Paper presented at Bellerive Conference, February 1979. He makes the additional point that distribution systems should be designed "so that different long-term technologies are adaptable."

96. The rationale and strategy for information and public education on conservation is set out in MIT Center for Policy Alternatives, *The Impact of Advertis-*

ing, Marketing, and Other Market Information on Consumer Energy Use: A Workshop Report, June 30, 1978, FTC Contract L0304. Another analysis that, like this chapter, points to the advantages of incentives for promoting conservation is Lind, *The Rationale for Federal End Use (Conservation and Solar) Programs.*

97. Incentives can, in many instances, be politically more acceptable than regulations in our society. If all homes at the time of sale were required to be brought up to certain conservation standards, the nation's realtors would oppose it. Tax credits for insulation would not call up such organized opposition. Public opinion analysts who carried out one of the major energy attitude surveys concluded: "Positive economic incentives—those that can help the consumer save money—may be more efficient than either behavioral regulation or price regulation." William H. Cunningham and Sally Cook Lopreato, *Energy Use and Conservation Incentives: A Study of the Southwestern United States* (New York: Praeger, 1977), p. 100.

Chapter 7
Solar America

1. Sun Day's Denis Hayes, however, is not optimistic about reaching these levels of solar contribution, since they would require "a World War II style mobilization"; Denis Hayes, personal communication. R. W. Scott, *World Oil,* June 1978.

2. For instance, 2 million solar water heaters are in operation in Japanese homes; Denis Hayes, "Short-Term Solar Prospects," in John Sawhill, ed., *Improving the Energy Efficiency of the American Economy* (New York: Prentice Hall, 1979), p. 6 of preliminary draft.

3. Satellite power systems and total energy systems are sometimes also considered in a separate category. See, for instance, U.S. Department of Energy, *Solar Energy, A Status Report,* DOE/ET-0062, June 1978, p. 9, and *Appendix A, Solar Technologies,* pp. 13–39.

4. This definition distinguishes solar energy from the products of "ancient sunshine" (oil, gas, and coal), but includes direct sunshine, the wind, falling water, and plant matter—except for some century-old trees.
Not surprisingly, the lack of an established definition for solar energy has tended to obfuscate press reports regarding solar energy and its true potential in this century. Recently, a story in *New York Times* was headlined "Solar Energy Held Still Decades Away" (*New York Times,* February 1, 1979, p. A7). The *Times* story, a summary of the report by H. Ehrenreich *et al., American Physical Society Study on Solar Photovoltaic Energy Conversion* (New York: American Physical Society, 1979), explained that "the group doubts that more than one percent of the nation's electricity can be generated from sunlight by the end of the century." This group's conclusions—which closely parallel our own—were actually limited to *photovoltaic* energy conversion, only one of five technologies proposed to generate electricity from the sun, and at present, certainly not the cheapest or most technically mature. As shown above, several non-electrical technologies may also be employed to obtain energy from the sun. Indeed, it is from these other technologies that the largest near- and middle-term contributions from solar energy can be expected. If, however, each of the eight individual solar technologies identified above were to contribute no more than "one percent of the nation's electricity" by the end of the century, the impact would be about 1.4 million barrels per day of oil equivalent, or about twice the level of U.S. oil imports from Iran in 1978.

5. The on-site technologies also include agricultural and process heat, small windmills, dams, and local plant matter usage. See, for instance, U.S. Congress, Office of Technology Assessment, *Application of Solar Technology to Today's Energy Needs,* vol. 1, June 1978, pp. 11 and 19.

6. See, for instance, Amory B. Lovins, *Soft Energy Paths: Towards a Durable Peace* (San Francisco: Friends of the Earth, 1977). See also Lovins, "Energy Strategy: The Road Not Taken?" *Foreign Affairs,* October 1976; E. F. Schumacher, *Small Is Beautiful: Economics As If People Mattered* (New York: Harper and Row, 1975).

7. *Biomass* is short for biological matter, and is generally used as a synonym for plant matter and animal waste.

8. For a comprehensive review of the economics and the technology of conversion processes for solar electricity, see Wolfgang Palz, *Solar Electricity: An Economic Approach to Solar Energy* (London: Butterworth, 1978).

9. Though I have chosen space and hot water heating to illustrate the potential—and the problems—of the on-site technologies, the remaining on-site technologies, such as small windmills, small dams, and, especially, agricultural and process heat, also have significant short-term potential. Some analysts believe that intermediate temperature systems for agricultural and process heat may well have the greatest short-term potential for solar energy utilization. See, for example, William D. Metz, "Solar Thermal Energy: Bringing the Pieces Together," *Science,* vol. 197, August 12, 1977.

10. R. G. Stein, *Architecture and Energy* (New York: Anchor Press/Doubleday, 1977). See also Wade Greene, "Solar Refractions," *New Times,* May 25, 1978, and Bruce Anderson, *Solar Energy: Fundamentals in Building Design* (New York: McGraw-Hill, 1977). The Roman bath data is from a personal discussion with Paul W. Cronin, president of Sunsav.

11. Bruce Anderson, *Solar Energy,* chaps. IIA, B, C. For an excellent analysis of passive design technologies, see John I. Meyer, "The Cost of Passive Solar Energy," M.S. thesis, Massachusetts Institute of Technology, June 1977. Quote is by Donald Watson (author of *Designing and Building a Solar Home* [Charlotte, Va.: Garden Way, 1977]), cited by Greene, "Solar Refractions."

12. The Texas house information is from *New York Times,* "Passive Approach Found to be Active Cost Saver in a Solar House in Texas," May 18, 1978, supplemented by personal communications from the owner of the house, Colonel Archie Erwin. The Maine building information is also from *New York Times,* "Solar Heated Shop in Maine is Called Most Innovative," May 9, 1978, p. A29.

13. See R. W. Bliss, "Why Not Just Build the House Right in the First Place?" *Bulletin of the Atomic Scientists,* March 1976; Owens Corning Fiberglass, *Energy-Saving Homes: The Arkansas Story,* June 1976. See also Reference 16.

14. For general background on active systems, see Anderson, *Solar Energy: Fundamentals in Building Design,* parts III, IV, V.

15. The estimate of the number of dwellings suitable for solar energy was derived from interviews with industry executives and academics. Hayes *et al.,* in *Blueprint for a Solar America* (p. 32), have estimated that by the turn of the century 30 percent of existing buildings could have the majority of their energy supplied with combined active space and hot water systems. The number of existing dwellings is from *Status Report on Solar Energy Domestic Policy Review,* Domestic Policy Review Integration Group, Washington, D.C., August 25, 1978. See also Fred S. Dubin, "Solar Energy Design for Existing Buildings," *ASHRAE Journal,* November 1975.

16. Market sizes for 1977 were estimated by assuming a solar heating system cost of $40 per square foot of *installed* collector for hot water and space heating, and $10 for swimming pool systems. The 1977 collector and unit sales volumes are from Sheldon H. Butt, president, Solar Energy Industries Association, Keynote Address, SEIA Convention, Phoenix, Arizona, February 1978. The installed col-

lector costs were derived from interviews with solar suppliers and installers and government energy officials. Earlier collector sales volumes are from Federal Energy Administration, *Solar Collector Manufacturing Activity Report, July through December 1975,* March 1976. For a review of industry performance in 1978, see *Solar Engineering Magazine,* "Industry Survives 1978 with Small Gains," February 1979, pp. 11–12. For more on federal tax credit, see Reference 38.

The number of swimming pools was estimated from Sheldon Butt's data by assuming that the average swimming pool required 150 square feet of collector area for heating.

Potential market was estimated by assuming an average system price of $6,000 per dwelling for hot-water and space heating. The $6,000 figure is a rough estimate. An active heating system costs $5,000 to $13,000, whereas a passive system can cost anywhere from nothing to $5,000. A hot water heating system ranges from $1,600 to $2,400. Source: interviews with manufacturers and government energy officials.

Other authors have given estimates for collector system shipments, exclusive of installation. Since installation accounts for about one half of overall cost, the numbers in the text should be reduced by one half (to about $150 million) to compare with their estimates. See, for instance, "The Coming Boom in Solar Energy," *Business Week,* October 9, 1978, p. 89. An estimate of $150 million, exclusive of installation, was also given by Martin Glensk of Arthur D. Little, a leading industry expert, at a presentation to the Energy and Resources Club of the Harvard Business School, Cambridge, Massachusetts, October 25, 1977.

17. Our interviews with solar heating manufacturers indicate that no manufacturer had solar heating sales of over $5 million in 1977. This compares with estimated manufacturer shipments of $100 million to $150 million in 1977.

18. M. A. Maidique and John Ince, *Grumman Energy Systems* (B) (Boston: Intercollegiate Case Clearinghouse, 1979).

19. In its March 1978 issue, *Solar Engineering,* the leading solar industry magazine, selected the first recipients of its Industry Leader Awards. Three of the six were Fortune 500 firms. Of the remaining three, only Solaron was a new firm.

	Firm	Main Business	1977 Sales (Millions)
1.	Honeywell	Electronic systems	$2911
2.	Grumman	Aerospace products	1552
3.	Revere Copper and Brass	Metal products	597
4.	Lennox	Heating & ventilation equipment	200 (est.)
5.	Grundfos	Pumps	10–20 (est.)
6.	Solaron	Solar air-heating systems	3 (est.)

The sales levels are from Standard and Poors' *New York Stock Exchange Reports* (Honeywell, vol. 45, no. 237, December 11, 1978; Grumman, vol. 45, no. 166, August 28, 1978; Revere Copper and Brass, vol. 45, no. 247, December 26, 1978), except for Lennox, Grundfos, and Solaron, which are author's estimates derived from employment data and interviews with industry sources.

20. William T. Schleyer and Donald M. Young, "Consumer Attitudes Towards Solar Energy," Harvard Business School Energy Project Report, May 20, 1977. Four thousand questionnaires were sent out to randomly selected residents of eight states, two from each of four regions: Northeast (Massachusetts, Connecticut), South (Florida, Georgia), Southwest (Arizona, Colorado), and West (California, Nevada). About 19 percent of those polled responded.

Other surveys have borne out the five-year payback requirement for major solar retrofit. See William H. Cunningham and Sally Cook Lopreato, *Energy Use and*

Conservation Incentives: a Study of the Southwestern United States (New York: Praeger, 1977), pp. 87–88.

For the sources of the estimates of solar heating system prices, see Reference 16.

21. For information on our nationwide consumer survey, see Reference 35. Also see *Massachusetts Solar Action Office Demographic Analysis of Applicants and Grant Recipients Under HUD Residential Solar Domestic Hot Water Initiative Program,* September 1977.

The provision for incorporation of solar heating in military housing and other military construction was made by Senator Gary Hart of Colorado. See Military Construction Act of 1979, Bill No. S3079.

22. Jacqueline Adams, "Industrial and Commercial Applications of Solar Heating Technology in Massachusetts," Harvard Business School Energy Project, Unpublished report, April 1978.

23. Jacqueline Adams, "Industrial and Commercial Applications."

24. The payback of our oil-heated example is bracketed by the payback that would result if electric heating or gas had been chosen. At current prices, the average residential user pays $2.40, $3.52, and $11.28 per million BTU's of gas, home heating oil, and electricity, respectively. The prices used in this calculation for oil and gas—$.486 per gallon for home heating oil and $2.45 for a million cubic feet of gas—are from the Department of Energy, *Monthly Energy Review,* November 1978, p. 74 (oil) and p. 82 (gas). In both cases the average price for the first four months of 1978 were used. For electric heating, an average price of $.385 per kilowatt hour was used. The average price per kilowatt-hour for the first four months of 1978 was obtained from John Damon, Edison Electric Institute, 90 Park Avenue, New York. Energy (BTU) content ratios for oil, gas, and electricity were taken from *The Encyclopedia of Energy* (New York: McGraw-Hill, 1976), p. 4.

Thus, relative to oil heat, payback in a gas-heated house—assuming similar furnace efficiencies of 60 percent—would be 1.5 times higher, while for an electric-heated house, assuming 90 percent efficient radiators, the payback would be 2.5 times lower. However, the gap between gas and oil heat is closing. Comparing the 1975 gas and oil prices to the 1978 prices given above, we find that oil prices increased 29 percent on the average, but during the same period, the price of natural gas sold to residential customers for heating purposes rose 58 percent. The 1975 data is also from the *Monthly Energy Review.*

All of the paybacks calculated above would have been 1.5 to 2 times larger had the example been space heating rather than water heating.

Several types of storage systems have been proposed to balance the inherent intermittency and cyclicality of technologies that depend on the sun's rays or the wind. However, though there is no dearth of possibilities, fly-wheel, gravity or battery storage is at present relatively costly. For these reasons, most solar heating systems provide only limited storage capacity. Typically an insulated oversized water tank or a rock bed provides a day or two of storage and a backup system is required for protection against longer periods of poor weather. For review of energy storage technologies see U.S. Congress, Office of Technology Assessment, *Energy Storage in Application of Solar Technology to Today's Energy Needs,* vol. I, June 1978, chap. XI.

For a detailed analysis of the economics and design of an actual solar home, see Mark Hyman, "Solar Economics Comes Home," *Technology Review,* February 1978, pp. 28–35.

25. For a review of the pitfalls of simple payback analysis, see Henry H. Leck, "Pitfalls of Payback Analysis," *Solar Engineering Magazine,* September 1978, pp. 22–25.

26. The following assumptions were made in this analysis:

1. System cost = $2,400
2. System life = 20 years
3. Fuel savings = $200 in the first year
4. Down payment = $480 (20% of $2,400)
5. Inflation rate = 6%
6. Mortgage:
   Life = 20 years, 20 payments
   Interest = 10%
7. Maintenance cost = 2% of purchase price per year.

Calculations were made for (a) a base case with no tax considerations, (b) a tax rate of 33 percent, and (c) a tax rate of 50 percent. Three cases of fuel cost escalation were also considered: (a) equal to inflation, (b) inflation plus 2 percent, and (c) inflation plus 4 percent. The resulting nine numbers are given in the table below. The numbers in case b were used in the text. The internal rate-of-return computations were made by John Kerr, as follows:

	Fuel Cost Escalation	Tax Rate 0%	33%	50%
Case a	Equal to inflation	5%	17%	30%
Case b	Inflation plus 2%	11%	24%	40%
Case c	Inflation plus 4%	16%	32%	48%

For a review of techniques of financial-return analysis, including the internal rate-of-return method used calculations above, see Eugene F. Brigham and Jay Fred Weston, *Managerial Finance* (Hinsdale, Ill.: Dryden Press, 1975), chap. 10.

27. Preliminary studies indicate that solar heating is significantly more labor intensive per BTU supplied than conventional power-generation technologies, such as nuclear or oil fired plants. Moreover, the jobs created would be primarily local jobs. One study in California compared the labor intensity of a nuclear plant to a "solar equivalent" and found an increase of over six times in labor hours for the solar heating case. Similar results have been found when solar heating has been compared to coal mining and power generation. One researcher has expressed skepticism about the reliability of such projections, but concludes that the use of solar heating will "probably require significantly more employment" than the conventional energy technologies. See Duane Chapman, "Taxation, Energy Use, and Employment," Testimony to the U.S. Congress, Joint Economic Committee, March 15, 1978, p. 4. See also Wilson Clark, "Creating Jobs Through Economic Policy," Testimony to the U.S. Congress, Joint Economic Committee, March 16, 1978. Also see Reference 2, Chapter 6.

Clearly the payback calculation with $35 oil is an oversimplification. First, the cost of home heating oil includes not only the price of crude oil but also refining and distribution costs, which do not necessarily rise in direct proportion to crude oil. Additionally, the analysis assumes that government regulations would allow the cost to be passed on to the consumer, whereas in reality the consumer—at present—would see a new price that is a combination of regulated domestic oil and imported oil. Also, the calculation is for *one* incremental barrel. If a substantial percentage of domestic consumption were imported at $35, plastics and energy-intensive components of solar systems, such as aluminum, would increase in price, making the payback less attractive.

28. For a review of the solar loan market, see "Solar Financing: Picture Brightens," *Solar Engineering Magazine*, January 1978, pp. 7–8. Some states, however, have passed laws excluding solar energy equipment from property taxes. See, for instance, *Solar State Legislation*, (Rockville, Md.: National Solar Heating and Cooling Information Center, January 1978).

29. President Carter's speech, delivered at the International Sun Day ceremonies, May 3, 1978, at the Solar Energy Research Institute, Golden, Colorado.

30. The estimate of the number of municipalities with their own building codes is from Richard Schoen, Alan Hirshberg, and Jerome Weingart, *New Energy Technologies for Buildings* (Cambridge, Mass.: Ballinger, 1975), p. 1975. For the current state of solar building code legislation, see *Solar State Legislation*. For further information on the Coral Gables ordinance, see Ordinance No. 2253, Planning Department, the City of Coral Gables, Florida.

31. The consultants who conducted the study, Robert O. Smith and Associates, found that only 15 percent of the one hundred New England utilities' solar domestic water heating experimental installations functioned well during the first year of operation. In addition to reliability problems, the consultants found that twenty-seven of the systems—when operating—produced very low savings of energy, or none. The average savings of energy used for hot water heating with the *best* fifteen systems was 37 percent. A variety of problems with pipes, valves, controllers, and undersizing of the panels and controllers—*primarily installation problems*—were cited as explanations of the poor results. The consultants concluded that the experiment had been a "learning experience" for the industry, and that with appropriate installation it was still reasonable to expect a target savings of 50 percent (in the New England area) of the energy consumed by water heating.

System Reliability	% of Installations
1. Functioned well	15
2. No *major* breakdowns	8
3. At least one major breakdown	57
4. Severely unreliable	20
TOTAL	100%

For further information on this study, see Robert O. Smith and Associates, "Summary of Performance Problems of 100 Residential Solar Water Heaters installed by the New England Electric Company subsidiaries in 1976–77," Boston, 1978. For a study of how utilities could help to alleviate installation problems, see Bruce M. Smackey, "Should Electric Utilities Market Solar Energy?" *Public Utilities Fortnightly,* September 28, 1978, pp. 37–43.

For the plight of one man and his solar installation, see John Rothchild, co-author with Frank Tenney of *The Home Energy Guide* (New York: Ballantine, 1977). Rothchild provides an informative and graphic description of how an inexperienced plumber can botch up the installation of a normally effective system. He also demonstrates how the homeowner can reinstall the system. His report, "Victory from Defeat: How One Man Saved His Solar System," can be obtained for a dollar handling charge per copy from Enersave Corp., P.O. Box 191, Everglades, Florida 33929.

32. See Joyce R. Swain, Oklahoma Solar Information Office, "States Focus on Tax Incentives, Lag Behind on Sun Rights, Codes," *Solar Engineering Magazine,* April 1978, p. 22, See also *Solar State Legislation;* W. R. Harris, "Is the Right to Light a California Necessity?" Rand Corporation, Paper 5558, 1975.

33. Letter from First American Bank for Savings, Brockton, Massachusetts, to Massachusetts Action Office, November 22, 1977.

34. As an example of the skepticism of the utility industry regarding solar energy, see Lelan F. Sillin, Jr., "Overview of Issues Facing the Electric Utility Industry," Paper presented to the 55th American Assembly, November 1978, pp. 10–14. Sillin cited a Booz-Allen study: "It is necessary for Northeast Utilities to continue to rely on conventional generation technologies to meet its base load capacity requirements in the late 1980s and early 1990s." Sillin also added that "no alternative technology . . . could justify delaying the decision to install a conventional—that is, nuclear or coal—base load generating facility in the 1990 time frame."

35. Smackey, "Should Electric Utilities Market Solar Energy," *Public Utilities Fortnightly.* For an alternative view, see Joseph G. Asburg and Ronald O. Mueller, "Solar Energy and Electric Utilities: Should They Be Interfaced?" *Science,* vol. 195, no. 4277, February, 1977, pp. 445–450. Eight of the major U.S. gas and electric utilities were also interviewed by one of the authors, Frank Schuller. For opposition to utility involvement, see Jane W. Stein, "Law to Force SDG&E to Stop Selling Solar Energy Systems," San Diego *Union,* December 24, 1978, p. A3.

36. Should the government intervene? Most analysts of this question have come to the same general conclusion: Government intervention is justified when it helps to "bring private decisions about costs and benefits more closely in line with total cost and benefits" (J. Herbert Hollomon and Michele Grenon, *U.S. Energy Research and Development Policy* [Cambridge, Mass.: Ballinger, 1975], p. 12). For instance, assume that a promising new home solar electric converter has been developed, but the cost of installing the first prototypes will be $100,000, or, say, fifty times the cost of a competitive conventional system. It may be in the interest of society to subsidize the first several users, because the knowledge gained in those first few installations can serve as a guide (to invest or not to invest) to tens of millions of potential buyers. The value of this knowledge for society will far exceed the cost of a 90 percent subsidy to a few pioneering users. Similarly, government investment in research and development can be justified when an industry, because of structural reasons and/or inappropriate incentives due to government regulation, is underinvesting in research and development. Housing, a fragmented industry, is a case in point. The housing industry is made up of many small firms, none of which are sufficiently large to carry out a major R&D program. See, for instance, Edwin Mansfield, *The Economics of Technological Change* (New York: W. W. Norton, 1968), p. 229. For an analysis of the proper role of the government in energy R&D, see Hollomon and Grenon, *U.S. Energy Research and Development Policy.*
  But the most compelling argument for government intervention is found in the marginal cost of energy. When the cost of a product is rising, as has been the case for energy, its *average* price is lower than the cost for the last additional BTU of energy—the marginal cost. Additionally, the market does not discriminate between the $15 per-barrel cost of imported oil, with its attendant risk-and-cost-of-supply interruption, and $10 per-barrel cost of domestic oil. Both are lumped into a gradually rising average price that disguises both the source uncertainty and the real cost of an additional BTU of energy. It is these latter factors that should be examined when comparing a BTU of solar energy to a BTU of petroleum energy. Although a system could be arranged to reflect the marginal cost of petroleum, gas, and so on, this would be politically difficult because of the income distribution consequences. The alternative—placing solar energy on an even basis with fossil fuels by means of subsidies—is politically much more realistic and palatable.

37. Case studies of the diffusion of a wide variety of innovations, from color TV to hybrid corn, show that market development follows a pattern characterized by an S-shaped logistic curve: Diffusion starts out very gradually, then rises rapidly, flattening off again in the last stage until the market is saturated. See, for instance, Edwin Mansfield, *Industrial Research and Technological Innovation* (New York: W. W. Norton, 1968), chaps. 7, 8, 9, pp. 133–191. For data on the actual diffusion of the color TV innovation over a twenty-year period, see, for instance, *Zenith C* (Boston: Intercollegiate Case Clearing House, 9–674–095, revised August 1977). For hybrid corn, see Zvi Grilliches, "Hybrid Corn and the Economics of Innovation," *The Economics of Technological Change,* ed. Nathan Rosenberg (Middlesex, Engl.: Penguin Books, 1971). For recent views on potential rates of diffusion of the solar energy technologies, see Dennis Schiffel, Dennis Costello, David Posner, and Robert Witholder, *The Market Penetration*

of Solar Energy: A Model Review Workshop Summary, Solar Energy Research Institute, Golden, Colo., January 1978. Similar diffusion curves, called product life cycles, are found by plotting sales of most products from birth to maturity. See, for example, Theodore Levitt, "Exploit the Product Life Cycle," *Harvard Business Review,* November–December 1965, p. 81.

38. The 1978 Federal Energy Policy Act tax-credit legislation allows a tax credit of 30 percent on the first $2,000 and a credit of 20 percent for the next $8,000. For a $10,000 hot-water-and-space-heating-system combination the tax credit would be $2,200 or a 22 percent credit. However, recommendations for significantly larger tax credits for consumers and industry ranging from 30 to 60 percent have been proposed by the solar policy review group created by President Carter on Sun Day, May 3, 1978. See, for instance, *Government R&D Report,* vol. XI, no. 4, February 15, 1979 (published in Ipswich, Massachusetts, by William Margetts), pp. 7–9. Not surprisingly, these proposals have met strong opposition from officials at the Office of the Management and Budget and the Department of the Treasury who are concerned about escalating budget deficits. By suggesting tax credits, it is not our intent to argue the merits of one policy instrument over another. Our primary intent here is to emphasize that solar heating presents a major opportunity for public policy. The potential perversities of tax credits, deductions, and direct grants or subsidies, and their possible disturbing effects on the economy, as well as the ranking of their political feasibility, are beyond the scope of this work. We will, however, in this essay use tax credits as our preferred policy instrument for three simple reasons: (1) Despite their potential for partial capture by suppliers, they enjoy wide political support; (2) the California experience indicates that a substantial tax credit can be a major stimulus to the adoption of solar technology; and (3) a tax credit is simple to administer, since the fund distribution system (the Internal Revenue Service in this case) is already operative. However, tax credits do not benefit low-income homeowners or those who pay little or no federal income tax. Clearly, a supplemental policy objective should be to reach this segment of the population, possibly by a system of direct grants. For the details of the California tax credit, which applies to certain "approved systems," see California Energy Commission, *Guidelines and Criteria for the California Solar Energy Tax Credit,* Sacramento, April 1978.

The impact of a 60 percent tax credit would be roughly about the same as a rise in oil prices to $30, which would make the economics about *three* times more favorable to solar heating. (See also Reference 26.) For the impact of the 55 percent California tax credit on the state's solar industry, see "The Coming Boom in Solar Energy," *Business Week.*

39. Nonetheless, there is a danger in the development of codes or standards too early in the industry's infancy, particularly if they are too restrictive. In a June 30, 1978 letter to the Division of Solar Energy, William B. Shurcliff argues that "standards should be worded in such a way that any company selling any queer kind of system, can claim it *meets* the standard unless it explicitly flunks. That is, I urge that you avoid placing a company under a dark shadow merely because you have not gotten around to writing a standard that applies to it. Write standards that are 'fail-safe', i.e., are not condemnatory by default."

40. The Energy Extension Service is now operating on an experimental scale in several states. See "Energy Extension Service," *Government R&D Report,* vol. IX, no. 6, published by William G. Margetts, Ipswich, Massachusetts.

41. Nobel laureate Melvin Calvin describes the sugar cane plant as "the best, most efficient solar energy device we have today on a large scale." Melvin Calvin, "The Sunny Side of the Future," *Chemtech,* June 1977, p. 353.

42. Denis Hayes, *Energy: The Solar Prospect,* Worldwatch Paper 11 (Washington, D.C.: Worldwatch Institute, March 1977).

43. Another 20 percent of the country's energy was derived indirectly from work animal feed. See, for instance, John C. Fisher, *Energy Crises in Perspective* (New York: John Wiley and Sons, 1974), p. 14. Only fifty years earlier, in 1850, solar energy—fuel wood, work animal feed, wind, and falling water—had accounted for 93 percent of U.S. energy consumption (Fisher, p. 13).

44. See E. L. Ellwood *et al.*, *The Potential of Lignocellulosic Materials for the Production of Chemicals, Fuels and Energy*, National Research Council (Springfield, Va.: National Technical Information Service, 1976), p. 11. For a similar estimate, including a 1970 breakdown into fuel wood and forest industry waste, see Thomas Gage, *Renewable Resources for Industrial Materials* (Washington, D.C.: National Academy of Sciences, 1976). For the estimates of Swedish and Finnish wood consumption, see Richard Merril and Thomas Gage, *Energy Primer, Solar, Water, Wind and Biofuels* (New York: Delta, 1978), a good basic source on biomass fuels, especially wood.

45. Biomass, or plant matter, is a byproduct of photosynthesis. Light (sunshine) is absorbed by chlorophyll, which in turn produces oxygen and hydrogen. The oxygen and hydrogen fuel the photosynthesis process by which carbon is converted to carbohydrates, the principal product of most green plants. For an excellent overview of the biomass alternative, see C. C. Burwell, "Solar Biomass Energy: An Overview of U.S. Potential," *Science*, vol. 199, March 10, 1978, pp. 1041–1948. For a compact classic article on biomass possibilities and processes, see Melvin Calvin, "The Sunny Side of the Future," *Chemtech*.

46. Alan D. Poole and Robert H. Williams, "Flower Power," *Bulletin of Atomic Scientists*, May 1976; John R. Benemann, *Biofuels: A Survey*, #ER–746–SR, Special Report (Palo Alto, Calif.: Electric Power Research Institute, June 1978).

47. Allen L. Hammond, "Alcohol: A Brazilian Answer to the Energy Crisis," *Science*, vol. 195, February 11, 1977. For more information on liquid and gaseous biomass, see Roberta Navicki, "Biomass," *Science News*, vol. 113, no. 16, April 22, 1978, p. 259.

48. Benemann, *Biofuels: A Survey*, sec. 1.

49. Allen L. Hammond, "Photosynthetic Solar Energy: Rediscovering Biomass Fuels," *Science*, vol. 197, August 19, 1977, pp. 745–746.

50. James S. Trefit, "Wood Stoves Glow Warmly Again in Millions of Homes," *Smithsonian Magazine*, October 1978, pp. 54–63. Our own interviews with leading wood stove and furnace manufacturers indicate that several of the largest manufacturers are selling 100,000 or more units a year.

51. The estimates for wood, dry crop wastes, and Western coals are from Allen L. Hammond, "Photosynthetic Solar Energy: Rediscovering Biomass Fuels," *Science*. The estimate for Eastern coal energy content is from C. Duane Schaub, Guarantee Fuels, Independence, Kansas.

52. Wood's sulfur content is .1 percent, compared to .5 to 1 percent for Western coal and 2 to 4 percent for Eastern coals. Source: C. Duane Schaub, Guarantee Fuels.

53. Analysis is based on sized, washed low-sulfur coal. Data for the economic analysis of pelletizing was provided by the engineering department of Guarantee Fuels.

54. E. A. Richards, "Alternate Energy Sources," *Brewer's Digest*, vol. 52, no. 8, August 1977, pp. 54–56, 58.

55. Benemann, *Biofuels: A Survey*, sec. 6–8.

56. See *New England Solar Energy Newsletter*, vol. 4, no. 4, published by the New England Solar Energy Association, Brattleboro, Vt., August 1978.

57. Hammond, "Photosynthetic Solar Energy: Rediscovering Biomass Fuels," *Science*. One firm in Canada estimates that yield of its forest land may be increased by seven times; others claim even higher yields. See "Growing Energy," *Science News*, March 11, 1978, p. 153. Claims of higher yield hybrid tree species are viewed with uneasiness by some environmentalists who are concerned with their impact on soil deterioration.

58. Burwell, "Solar Biomass Energy: An Overview of U.S. Potential," *Science*. See also Navicki, "Biomass," *Science News*. See also Stephen H. Spurr, "Silviculture," *Scientific American*, vol. 240, no. 2, February 1979, pp. 76–91.

59. A thousand-acre pilot energy plantation on the Savannah River in Aiken, South Carolina, has been already funded by the Department of Energy; see Navicki, "Biomass," *Science News*, p. 258.

60. Benemann, *Biofuels: A Survey*, sec. 6–8. Personal communication from Shelley Don, chairman of Bio-gas of Colorado.

61. The Energy Extension Service was created for such a purpose. See "Energy Extension Service," *Government R&D Report*.

62. Hammond, "Photosynthetic Solar Energy: Rediscovering Biomass Fuels," *Science*, p.745.

63. Daniel Behrman, *Solar Energy: The Awakening Science* (Boston: Little, Brown, 1976), pp. 32–33, 36.

64. This quote is from interviews conducted at Font-Romeu and Paris with French solar energy officials. Felix Trombe has expressed similar views: "America will bring out a solar power system for half the price, then they will sell it to the rest of the world" (Trombe as quoted by Behrman, *Solar Energy: The Awakening Science*, p. 41). For a description of the French solar thermal program, see *Solar Energy from France* (Paris: Délégation Aux Energies Nouvelles, Ministère de L'Industrie et de la Recherche, 1976), pp. 10–33. For a popular account, see Alain Ledoux, "Them I, Première Centrale Solaire," *Science & Vie* (June 1977), pp. 108–112.

65. In fact, most of the funds under the solar thermal category are absorbed by the power-tower program. See, for instance, William D. Metz, "Solar Thermal Electricity: Power Tower Dominates Research," *Science*, July 22, 1977, p. 353.

66. Palz, *Solar Electricity: An Economic Approach to Solar Energy*, p. 226. For heliostat technology and economics, see pp. 129–133 and 166–174.

67. For a summary of the U.S. power-tower program, see Department of Energy, *Solar Thermal Energy Conversion, Program Summary*, EROA 76-159, October 1976, pp. 1–4. See also Energy Research and Development Administration, *Central Receiver Solar Thermal Power System, Phase 1, 10MW Electric Pilot Plant* (Washington, D.C.: Government Printing Office, 1976). For cost-sharing information, see "Sun Power as a Peak Source for the Electric Utilities," *Solar Engineering Magazine*, June 1978, pp. 21–22. For a popular account, see Walter Sullivan, "Solar Test Facility is Near Completion," *New York Times*, November 13, 1977, p. 59. For local attitudes toward the Barstow power tower, see Robert Lindsey, "Desert Town Puts Hopes in Solar Plant," *New York Times*, October 14, 1977, p. A.14.

68. Metz, "Solar Thermal Electricity: Power Tower Dominates Research," *Science*, pp. 353–356. For an alternative view of power-tower economics, see Alvin F. Hildebrand and Lovin L. Vant-Hull, "Power with Heliostats," *Science*, September 16, 1977, pp. 1139–1146. For additional views, see Palz, *Solar Electricity: An Economic Approach to Solar Energy*, pp. 170–174.

69. Estimates for power-tower maximum sizes are ꞏfrom Allen L. Hammond and William D. Metz, "Solar Energy Research: Making Solar After the Nuclear

Model?" *Science,* vol. 197, July 15, 1977, pp. 241–244. By comparison, the typical nuclear or coal-fired power plant generates 1,000 to 2,000 megawatts of power. See, for example, Chapter 4.

For estimates of dates of commercial feasibility of power towers, see Walter F. Morrow, Jr., "Solar Energy: Its Time is Near," *Technology Review,* December 1973, pp. 40.

70. See, for instance, Morrow, "Solar Energy: Its Time is Near," *Technology Review,* pp. 31–43.

71. For a review of the environmental risks of microwaves, see Paul Brodeur, *The Zapping of America* (New York: W. W. Norton, 1978). See also *Science News,* vol. 113, no. 3, January 21, 1978, p. 40, for a listing of potential environmental hazards of solar microwave power satellites.

Reservations about the feasibility of solar satellites go far beyond environmental and counterculture groups. An economist at the nation's eighth largest oil company who believes that payoffs from Small Solar are "more quickly achievable than most people believe," sized up solar satellites thus: "A colleague was recently approached by NASA about solar satellites. They sent him about five pounds of paper to read; when he read it all, he thought about it and discovered that there were about a hundred specific events that all had to happen for the solar satellite to go at all—about a hundred things that had to go right for the thing to work—and then, if all hundred went right, what you got was electricity at a hundred mills [ten cents] at the bus bar. The most expensive alternative he could think of for electricity cost sixty mills [6 cents] at the bus bar, and he had to work to get the price up that high. Perhaps that makes us think a little more about solar satellites." David Sternlight, "Making Energy Policy Analysis Relevant," Keynote speech, Hawaii Conference on Jobs and the Environment, November 14, 1978. See also Kit Smith, "Oil Economist Bullish on Solar Energy Hopes," Honolulu *Advertiser,* November 18, 1978, p. E3.

72. For a small classic on photovoltaic principles and technology, see Bruce Chalmers, "The Photovoltaic Generation of Electricity," *Scientific American,* November 1976, pp. 34–43. For an excellent review of the photovoltaic field, see John C. C. Fan, "Solar Cells: Plugging into the Sun," *Technology Review,* August/September 1978, pp. 14–35.

73. See, for instance, Isaac Asimov, *Isaac Asimov's Biographic Encyclopedia of Science and Technology* (New York: Avon, 1972), p. 588.

74. Allen L. Hammond, "Photovoltaics: The Semiconductor Revolution Comes to Solar," *Science,* vol. 197, July 29, 1977, pp. 445–447.

75. For a review of government's role in the civilian electronics industry (including semiconductors), see J. M. Utterback and Albert E. Murray, "The Influence of Defense Procurement and Sponsorship of Defense and Development on the Development of the Civilian Electronics Industry," M.I.T. Center for Policy Alternatives, CPA–77–5, June 30, 1977.

76. *Solar Engineering Magazine,* November 1977, p. 12, gives a range of $13 to $27 per peak watt for 1977, and estimates an average of $16. My own interviews with industry executives indicated that average prices were $2 to $3 below this level by late 1977. For an overview issue on the photovoltaic industry, see "Watts Ahead for Solar Cells," *Solar Engineering Magazine,* November 1977; for price estimates, see same issue, pp. 12, 14.

Bids for a 250 kilowatt system for an Arkansas school were as low as $3 per watt. "Solar Power, Now or Never," *Science News,* vol. 113, no. 1, January 7, 1978, p. 8.

77. "Commercial Markets Increasing for Photovoltaic Industry," *Solar Engineering Magazine,* November 1977, pp. 12–16. The growth estimate was prepared

by Anthony Adler, Photovoltaics Division director of the Solar Energy Industries Association, p. 12. The estimate of market size, which coincides with the results of the author's interviews with industry executives, was given in the same article, also on p. 12.

78. Personal communication, June 1977, Robert C. Peterson, vice-president, Operations, Solar Power Corporation.

79. For the RCA work, see Charles A. Miller, "A Low-Cost Cell is Here," *Mechanix Illustrated*, April 1978. For a review of Ovshinsky's work, see Jerry E. Bishop, "Ovshinsky's Theories Finally Win Approval In the World of Science," *Wall Street Journal*, July 7, 1977, p. 1. See also "Materials for Converting Sunlight, Heat Into Electricity Announced by Inventor," *Wall Street Journal*, July 6, 1977, p. 10. The price estimate for amorphous semiconductors is from "A Promise of Cheap Solar Energy," *Business Week*, July 18, 1977, p. 20.

80. Personal communication, Paul Maycock, assistant director for Photovoltaics, Department of Energy.

81. The Department of Energy has set photovoltaic system prices for the near term (1982), midterm (1986), and far term (1990) of $2.00, $.50, and $.30 per peak watt respectively. See *National Photovoltaic Program Plan*, DOE/ET–0035, February 3, 1978, published in March 1978.
Interviews with the marketing managers of four of the leading photovoltaic suppliers indicated that prices are expected to drop to $5 or $7 per peak watt within the next year. All agreed that a large—$100 million to $200 million—purchase order would drive them down several times further but were not willing to officially commit their firm to a specific price and volume.
For the Clinch River estimate, see Henry Kelly, "Photovoltaic Power Systems: A Tour Through the Alternatives," *Science*, vol. 199, February 10, 1978, p. 643.
A major study of photovoltaic conversion published as this book went into print reaches similar conclusions regarding the desirability of pursuing a diversity of technological approaches for photovoltaic conversion; H. Ehrenreich *et al.*, *American Physical Society Study on Solar Photovoltaic Energy Conversion.*

82. In a June 1978 Department of Energy study (*Solar Energy, A Status Report*, DOE/ET–0062), the authors examined sixteen projections and found a range of from 3 to 39 quads of solar energy for the year 2000 and from 11 to 109 quads for 2020, not including hydropower.

83. For instance, to Denis Hayes the solar options are "wind, falling water, biomass and direct sunlight." Hayes includes solar heating and cooling and photovoltaics under direct sunlight; "falling water" is his term for hydropower. See Denis Hayes, *Energy: The Solar Prospect.* This is a classic primer on solar energy. For a more extensive treatment of the subject by the same author, see *Rays of Hope: The Transition to a Post-Petroleum World* (New York: W. W. Norton, 1977).

84. Council on Environmental Quality, Executive Office of the President, *Solar Energy, Progress and Promise*, April 1978.

85. Morrow, "Solar Energy: Its Time is Near," *Technology Review*, pp. 31–43.

86. Division of Solar Energy, Energy Research and Development Administration, *Solar Energy in America's Future: A Preliminary Assessment* (Washington, D.C.: ERDA, March 1977). This report documents a Stanford Research Institute study.

87. Other countries have set similar goals for the year 2000. In a study by the Australian Academy of Science, a target of one quarter of the country's energy from solar was set for the year 2000. See *Report of the Committee on Solar Energy Research in Australia*, September, 1973. Plans for virtually complete solar

self-efficiency by 2020 have already been developed by both Sweden and California. See, for instance, Thomas B. Johannson and Peter Steen, "Solar Sweden," *AMBIO*, vol. 7, no. 2, June 1978; Mark Christiensen *et al.*, *Distributed Energy Systems in California Energy Future: A Preliminary Report* (Springfield, Va.: National Technical Information Service, September 1977).

88. Unlike the federal tax credit program, any such program should include passive solar energy designs. A key difficulty here is qualifying the cost of the passive "system" and distinguishing it from the remainder of the structure. See also Reference 37.

The funds required for such a program and the amount of solar energy produced could vary widely depending on how much of an investment is made in conservation and which solar technology is chosen. For instance, an active system to heat a conventionally insulated residence may cost $8,000, while a passive system incorporated into a highly insulated structure may only cost a few hundred dollars. In general, the best policy is first to insulate the structure very well and only then to consider solar energy. If, as an example, an average installation cost of $3,000 is assumed (averaged over a mix of old and new homes, active and passive systems, wall-insulated and poorly insulated structures, and hot water and space heating systems) for each of 10 million installations over the next ten years, the total cost would be $30 billion. If commercial installations are included, the total could rise to perhaps $40 billion, or $4 billion yearly. With a 50 percent tax credit, this would result in a subsidy of $2 billion a year.

89. With the exceptions noted below, background information and statistical data regarding solar energy in California is from *Towards a Solar California: The Solar Cal Action Program,* Solar Cal Council, State Capitol, Sacramento, California, January 1979. This is the most outstanding solar energy program this author has yet encountered. It is worthy of study by every state energy commission and by the Department of Energy. The data on the Davis homes are from *Solar Engineering Magazine,* January 1979, p. 12, supplemented with a telephone interview of representatives of the builder, Corbett Homes. Data on the Davis ordinance are from *Energy Conservation Building Code Handbook,* July 1976, Davis, California. Additional information was provided by city building department officials in Davis, California. For the involvement of—and controversy over—utilities in California's solar industry, see, for instance, *Solar Energy Intelligence Report,* published weekly by Business Publishers, Inc., Silver Spring, Md., July 11, 1977, p. 149, and August 28, 1978, p. 261. Detailed references regarding other California data are provided in *Towards a Solar California.* The quote is from a February 7, 1979, letter from Jerry Yudelson, the director of the Solar Cal office, Business and Transportation Agency, State of California, Sacramento, California. For sources regarding the 55 percent tax credit in California, see Reference 38. For information on the San Diego ordinance, see Will Corry, "Solar Mandate for San Diego," *Sun Up: Energy News Digest,* vol. 3, no. 2, February 1979, p. 1; Jane Weisman Stein, "Solar Heating: Don't Be Burnt," San Diego *Union,* December 24, 1978, p. A-3. For a general California survey, see Jane Weisman Stein, "State Builds Bridge Over Energy Gaps," San Diego *Union,* January 14, 1979, p. 1.

90. Predictions regarding total U.S. energy in the year 2000 vary widely. The projections cited in Table 7–2 vary from 45 to 79 mbd, or a 2 to 1 range. By choosing a 50 mbd scenario my aim is not to introduce an additional projection, but to emphasize that a strong conservation program, which a 50 mbd scenario would require, should be an integral part of a shift toward solar energy.

91. Relative to the remainder of the world, hydropower in the United States is highly developed. While North America includes only about 13 percent of the world's hydropower potential it produces 40 percent of the total power in the world. But according to at least one estimate, the total potential hydroelectric

capacity of the United States is about three times present output, or about 170,000 megawatts. See, for instance, Federal Power Commission, *Hydroelectric Power Resources of the United States* (Washington, D.C.: Government Printing Office, November 1976), p. vii. Yet, due to environmental opposition and economic constraints, it is not likely that much of this potential will be realized. For these reasons, I have used an estimate of 1 to 2 percent growth for major hydroelectric installations, instead of the historic average of 4 percent growth, and rounded to the nearest quad. For an estimate of hydroelectric growth, see John C. Fisher, *Energy Crises in Perspective* (New York: John Wiley and Sons, 1974), p. 28.

However, the potential for low-head (*head* is the height of fall of the water) hydropower, or small installations typically under one megawatt of power, is considerable. According to the Federal Power Commission, present hydroelectric generating capacity could be doubled by developing the nation's 50,000 existing small dams. Two key advantages of the small dams are reduced environmental opposition and the speed with which they can be developed, in comparison to the decade or more required for the large nuclear or coal burning plants. The village of Lyndonville, Vermont, is an example of what is possible. It earns a profit from its beautiful 600-kilowatt plant on the Passumpsic River. For an analysis of the potential of low-head hydro, see David E. Lilienthal, "Lost Megawatts Flow Over Nation's Myriad Spillways," *Smithsonian Magazine*, vol. 8, no. 6, September 1977, pp. 83–88. Mr. Lilienthal is former chairman of the Atomic Energy Commission and Tennessee Valley Authority. See also J. R. McDonald, "Estimate of National Hydroelectric Power Potential at Existing Dams," U.S. Army Corps of Engineers, Institute for Water Resources, Mimeographed paper, July 20, 1977.

92. If the United States had planned ahead, the solar energy industry might have been expected to be supplying one fifth of U.S. energy and to have revenues of $50 billion a year today. Such an industry could be expected to spend $2 billion a year (4 percent of sales) on research and development. Only about a billion—or half of these research-and-development funds—might be expected to be invested in new products and technologies, and the remainder would go to sustaining existing products. My suggestion is simply to assume that the new product and process research that might have been done independently by a strong industry be done with the help of government intervention.

93. Ellis Rubinstein, "Technology and Society," *IEEE Spectrum*, January 1978, pp. 73–74. See also *Science News*, vol. 113, no. 4, January 28, 1978, p. 52.

Chapter 8
Conclusion: Toward a Balanced Energy Program

1. U.S. Department of Energy, Energy Information Administration, *Annual Report to Congress, Volume III, 1977* (Washington, D.C.: Superintendent of Documents, May 1978), p. 61. This represents an increase, in 1978 dollars, of $10 per barrel of the 84 billion barrels of crude oil equivalent that were in U.S. proved reserves of crude oil, natural gas, and natural gas liquids at the end of 1973. This estimate overstates the value, because future income was not discounted to obtain a present value. It understates the increase in value of all oil and gas in the ground because additions will be made to proved reserves. We benefited from discussions with Kenneth Arrow on welfare economics as well as on a number of other issues in this chapter.

2. By any measure, exploration and development activity for both oil and gas has increased substantially; the number of wells drilled, as one example, increased 69 percent between 1973 and 1977; *Monthly Energy Review*, January 1979, pp. 46–48. From 1973 to 1977, domestic production of crude oil and natural gas liquids

(including condensate) declined from 10.95 million barrels daily to 9.83 mbd; domestic production of natural gas declined from 24.9 trillion cubic feet to 22.7 tcf. (These natural gas figures include extraction losses, fuel for leases, plants and pipelines, and other adjustments; hence, they differ from the figures for natural gas delivered to consumers.) Domestic reserves of oil and gas declined from 83.8 billion barrels of crude oil equivalent to 73.3 billion; and imports of petroleum climbed from 6.3 million barrels daily to 8.7 million mbd. See Department of Energy, *Annual Report, Volume III*, pp. 23, 51, 61. Although Alaskan production reversed the adverse production trends in oil production and imports in 1978, the adverse trends will be reestablished in 1979.

3. A number of examples are listed in the Appendix.

4. Natural gas liquids are included as "oil" for consistency with most statistical sources. Such liquids are about 16 percent of total oil.

5. See projections in Appendix.

6. These costs are a result of "market failure." For an excellent discussion of this general subject, see Francis M. Bator, "The Anatomy of Market Failure," *Quarterly Journal of Economics,* LXXII (August 1958), pp. 351–379. Authors have referred to such costs by a variety of names. For examples, see "Externalities," in Robert M. Solow, "The Economics of Resources or the Resources of Economics," *The American Economic Review,* 64 (May 1974), pp. 1–14; "side effects," in Fred Hirsch, *The Social Limits of Growth* (Cambridge, Mass.: Harvard University Press, 1977), p. 3; "uncounted costs and benefits," in Charles E. Lindblom, *Politics and Markets* (New York: Basic Books, 1977), pp. 79, 80; "byproducts" in Thomas C. Schelling, "On the Ecology of Micromotives," in Robbin Marris, ed., *The Corporate Society* (London: Macmillan, 1974), p. 39. This category of costs, together with the costs paid for directly by the users of the energy, is called "social costs"; see Robert Dorfman and Nancy S. Dorfman, *Economics of the Environment* (New York: W. W. Norton, 1977).

7. Executive Office of the President, Energy Policy and Planning, *National Energy Plan* (Washington, D.C.: Government Printing Office, April 29, 1977), p. 95.

8. Although carbon dioxide emissions from burning hydrocarbons could eventually prove to be a large cost for future generations.

9. 1977 figures, see Table 8–1.

10. They are discussed at greater length in Chapter 2.

11. The increased costs to the United States flowing from higher oil prices can be thought of as an externality in the sense that a user does not encounter, at the time the consumption decision is made, the subsequent higher costs resulting from his (or her) actions. Such costs are often called "pecuniary costs." This example is but one of a class of actions that causes market failure in the economic sense. See Alfred Kahn, "The Tyranny of Small Decisions: Market Failures, Imperfections, and the Limits of Economics," *Kyklos* (1966), pp. 23–46; Thomas C. Schelling, "On the Ecology of Micromotives," pp. 19–64; and Schelling, *Micromotives and Macrobehavior* (New York: W. W. Norton, 1978).

12. Although recently the Saudis have moved to reduce dependence on the United States. See Chapter 2. Some authors place a heavier emphasis on "profit maximization" than do we. These authors also expect higher oil prices and recognize the large amount of uncertainty associated with predicting OPEC oil prices. For example, see M. A. Adelman, "Constraints on the World Oil Monopoly Price," *Resources and Energy,* 1 (1978).

13. For a list of methods available to correct distortions, see Schelling, "On the Ecology of Micromotives," p. 28. For an indication of the preference of business

executives (as well as of the chairman of the Council of Economic Advisors, Charles Schultz) for incentives rather than sanctions, see *Chemical Week,* September 6, 1978, p. 15.

14. The subjective nature of costs is developed at length in James M. Buchanan, *Costs and Choice* (Chicago: Markham Publishing, 1969). For a discussion of the difficulty in using the Gross National Product to assess economic welfare, see William D. Nordhaus and James Tobin, "Is Growth Obsolete?" in *Economic Growth* (New York: National Bureau of Economic Research, 1972). For an indication of the difficulty of determining the relationship of higher energy prices to GNP, see Energy Modeling Forum, *Energy and the Economy, EMF Report 1, Volume 1* (Stanford: Stanford University, Institute of Energy Studies, September 1977).

15. For general discussions, see Bator, "Anatomy of Market Failure"; Dorfman and Dorfman, *Economics of the Environment;* Kahn, "The Tyranny of Small Decisions"; and Schelling, *Micromotives and Macrobehavior.* But something is needed to force action, as volunteerism is not suited to solve inefficiencies in energy markets. Schelling's observations about conditions in which volunteerism is unlikely to work apply to energy. There is "nothing heroic in the occasion"; what is required is often a "protracted nuisance." The individual feels no particular community with many others who would benefit from his sacrifices, and he likely would suspect that large numbers of people were not cooperating. Thus, it is not surprising that President Carter discovered after he launched his "moral equivalent of war" that the nation needs quite a bit more than moral exhortation from the White House to restrain oil imports. Perhaps the classic illustration of individual actions which result in suboptimal economic results is the "tragedy of the commons," in which everyone benefits from the upkeep of the common, but no one has the motivation to tend it himself; thus, there develops the opposite and fatal incentive to overgraze it before others complete its ruin. See G. Hardin, "The Tragedy of the Commons," *Science,* December 13, 1968, pp. 1243–1248. Perhaps the most dramatic case was the killing of 20 million to 30 million buffalo within half a dozen years, just for the hides. For every penny of hide, five pounds of meat rotted. Fifteen years later, this meat could have been transported to market. This example is given in Thomas Schelling, "On the Ecology of Micromotives."

16. For estimates of price subsidies received by U.S. consumers, see Reference 75 of Chapter 2 (for oil) and Reference 2 of Chapter 3 (for natural gas). Consumers are also receiving a large price subsidy by paying much less for electricity —sometimes less than half—than for electricity from new coal, nuclear, and hydroelectric plants. Estimates shown elsewhere, although not complete for all consuming sectors and based on different data, are consistent with the $50 billion figure. See Sant in Reference 19. In addition to price subsidies granted consumers (principally at the expense of the owners of "old" oil, gas sold into the interstate market, and electrical generating plants), producers of conventional energy sources have received large subsidies from the government, and are continuing to receive subsidies. One study showed past subsidies to be $120 billion. See Battelle Memorial Institute, *An Analysis of Federal Incentives Used to Stimulate Energy Production* (Springfield, Va.: National Technical Information Service, March 1978).

We recognize that some forms of pollution could be caused by certain forms of solar energy and cogeneration, but our statement is generally correct. Of course, all oil, domestic and imported, has external costs because of environmental effects. For example, even so-called low-sulfur oil, whether found in nature or refined from crude oils containing sulfur, still have sulfur in them; thus sulfur dioxide is released into the atmosphere when the oil is burned. And burning oil (and gas), of course, also releases carbon dioxide, thereby contributing to a possible social cost for some future generation.

17. This conclusion is consistent with the so-called theory of the second best, which establishes that when the prices of some products (such as oil) are held below free-market levels, it may be economically beneficial for the nation to hold the prices of competing prices (such as conservation and solar energies) below their free-market levels. See R. G. Lipsey and Kelvin Lancaster, "The General Theory of Second Best," *Review of Economic Studies,* 24 (1956), pp. 11–32; E. J. Mishan, "Second Thoughts on Second Best," *Oxford Economic Papers,* 14 (October 1962), pp. 205–17; R. Rees, "Second-Best Rules for Public Enterprise Pricing," *Economica,* 35 (August 1968), pp. 260–73; William J. Baumol and David F. Bradford, "Optimal Departures from Marginal Cost Pricing," *American Economic Review,* 60 (June 1970), pp. 265–83.

18. Subsidizing the development of solar energy and the greater use of conservation is far more politically acceptable than decontrolling prices or spending money to develop additional fossil fuel resources. One nationwide poll showed that only 9 percent of the general public favored decontrol of prices on oil and natural gas to encourage new production; Yankelovich, Skelly and White, Inc., No. 7, 1978. Another poll (in California) showed that "energy development" was first priority for additional government expenditures, and among those favoring energy development, 72 percent favored expenditures for alternative energy sources (such as solar), 16 percent for conservation, and only 5 percent for fossil fuels; Boston *Herald American,* June 25, 1978, p. A17.

19. In fact, there is no way of knowing for sure what the return on investment would be for the nation in giving financial incentives for conservation and solar energy. But there are powerful reasons for trying fairly large payments and for believing that the nation would be justified in spending tens of billions of dollars on such a program. Certainly there are formidable political and economic constraints in providing supplies of energy from conventional sources in sufficient quantities to stop the growth of oil imports. Furthermore, although more analyses are required, it appears that a saving of the equivalent of 10 million barrels of oil per day, estimated to require an investment of some $220 billion, is cheaper than providing the equivalent of an additional 10 million barrels per day of oil from conventional energy sources—and certainly cheaper than the cost of marginal oil imports. For the comparison of conservation and conventional energy sources, see Roger Sant, "A Preliminary Assessment of the Potential to Adjust Capital Stock to Higher Energy Using Efficiencies," Paper presented at the American Association for the Advancement of Science, Washington, D.C., February 14, 1978. A similar analysis is not available for solar energy, but data in Chapter 7 suggest that the cost of solar energy varies over a wide range and some forms are quite cost effective when compared with the cost of incremental supplies of conventional energy sources, as reported by Sant. Note that the analysis by Sant does not include external costs. And it apparently does not consider a possible upward bias in econometric and technological forecasts of supplies of conventional energy sources, as discussed in our Appendix.

Information in Reference 2 of Chapter 6 indicates that conservation and solar energy are more cost effective than nuclear power.

20. Ideally, it would be desirable for the cost of marginal BTUs of energy from conservation and solar to equal the cost of marginal BTUs from each of the other energy sources. No adequate information exists to fine-tune financial incentives in such a manner. But our contention is that for the next five million or so barrels of oil daily above current levels, conservation (with some solar energy, also) will be at lower cost than the main alternative: oil imports. *In addition, financial incentives for conservation and solar have only a nominal impact on the consumer price index compared with the major impact of imported oil.*

21. Separate economic units, because of lack of risk-pooling, are more risk-averse than all economic units taken together. As discussed in Chapter 7, a

financial incentive of 55 percent in California so far has proven quite effective. Reference 36, Chapter 7, contains references justifying payments to speed the diffusion of innovation. In addition, because investors do not capture all the beneficial effects of experience, there is a natural tendency for society to underinvest; see Kenneth Arrow, "The Economic Implications of Learning by Doing," *Review of Economic Studies*, 29 (June 1962), pp. 155–173; and because learning-curve effects are greater during the early stages of an industry's life than during the later stages, the tendency to underinvest is greater during the early stages.

22. In order to partially correct such an obvious underestimation of the cost of sickness, one U.S. government study gave all housewives the value of a domestic servant. See Dorothy P. Rice, *Estimating the Cost of Illness* (Health Economics Series No. 6), U.S. Public Health Service, 1966. A more recent approach has given values to each duty a housewife performs, an approach that resulted in a value (in 1972) of about $6,000 yearly. The psychic value of a housewife to her family or society was not considered. Also, pain and suffering are not included in the cost of illness. See Barbara S. Cooper and Dorothy P. Rice, "The Economic Cost of Illness Revisited," Social Security Administration, DHEW Publication No. (55A) 76–11703, reprinted from the *Social Security Bulletin*, February 1976, U.S. Department of Health, Education, and Welfare. Also see Lester B. Lave and Eugene P. Seskin, *Air Pollution and Human Health* (Baltimore: Johns Hopkins University Press, 1977), pp. 225, 348, 349. Another complicating factor in estimating the cost of pollution is the impossibility of proving empirically that it causes illness. There is controversy surrounding, especially, the issue of air pollution, but two leading investigators conclude that it does. See Lester B. Lave and Eugene P. Seskin, "Does Air Pollution Cause Mortality?" *Statistics and the Environment*, Proceedings of the Fourth Symposium, March 3–5, 1976 (Washington, D.C.: American Statistical Association). Their conclusion is supported by other observations. For example, more than 4,000 citizens of London died from an extended period of severe air pollution there in 1952; see *Resources* (Washington, D.C.: Resources for the Future, April–July 1978), p. 1.

For an example of an oil-company publication that ignores the fact that conservation and solar energy are competing against oil and gas priced below their economic values, see *Our Energy Dilemma . . . Fact and Commentary*, Report by Marathon Oil Company, Findlay, Ohio, July 1977.

23. We call for considerably larger financial incentives than those contained in the National Energy Act of 1978, which specified tax credits of 15 percent (with a $300 ceiling) on conservation, and 15 to 30 percent on solar energy (with a $2,200 limit). The taxes collected by the government from a "windfall profits" tax on crude oil would be in the general range of tax credits (or other incentives) likely to be given for conservation and solar energy subsidies under our recommendations. Of course, the rationalization of a price reflecting its true economic view would also require the deregulation of old gas, again presumably with some kind of windfall tax on the profits. The same argument would be made for "old" electricity. But we shall leave those topics for another day. In the calculation below, we deal only with oil. Our oil tax calculations are meant to illustrate a reasonable program without necessarily being exactly what, after further analysis and debate, we would recommend.

The revenues that would be collected by the government from a tax on crude oil would depend on the level of world oil prices, the speed with which prices were allowed to move to world levels, the categories of oil covered, and the amount of the windfall retained by the companies rather than being collected by the government. How much extra income would accrue from deregulation? In late 1978, two well-known consulting firms each estimated that the amount to be split between producers and the government would be $15 billion yearly at the initial level, and then would gradually decline as the flow of old oil declined.

Our own calculations are in the same range. However, the rise in imported oil prices in early 1979 may have added another $5 billion—depending on the government's response in changing controlled oil prices—bringing the total pie up to $20 billion. The potential revenue, of course, would be considerably higher if "old" natural gas were included in the program as well.

There is greater uncertainty surrounding the estimate of the cost of financial incentives for conservation and solar energy. But an estimate of the order of magnitude can be made. Some energy savings can be achieved by direct regulation. Also, some energy can be saved by the removal of institutional and educational barriers. Let us assume that, of the 7 million barrels daily of oil equivalent to be saved in the late 1980's beyond that shown in the conventional program, 3 million could be saved by a combination of regulation and the removal of barriers. Assume further that all of the remaining 4 million would require financial incentives.

A simple projection from Sant's figures (Reference 19) suggests that the cost of saving four mbdoe would require an investment of $88 billion (.4 times $220 billion). On the one hand, the $88 billion could be on the low side, because inflation has increased costs beyond Sant's estimate, which was based on 1978 dollars. On the other hand, the $88 billion could be on the high side, because the cost per barrel of the first four mbd of oil equivalent saved would be less than that for the 10 mbd. There are two reasons: Lower cost barrels would be saved first; and the cost of some solar in energy installations would be lower than the average cost of conservation estimated by Sant.

The cost of the program to the government (assuming an average financial incentive of 50 percent) would be $44 billion spread over ten years, or $4.4 billion per year. Even if payments in the initial years were higher than in later years because of a gradual elimination of the subsidy (less necessary in later years because energy prices would be nearer their true economic values), there would be ample funds available from a windfall tax to fund the program and still allow the oil companies to share, along with consumers, in the windfall.

Furthermore, it is possible that most or even all of the subsidies could be paid by the additional tax revenues collected because of the increased domestic economic activity that would exist with stable, as opposed to higher, world oil prices. The following should be noted: *The tax revenues lost because of the 1973–74 rise in the price of oil would certainly have paid for all the government's share of the conservation and solar energy called for in the balanced program. And the tax revenues lost as a result of the oil price rises of 1979 will also be substantial.*

24. Adequate data are not available to indicate what the mix of conservation and solar energy would be, except that expenditures and results for conservation will be considerably larger than for solar energy, especially over the next decade. For one thing, in some solar applications—home heating, for example—a reasonably high level of conservation often is needed to make solar energy cost effective.

We comprehend that energy is not one homogeneous commodity, and that a saving in total energy use might not result in an equal saving in imported oil, especially if the total saving were due to a reduction in GNP (see Reference 30, Appendix).

25. Some might think that the consumption figures for the conventional program for the late 1980's are exaggerated. But the rate of annual growth in the consumption level shown for the conventional program without the extra conservation savings—54 million barrels of oil equivalent per day—would represent a growth rate from 1977 of only 3.2 percent. If the annual growth rate of energy should be 3.5 percent, as two leading energy-consulting firms indicated in 1978 in reports to their clients, then U.S. energy consumption in the late 1980's would be 56 million barrels of oil equivalent per day. This would imply oil imports of 19 million barrels per day if production of domestic energy sources were those shown in the

conventional program! In our judgment, higher levels of domestic production are quite unlikely. Almost surely, world oil output would not permit such high American petroleum imports. Instead, there likely would be an even greater jump in world oil prices than any mentioned here. It is such a possible consequence that requires that conservation and solar energy be given their fair chance.

Appendix: Limits to Models

This Appendix is based heavily on the analyses in Sergio Koreisha, "A Survey on the Limitations of Three Energy Policy Models," Working paper, Energy Project at the Harvard Business School, 1979. We wish to express our appreciation to William W. Hogan, Hendrik Houthakker, and Henry Jacoby for their efforts in facilitating our review of their work, and to Hogan in particular for general discussions on the roles of models and the importance of lags. We, of course, bear full responsibility for this appendix.

1. "Rear-view mirror" analogy from Arthur Schlaifer, Jr., professor at Harvard Business School, to whom we express appreciation for reviewing the manuscript. Of course, in some cases econometric models are used to forecast *new* energy sources, and technological models are used to forecast *existing* energy sources. The use of judgment represents the use of an "implicit mental model," a term used by William W. Hogan, "Energy Modeling: Building Understanding for Better Use," Paper presented at the Second Lawrence Symposium on the Systems and Decision Sciences, Berkeley, California, October 3, 1978. In using their judgment, modelers sometimes change the results directly and sometimes change the assumptions in the model in order to obtain different results.

2. "The Economic Modelers Vie for Washington's Ear," *Fortune,* November 20, 1978; quote on "public officials," p. 105, and "when the history," p. 102. Some authors have cited examples in which U.S. political leaders have ceded "large tracts of traditionally political territory to the custody of experts"; Martin Greenberger *et al., Models in the Policy Process* (New York: Russell Sage Foundation, 1976), p. 28. Also see Robert E. Lance, "The Decline of Politics and Ideology in a Knowledgeable Society," *American Journal of Sociology* 31 (1968), pp. 657–658. In addition to the information in this appendix, see especially Chapters 2, 3, and 6 for support of this conclusion. For further discussion of problems in using models, see Lester C. Thurow, "Economics 1977," *Daedalus,* 106 (Fall 1977), pp. 79–94. For an essay on the misuse of macroeconomic models by economists, see James B. Ramsey, *Economic Forecasting—Models or Markets* (London: The Institute of Economic Affairs, 1977). Also see "The Economist as Prophet," *Morgan Guaranty Survey,* August 1978, pp. 9–14; and Leonard Silk, "The Highly Inexact Science of Economics," *New York Times,* December 8, 1977, p. 65. Our judgment is that reliance on economic and technology studies is more likely by the executive branch than by members of Congress, who are elected by constituents, who in turn feel little need for such studies to form an opinion about their interests.

3. *New York Times,* November 13, 1974, p. 1; *Business Week,* January 13, 1975, p. 67. For information about oil prices in 1974, see Reference 26.

4. See article by Assistant Secretary of State Thomas O. Enders about the need for a floor price in the industrial nations, in "OPEC and the Industrial Countries: The Next Ten Years," *Foreign Affairs,* July 1975, pp. 625–637.

5. "A Conversation with Henry Ford," Los Angeles *Times,* January 21, 1979.

6. For an excellent example of a model builder's description of limitations, see the writings by William W. Hogan in Energy Modeling Forum, *Energy and the Economy, EMF Report 1,* vol. 1 (Stanford: Stanford University, Institute for Energy Studies, September 1977), p. iv. Within the model-building profession, there has been concern for some time about the use and misuse of models; see Hogan, "Energy Modeling: Building Understanding for Better Use." For a recent

review of energy models, see Alan S. Manne, Richard G. Richels, and John P. Weyant, "Energy Policy Modeling: A Survey," *Operations Research*, forthcoming. Also, in recent years much greater emphasis has been placed on assessing models; the Energy Modeling Forum at Stanford has an active program underway. For an example of their work, see Energy Modeling Forum, *Coal in Transition: 1980–2000, EMF Report 2*, vol. 1 (Stanford: Stanford University, Energy Modeling Forum, July 1978). A group at MIT also has been active in this field; for example, see MIT Energy Laboratory Policy Study Group, "The FEA Project Independence Report: An Analytical Review and Evaluation," Energy Laboratory Report MIT–EL–75–017, Mimeographed paper, May 1975.

7. The requirement of relevant historical experience, of course, has been recognized. In January 1974, for example, William Fellner, then a member of the President's Council of Economic Advisors, stated that "it is definitely impossible" to specify what the price elasticity of demand for gasoline would be with the drastically higher gasoline prices; quoted in *New York Times*, January 11, 1974, pp. 41, 48. Quotes by others in this article reflect the same sentiment.

The problem of extrapolating far outside the range of prior experience often is compounded by the assumption (for convenience) that one elasticity equation exists over the entire range. For example, for convenience and because of lack of a more precise theory, model builders often assume that over the entire range of forecasts, quantity and price have either a linear relationship or constant elasticity. If one assumes a linear relationship, then a change in price from $10.00 to $11.00 is assumed to have the same effect on quantity demanded (or supplied) as a change in price from $1.00 to $2.00; in other words, a 10-percent price increase at the upper level would have the same effect on quantity as a 100-percent price increase at the lower level. Thus, at high price levels, demand and supply are assumed to be very responsive to price changes; so, as the price rises, demand is forecast to decrease very rapidly and supply to increase very rapidly. Therefore, with upward price movements, demand and supply are forecast to come into equilibrium very quickly. This condition existed with the Kennedy-Houthakker model, which used linear equations for demand and supply. (The demand equations were linear approximations of the double-logarithmic equations used to estimate the demand function; see Reference 9.)

But what if the model builder assumes a constant elasticity over the entire range? The practical implication of this is that if the price elasticity of demand is less than one—as was the case for OPEC oil in 1974 and likely still is the case in 1979—then the higher the price, the greater the income of the suppliers (see Adelman in Reference 45); thus, in this formulation, there is no practical cut-off in price increases. Therefore, using a constant elasticity over a very broad range has the potential to create unrealistic results.

8. Gary C. Hufbauer, *Synthetic Materials and the Theory of International Trade* (Cambridge: Harvard University Press, 1966), pp. 46–57, and Robert B. Stobaugh and Phillip L. Townsend, "Price Forecasting and Strategic Planning: The Case of Petrochemicals," *Journal of Marketing Research*, XII (February 1975), pp. 19–29.

9. For a description of the model, which has been documented more than most models, see Michael Kennedy, "An Economic Model of the World Oil Market," *The Bell Journal of Economics and Management Science*, 5 (Autumn 1974), pp. 540–577. For more detailed descriptions of the model, see Michael Kennedy, "A World Oil Model," in Dale Jorgenson, ed., *Econometric Studies of U.S. Energy Policy* (Amsterdam: North-Holland, 1976), pp. 95–175, and Michael Kennedy, "An Economic Model of the World Oil Market," Ph.D. thesis, Harvard University, November 1974. Also, an important contribution of this model is its solution algorithm, which is much more efficient than most models. For an earlier paper on a similar subject, see H. S. Houthakker, "The Capacity Method of Quadratic Programming," *Econometrica*, 28 (January 1960), pp. 62–87.

10. The host-government "revenue" officially is a combination of royalty and income tax, but often is referred to as a royalty or export duty—the terminology of Kennedy and Houthakker. For convenience, we refer to it as a host-government tax. What the U.S. government officially deems it to be—royalty or income tax—could make a substantial difference in U.S. taxation of U.S. companies.

11. Parts of the World Oil Model have been incorporated into a new model, the World Energy Model, that is, a dynamic long-run model that contains natural gas, coal, nuclear, hydro, and electrical power, in addition to oil. For the results of the model, see H. S. Houthakker and Michael Kennedy, "Long-Range Energy Prospects," in Ragaei El Mallakh and Dorothea H. El Mallakh, *Energy Options and Conservation* (Boulder, Colo.: The International Research Center for Energy and Economic Development, 1978), pp. 11–41. For a description of the methodology, see Houthakker and Kennedy, "A Long-Run Model of World Energy Demands, Supplies and Prices," Mimeographed paper, prepared in 1978.

12. "The equation for gasoline demand was run on pooled time series observations from 12 countries (Portugal, Italy, Austria, Belgium, Denmark, France, West Germany, Netherlands, Norway, Sweden, United Kingdom, and United States) which covered the years 1962–1972. The remaining equations (kerosene, distillate fuel, residual oil fuel) were fitted over a smaller sample of 9 countries (Japan, Italy, Belgium, Denmark, France, West Germany, Netherlands, Sweden, and United Kingdom), for the years 1965–1970. The U.S. was not included due to data deficiency"; Kennedy, "An Economic Model," Ph.D. thesis, pp. IV–11 and IV–12.

13. Some of the standard errors for the regression coefficients for certain product-demand equations were larger or as large as the coefficients themselves; Kennedy, "An Economic Model," *Bell Journal*. It is quite common for model builders to modify their elasticities in order to ensure that the models produce what the modelers consider to be reasonable results. "The Economic Modelers Vie for Washington's Ear," *Fortune*, November 20, 1978, p. 105. At times, modelers handle some of the potential problems raised by our "red flags" by running several scenarios and choosing one that reflects their judgment. It is desirable, of course, to show estimates obtained with elasticities based upon data and those obtained with elasticities based upon judgment.

14. Kennedy, "An Economic Model," *Bell Journal*, p. 571. See Reference 26 for a discussion of world oil prices in 1974. They assumed, of course, that U.S. price controls on oil would be removed, but even with controls the average prices paid by U.S. refiners and prices paid for new oil were higher than some of the equilibrium prices estimated by Kennedy and Houthakker. See discussion later in text about possible nonmonetary effects of price controls. Effective January 1, 1979, the OPEC price for market crude had been raised to $13.34 and the host-government tax is a little lower, generally 10 to 30 cents, depending on operating costs, whether the company operated the oil fields, and the reinvestment rate of the companies within the host-nation; *Wall Street Journal*, December 18, 1978, p. 2.

15. Kennedy and Houthakker used a nonlinear model, which made it difficult to calculate likely boundaries; see Reference 9.

16. For information about world oil prices during 1974, see Reference 26.

17. First quote from Hendrik S. Houthakker, "The Oil Problem and the International Monetary System," Talk before the Conference Board on United States–Japanese Economic Policy, Tokyo, Japan, April 1, 1974, p. 10. Other quotes by Vermont Royster, a writer for *Wall Street Journal*, in reporting on a conference on the world oil situation at the American Enterprise Institute, in "Glimmer Amid the Gloom," *Wall Street Journal*, October 9, 1974, p. 26.

18. Quote from Hendrik S. Houthakker, *The World Price of Oil: A Medium-*

Term Analysis (Washington, D.C.: American Enterprise Institute for Public Policy Research, 1976), p. 27. This publication also contains estimates for 1985. Information about reason for using judgment in estimating output of Middle East and North Africa, from interview with Houthakker.

19. The Policy Study Group of the MIT Energy Laboratory, "Energy Self-Sufficiency: An Economic Evaluation," *Technology Review*, May 1974, pp. 23–58; quote is from p. 24. Note that this was an *assumption* used for analytical purposes, not a *prediction*. For their economic models, they used long-term elasticities and assumed that they applied for 1980.

20. The group recognized the explicit problem of aggregating onshore and offshore drilling activity; "Energy Self-Sufficiency," p. 34 (quote is on p. 29). In fact, U.S. production of crude oil in 1980 almost surely will not be as high as that indicated by the MIT econometric study for the prices actually received for newly found oil. But also see later discussion in text about possible nonmonetary effects of price controls.

21. The group's report explicitly recognized these two problems; "Energy Self-Sufficiency," p. 34.

22. See Chapter 2, Reference 69.

23. The Department of Energy estimate was $24 in 1975 dollars; *Chemical Week*, November 8, 1978, p. 36.

24. For example, refer to the shale-oil story in Chapter 2.

25. For starts toward developing a conceptual model describing the progress of new technologies, see John Diffenbach, "New Energy Technologies: Evolution and Cross-Impact," Harvard Business School doctoral thesis, and David Bodde, "The Influence of Regulation on Technical Evolution of the Electric Generating Industry," Harvard Business School doctoral thesis, 1976.

26. The estimates of natural gas supply in the MIT study seem especially high, presumably because of the elasticities in the MacAvoy-Pindyck model used by MIT. See Chapter 3 of this book for prices of newly found natural gas. In addition to assuming price decontrol of oil in all cases and gas in some cases, the MIT group also assumed that offshore territories would be leased. The price of OPEC oil at the time of the MIT study was not clear to anyone, not even the oil companies. The OPEC members had agreed on a host-government tax of $7.00, effective January 1, 1974. But demand for oil continued strong and in March 1974, the oil companies in anticipation of retroactive tax action by OPEC members began transferring crude oil from their producing subsidiaries to their other subsidiaries at about $9.50 per barrel, which would imply a government tax of about $9.00. By the end of 1974, through retroactive action and OPEC "participation," that is, the OPEC members' takeover of all or part of the operating properties from the companies, the government tax on market crude was raised to $10.12 a barrel, effective January 1, 1975. See Edith Penrose, "The Development of Crisis," in Raymond Vernon, ed., *The Oil Crisis* (New York: W. W. Norton, 1976), p. 52; and "Energy Self-Sufficiency," pp. 48–49. For more details on oil prices during 1974, see Reference 32, Chapter 2. We interpret a host-government tax of $7.00 a barrel to be the basis for the MIT group's world capacity estimates as well as to be the "current" level from which they based their prediction that price was more likely to fall than rise. After describing the uncertainty surrounding OPEC price at that particular time, they state, "Our assumption at other points in this study has been that the average payment will converge in the short run on a figure close to $7.00 per barrel"; "Energy Self-Sufficiency," p. 49. They mention a $9.00 level delivered to the United States, which equates to about $7.00 host-government tax; see p. 54.

Because of the limitations inherent in using a single econometric model to predict world oil supplies, the Supply Analysis Group of the MIT World Oil Project is developing a forecasting method that incorporates explicit representations of the different steps in the oil-supply process for different major oil fields. See Paul L. Eckbo, Henry D. Jacoby, and James L Smith, "Oil Supply Forecasting: A Disaggregated Process Aproach," *The Bell Journal of Economics,* 9 (Spring 1978), pp. 218–235; and M. A. Adelman and H. D. Jacoby, "Alternative Methods of Oil Supply Forecasting," in Robert S. Pindyck, ed., *Advances in the Economics of Energy Resources, Volume II: The Production and Pricing of Energy Resources* (Greenwich, Conn.: J.A.I. Press, forthcoming).

27. "Energy Self-Sufficiency," pp. 28–30.

28. "Energy Self-Sufficiency," p. 28.

29. The National Petroleum Council studies on which the MIT group relied heavily for their judgmental supply forecasts consist of a series of reports published in 1972 and 1973 on the U.S. energy outlook from then until the end of the century. The studies' participants included energy experts not only from the oil and gas industries, but also from government and the coal, nuclear, and electric utilities industries. The NPC supply forecasts were relatively insensitive to price, being 38.4, 39.3, and 39.4 million barrels of oil equivalent per day, respectively, at the three price levels ($7, $9, and $11 in 1973 dollars). "Energy Self-Sufficiency," p. 28. The MIT group's judgmental demand estimates relied most heavily on the NPC studies and a study by the National Economic Research Associates; p. 30. We made no attempt to evaluate the judgmental forecasts. First, evaluating such forecasts is more difficult than evaluating econometric estimates—a factor that makes econometric models easier to criticize than judgmental forecasts—for the assumptions underlying judgmental assessments are seldom as clearly identified. Second, we are not in a position to question the expertise of those making the assessments. In fact, we know personally a number of the experts who participated in the National Petroleum Council work used by MIT and doubt whether it would have been possible to obtain a better qualified panel. But the ability of any expert to make accurate predictions of supply-and-demand relationships under conditions quite different from any historical experience is very limited. See Reference 30, this chapter, about the disadvantage of considering energy as one homogeneous source.

30. Federal Energy Administration, *Project Independence Report* (Washington, D.C.: Government Printing Office, November 1974), app. AII. The demand forecasts do not reflect economic and resource restrictions, such as levels of domestic supplies of energy sources, financial ability to build new facilities, or import levels of energy products. These restrictions are incorporated into the model at a later stage. The demand forecasts, however, do reflect effects of non-price initiatives taken to conserve energy. Regional forecasts were obtained by disaggregating the national forecasts according to census regions. Note that there obviously is an important advantage in a model that determines demand by fuel type and allows substitution of one fuel for another only when this is possible. In contrast, relatively simple energy models in which energy is considered as one homogeneous product can be misleading. For example, GNP affects electricity demand, which in turn affects the demand for coal. Thus, as long as coal is available, a rise in energy usage caused by a rise in GNP would not be met entirely by imported oil. In two computer runs done by the PIES–74 group, for example, GNP was reduced, thereby lowering demand for energy by 10 quads, but oil imports were reduced only 2 quads; this example from interview with William Hogan.

31. If the prices derived by the linear program were not significantly different from the ones estimated by the econometric model, then the linear program prices and quantities were considered to be the equilibrium solution. If the prices were

significantly different, then trial and error was used until there was a convergence of the prices from the linear program and econometric model.

32. A good critique is J. A. Hausman, *"Project Independence Report:* An Appraisal of U.S. Energy Needs up to 1985," *Bell Journal of Economics,* 6 (Autumn 1975), pp. 517–551; and a related study, MIT Energy Laboratory Policy Study Group, "The FEA Project Independence Report." Also see U.S. General Accounting Office, *Review of the 1974 Project Independence Evaluation System,* OPA–76–20, Washington, D.C., April 1976. A later version of PIES was critiqued in Hans Landsberg, "Review of the Federal Energy Administration *National Energy Outlook, 1976,"* Report to the National Science Foundation, Resources for the Future, Washington, D.C., March, 1977.

33. Federal Energy Administration, *Project Independence Report,* pp. 414–416.

34. Martin Greenberger *et al., Models in the Policy Process,* p. 45.

35. For another example, see Reference 6, this appendix.

36. In technical terms, substitute goods theoretically should have a positive cross elasticity, whereas the PIES authors obtained a number of negative cross elasticities; *Project Independence Report,* app. A–11, p. 62.

37. U.S. Comptroller General Report to Congress, *Review of the 1974 Project Independence Evaluation System* (Washington, D.C.: Government Printing Office, November 1974), p. 38.

38. *Project Independence Report,* pp. 82–83.

39. *Project Independence Report,* p. 49.

40. Hogan, "Energy Modeling: Building Understanding for Better Use"; and "The Economist as Prophet," *The Morgan Guaranty Survey,* August 1978, p. 13.

41. Hogan, "Energy Modeling," and A. M. Geoffrion, "The Purpose of Mathematical Programming Is Insight, Not Numbers," *Interface,* November 1976, pp. 81–92. Some model builders, however, disagree with this.

42. This is even ignoring the deliberate misuse of model results for political purposes. In this appendix, we are not suggesting that there is something better than models, but that sometimes formal models are not adequate to provide predictions within a range sufficiently narrow to be useful. Neither are judgmental models, sometimes.

43. Robert Pindyck, "The Economics of Oil Pricing," *Wall Street Journal,* December 20, 1977, p. 16. Also, see his "OPEC's Threat to the West," *Foreign Policy,* Spring 1978, pp. 36–52, and "Gains to Producers from the Cartelization of Exhaustible Resources," in *Review of Economics and Statistics,* LX (May 1978), pp. 238–251.

44. Pindyck, "The Economics of Oil Pricing."

45. M. A. Adelman, "Need for Caution over Prices," *Petroleum Economist,* September 1977, p. 359.

46. Hogan, "Energy Modeling."

47. Note that on models linking energy consumption to GNP, the elasticity can be 1.0 and the ratio of energy to GNP can change. For evidence on the conclusion that too much attention was focused on the potential of traditional energy sources at the expense of conservation and solar energy, see especially Chapters 6 and 7, this book.

48. *Chemical Week,* November 8, 1978, p. 36.

49. "Energy Self-Sufficiency," p. 24.

Index

oil fields in, 10; oil production in, 32, 33

Chirac, Jacques, 108

Citibank, 180

Clean Air Act, 1977 Amendments to, 90, 92

Clinch River breeder reactor, 210; construction of, 131

Coal, 8, 9; Carter's National Energy Plan on production of, 10, 80–82, 90, 93; conflicts over, 219; cost of, compared with wood, 199–200; decline in, after WWII, 79; difficulties posed in transition to, 81; disappointment of, 105–106; environmental barriers to short-term utilization of, 91–94; geographic distribution of, 87; increasing dependence on railroad of, 88–91; industry research and development, 104–105; long-term prospects for utilization of, 81, 97–105 *passim;* and National Coal Policy Project, 106–107; new companies participating in, 97–102; new technologies in, 102–105; outlook for, 219–220; people as barriers to short-term utilization of, 94–97; position of, in electric utility industry, 82, 84–86; rediscovery of, 79–80; short-term utilization of, 81–97 *passim;* as source of gas, 68; synthetic gas manufactured from, 68–69; systemic barriers to short-term utilization of, 82–91; U.S. abundance of, 81

Coalcon project, 104

"Coal conversion," 84

Coal strike (1977–1978), 86, 94, 95, 169; Carter's threat to use Taft-Hartley injunction in, 79;

impact on coal's future of, 107; issues not settled by, 97, 106

Cogeneration, 157–158; advantages of, 159–160; obstacles to, 160–161; two types of, 158–159

Combined heat and power, *see* Cogeneration

Combustion Engineering, 114

Commonwealth Edison Company, 86

Compagnie Française de Pétroles, 23

Conference Board, 143

ConPaso project, 101–102

Conservation, energy, 176; conclusions on, 179–182; and Los Angeles' Emergency Energy Curtailment Plan, 144–146; as major alternative to imported oil, 11–12, 54, 136–138; need for public policy on, 178–179; obstacles to, 139–144; productive, 138–139; projections, inventory, and observation of potential for, 176–178; recommendations for, 216, 226–233 *passim;* three types of, 138–139. *See also* Buildings, energy conservation in; Industrial energy conservation; Transportation, energy conservation in

Consolidated Edison, 158

Consolidation Coal Company, 98, 103

Continental Oil Company, 98

Cuba, military aid for Qaddafi from, 35

Cubbage, Ben, 72

Curtailment, as category of conservation, 138

Darmstadter, Joel, 139

Daugherty, Jim, 72

About the Authors

ROBERT STOBAUGH is a professor of business administration at the Harvard Business School and director of the Energy Project. He has a doctorate from Harvard Business School and also has a degree in chemical engineering. Author or coauthor of six books, he is president of the Academy of International Business and has been a consultant to a number of Cabinet-level departments and congressional committees and to the United Nations and several foreign countries.

DANIEL YERGIN is a lecturer at the Kennedy School at Harvard, and directs the International Energy Seminar at Harvard's Center for International Affairs. He formerly was a lecturer at the Harvard Business School. Author of *Shattered Peace: The Origins of the Cold War and the National Security State,* he has a Ph.D. in international relations from Cambridge University, where he was a Marshall Scholar. He is a consultant to the Energy Directorate of the Commission of the European Community.

I. C. BUPP is a lecturer at the Harvard Business School and coauthor of *Light Water: How the Nuclear Dream Dissolved.* He holds a Ph.D. from the Harvard Government Department.

MEL HORWITCH is an assistant professor at the Harvard Business School, where he received his doctorate, and author of the forthcoming book, *Clipped Wings: The SST Controversy and a Changing America.*

SERGIO KOREISHA has a master's degree in engineering from the University of California and is a doctoral candidate at the Harvard Business School.

MODESTO A. MAIDIQUE is an assistant professor at the Harvard Business School. He holds a Ph.D. in physics from MIT. He was a founder and for seven years a vice-president of an electronics firm, and holds several patents.

FRANK SCHULLER has an M.B.A. from the Harvard Business School and is a candidate for the degree of doctor of business administration.